The Economics of Global Allocations of the Green Climate Fund

S. Niggol Seo

The Economics of Global Allocations of the Green Climate Fund

An Assessment from Four Scientific Traditions of Modeling Adaptation Strategies

Springer

S. Niggol Seo
Muaebak Institute of Global Warming Studies
Seorim-dong, Gwanak-gu, Seoul, Korea

ISBN 978-3-030-18273-1 ISBN 978-3-030-18274-8 (eBook)
https://doi.org/10.1007/978-3-030-18274-8

This Springer imprint is published by the registered company Springer Nature Switzerland AG.
The registered company address is: Gewerbestrasse 11, 6330 Cham, Switzerland

To Moonsoo and Jijang

Preface

The Green Climate Fund (GCF) is a transformational global fund because of its financial size and a global scale of monetary transfers. The size of the GCF was agreed to be about ten times larger than the annual United Nations (UN) budget, both the general budget and the peacekeeping/security budget combined. Ardent supporters of the UN climate conferences have even pushed to increase the size of the GCF in the near future to about 100 times of the UN annual combined budget.

First announced at the Copenhagen Conference in 2009, the Fund started its initial allocations to a batch of selected projects in November 2015 in time for the Paris Conference. As of January 2019, a total of 93+ projects received a grant from the Fund. The GCF has quickly become the ground zero of the global climate change policy negotiations coordinated by the United Nations since 1992.

Notwithstanding the vital importance of the GCF's works reverberating vibrantly at each global climate change roundtable, little have been written by academics about the rationale behind the Fund and the funding allocation decisions it has been issuing. This is owing largely to the lack of expertise among the academics on the theory of global funds, but more critically to the lack of scientific information on the range of possible GCF-funded projects and their potential economic consequences.

In this book, I have taken an ambitious step toward disentangling the complexities of the works of the GCF by putting together an ensemble of essential scientific literatures that are primarily concerned on the central question of the relationships between economic activities and climate consequences. The four empirical traditions explained in this book in the contexts of the GCF funding decisions are a microbehavioral economic model of behavioral adaptations, an agro-economic climate change impact model, a statistical model of yield changes, and an ecosystem model of climate change impacts.

The book begins with a comprehensive review of the GCF, which includes its historical developments, the organization, institutional relationships with the UN and climate conferences, governing principles, investment frameworks, pledges/contributions, and awarded projects. In providing an evaluation of each of the 93+ GCF-funded projects, I am founded on three essential economic theories: a theory of public expenditures initiated by Paul Samuelson, a theory of Ricardo-Pareto efficient resource allocations from the eponymous economists, and a theory of public and group adaptation.

Rephrased in plain terms, the project evaluations are performed on the following three criteria: (1) Is the GCF funding allocated to the project that yields the largest benefit pertinent to climate change policy, for example, the carbon dioxide equivalent emissions? (2) Does the GCF funding induce a recipient to make the best, that is, most profitable, decision given the changed climate? (3) Is the GCF funding given to the group can provide the proposed or needed group action at the lowest cost?

This book goes further on to elucidate the relationships between the GCF and other global funds whose funding areas often overlap those of the GCF: UN specialized funds and programs, crowdfunding, cryptocurrency, and global environmental funds such as the CO2 emissions allowance trading. I explore the possibility of tying the GCF funding decisions to the prices that arise from these global funds.

I am convinced that the book will provide a fascinating, at times provocative, alternative perspective on the current global warming economics as well as on the current global policy negotiations on global warming coordinated by the United Nations. At the end of this book, readers will be able to appreciate and evaluate the two fields from a fresh vantage point which holds much promise.

That said, I feel that the book can be relished by non-climate change experts as well and should be interpreted as a general contribution to the economics literature. To give you an idea, the monetary size of the global scale funds discussed in this book will be larger than the gross domestic product (GDP) of any nation in the world, with a dozen exceptions, at their full realizations. As such, a sound theory and analysis of the long list of global funds will become increasingly integral to the economic thinkers of the future.

If you belong to the so-called Millennials and especially Generation Z, I strongly recommend this book for your reading during your long train travel to a summer vacation resort or your hometown. You will run into a starkly different planet from the drastic and catastrophic world emotionally depicted in the Green New Deal by politicians or *The Day After Tomorrow* by filmmakers. You will find it much cooler and more brain-stimulating.

This book could not have been completed without the contributions of many outstanding researchers, most of whom are my contemporaries, who are, justly enough, frequently cited throughout this book as a key contributor to their respective literature. I would like to express my gratitude to them for their brilliant and enduring scientific works. I feel particularly thankful to Professor Robert Mendelsohn at Yale University with whom I started my research on climate change adaptation behaviors 18 years ago and to late Professor Robert Evenson by whom I was introduced to the economics of green revolution and the research and development in agriculture.

Lastly, I would like to express my appreciation to Margaret Deignan and the editorial team at Springer Nature for their masterful editorial works in producing this book.

Seoul, South Korea S. Niggol Seo

Contents

About the Author

S. Niggol Seo is a natural resource economist who specializes in the study of global warming. Born in a rural village in South Korea in 1972, he received a Ph.D. degree in Environmental and Natural Resource Economics from Yale University in May 2006 with a dissertation on microbehavioral models of global warming. While at Yale, he learned from Professors Robert Mendelsohn and William Nordhaus (Nobel Prize in 2018) on the economics of global warming. Since 2003, he has worked on various World Bank projects on climate change in Africa, Latin America, and Asia. He held professor positions in the UK, Spain, and Australia from 2006 to 2015.

Professor Seo has published over a hundred (peer reviewed) articles on global warming economics, which include six books. He has been on the editorial boards of the three journals, *Climatic Change* (Stanford Univ.), *Food Policy* (Univ. of Bologna), and *Applied Economic Perspectives and Policy* (Oxford Univ.) and received an Outstanding Applied Economic Perspectives and Policy Article Award from the Agricultural and Applied Economics Association (AAEA) in Pittsburgh in June 2011 for developing a behavioral economic model of adaptations to climate change.

List of Figures

List of Tables

Chapter 1
Economics of the Green Climate Fund, Paris Agreements, and Global Funds and Currencies: An Overview

Abstract This introductory chapter poses the central question of this book and provides an overview of the chapters that follow. The challenges faced by global decision-makers of allocating the Green Climate Fund (GCF) funds justifiably to a set of selected projects are very high but underappreciated. This book elucidates these challenges and addresses them by laying out an analytical framework for the GCF and providing an evaluation of each of the GCF-funded projects, which totals 93 projects and US$ 4.6 billion as of November 2018. For empirical evaluations, this book takes a novel step by putting together four scientific traditions of climate change that have produced quantitative results and predictions that are highly pertinent to the GCF funding decisions. This book extends the analysis of the GCF to other global funds whose areas of funding may overlap those of the GCF such as the United Nations specialized funds and programs, the European Union's carbon dioxide emissions allowance trading.

Keywords Green climate fund · Paris agreements · Four scientific traditions · UN specialized funds and programs

1.1 Challenges of the Green Climate Fund

The global community is facing an unprecedented challenge in allocating the Green Climate Fund (GCF), the fund which was created by the UN climate conferences of the parties to assist the global efforts to fight global and dangerous climate changes (UNFCCC 2009, 2010). At the signing of the Paris Agreements, praised by many as a turning point or a landmark climate agreement (NPR 2016; CBS 2016), the GCF became the primary financial instrument of the United Nations Framework Convention on Climate Change (UNFCCC), the UN agency for climate change policy negotiations and implementations (GCF 2011; UNFCCC 2015; Seo 2017a, b).

The sheer size of the GCF fund would pose a big challenge, which is over 10 times larger than the annual United Nations regular budget and the annual UN

S. N. Seo, *The Economics of Global Allocations of the Green Climate Fund*,
https://doi.org/10.1007/978-3-030-18274-8_1

peacekeeping budget combined, to the group which is responsible for allocating the money to someone somewhere who is taking climate-related actions (UN 2017, 2018). The annual allocation of the GCF funding was set at US$100 billion per year (UNFCCC 2009, 2010), but climate enthusiasts and activists have argued for the increase of the size of the GCF funding to US$ 1 trillion, for example, the Economic Times of India and the Global Environmental Facility's evaluation (Economic Times 2015; GEF 2017).

Will countries pledge and contribute as much as they gave promises at numerous conferences? Who should be responsible for allocating the fund globally to different climate actors and actions? Is there a guideline or rule that should dictate the allocation decisions by the GCF Board? How should the outcomes from funded projects be evaluated? Who should monitor the progresses in each of the funded projects? There are many intellectual questions unanswered yet.

Despite the great challenges in terms of conceptualizations, legal issues, and practical matters, there has been little written about the GCF by climate change researchers. This is in a huge contrast to the climate activists who vehemently argue for the moral responsibility of the rich nations to increase the size of the funding as well as to the high tensions around the GCF aids among the negotiators at every United Nations conference (NYT 2015; Carbon Brief 2018).

The present author argued, just before the Paris Conference in late 2015, that a climate foreign aid to the poor in developing countries dressed up as the GCF funding could end up hurting the poor by failing to encourage efficient adaptations to unfolding climatic changes (Seo 2015b). The present author provided a review of the Paris Agreements from the economic and policy standpoints (Seo 2017b).

Analysts of the Paris Agreements have concentrated on the emissions reduction commitments made by the parties of the conferences and asked whether such commitments are sufficient, for example, the United Nations Enviroment Programme's emissions gap report or the Intergovernmental Panel of Climate Change (IPCC)'s special report on global warming of 1.5 °C (UNEP 2017; IPCC 2018). A group of legal experts from the UNFCCC Secretariat also provided a legal overview of the Paris Agreements as an international agreement (Klein et al. 2017).

As of October 2018, the international fund is securely kept at a vault in Songdo City, South Korea where the GCF Secretariat is located or at the World Bank which is designated as the trustee of the GCF (UNFCCC 2012, GCF 2018a). The international climate fund was raised by voluntary pledges and contributions by some of the richest countries in the world to help developing countries to participate in a global climate agreement or protocol which eventually turned out to be the Paris Agreements. The contributions up to this point came mostly from five richest countries that were vying, at that time, for a global leadership on climate politics: the US, Japan, Germany, the UK, and France (GCF 2018d).

First conceived at the Copenhagen Conference in Denmark at the end of 2009, the GCF quickly became the top battle ground between developing countries and developed countries, that is, between the Annex A countries that are demanding monetary aids and the Annex B countries that have promised such aids (UNFCCC 2009; Nordhaus 2010). Soon afterwards, it became a primary instrument for the

Table 1.1 Green climate fund and the conferences of the parties (COP)

Conference of the parties (COP)	Place and year	Outcomes	Contents
	Rio de Janeiro, Brazil; 1992	United Nations Framework Convention on Climate Change	Common but differentiated responsibilities (UNFCCC 1992)
COP 3	Kyoto, Japan; 1997	Kyoto Protocol	Developing and transitional countries are excluded from mitigation responsibilities (UNFCCC 1997)
COP 13	Bali, Indonesia; 2007	Bali Action Plan	Decides to launch a comprehensive process for an agreement on the 2nd phase of the Kyoto Protocol after 2012 by Copenhagen COP in 2009 (UNFCCC 2007)
COP 15	Copenhagen, Denmark; 2009	Copenhagen Accord	The Copenhagen Green Climate Fund is announced. Fails to agree on the 2nd phase of the Kyoto Protocol (UNFCCC 2009)
COP 16	Cancun, Mexico; 2010		Decides to establish the Green Climate Fund as an operating entity of the financial mechanism of the Convention (UNFCCC 2010)
COP 17	Durban, South Africa; 2011	Durban Platform for Enhanced Action	Decides to launch a process to develop a protocol, another legal instrument or an agreed outcome with legal force under the Convention applicable to all Parties; Approves the governing instrument for the GCF (UNFCCC 2011a, b)
COP 18	Doha, Qatar; 2012		An independent GCF Secretariat is to be established in Songdo city, South Korea (UNFCCC 2012)
COP 21	Paris, France; 2015	Paris Agreement	Requests the Green Climate Fund to expedite support for the least developed countries and other developing country Parties for the formulation of national adaptation plans and for the subsequent implementation of policies, projects and programmes identified by them (UNFCCC 2015)
COP 24	Katowice, Poland; 2018	Decisions adopted at Katowice	Developed country parties shall biennially communicate projected financial resources; Both grants and loans allowed for counting (UNFCCC 2018)

implementation of a global climate landmark deal called the Paris Agreements (UNFCCC 2015). The timeline of the GCF in the context of the United Nations conferences is summarized in Table 1.1.

To put this in a perspective, the widely hailed Paris Agreements would have never been achieved, had it not been for the promise of international aids through the Green Climate Fund by the Annex B countries. A careful examination of the Intended Nationally Determined Contributions (INDCs) by developing countries submitted to the document of Paris Agreements reveals that their intended contributions were stated in dependence upon the actualization of the GCF aids to their countries (Seo 2017a).

Despite the pivotal and decisive role thrust upon the GCF, there have been few scientific attempts to clarify a host of issues related to the GCF conceptually, including how and on what grounds the GCF fund can be generated, how the funds should be allocated to which projects, what the GCF grants should achieve in the end for climate change policy purposes (Seo 2015b; 2017a). The gigantic knowledge and analytical gap is turning out to lead to endless disputes over the GCF funding, even endangering the existence of the Paris Agreements itself (Roberts and Weikmans 2015; Politico 2018; Carbon Brief 2018).

The first allocations of the GCF funding began in November 2015, timed for the kick-off of the Paris Conference in France (GCF 2018e). As of November 2018, 93+ international projects received a grant or a loan from the GCF Board. Further, the GCF Secretariat was established in December 2013 in Songdo City, South Korea and, as of April 2018, the Secretariat had 172 staff members arriving from 55+ countries and a large number of consultants.

This book provides, first and foremost, a scientific undertaking to fill the aforementioned knowledge and analytical gaps in this important global green fund established for climate policy actions, a successful completion of which will no doubt contribute immensely to climate change policy directions as well as to resolving international disputes over climate change and climate aids.

The portfolio of the GCF-funded projects up to this point gives the present author additional but essential data that can be analyzed for the purposes of this book. The GCF Secretariat makes available a project document for each of the funded projects which contains key information pertinent to this book such as the recipient's geographical origin, the project's category in the results areas defined by the GCF, monetary size of the funding, duration of the project, specific activities to be performed, and expected contributions from the project (GCF 2018e).

An almost insurmountable task in the climate change literature has been how the international climate community should design the GCF framework for investment decisions. In particular, it has been extremely difficult to develop an analytical framework for evaluating each project proposal against the other proposals as well as evaluating each project outcome against the other projects' outcomes.

This book takes on this important but difficult task, for which the present author draws on the lessons from the four scientific traditions of climate change developed during the past three decades by different groups of researchers and applies major results from these traditions for the purpose of evaluating each GCF project proposal and providing the guidelines for the GCF decisions.

The four traditions are a microbehavioral economic model of adaptation to climate change (Seo 2006, 2016a, b, c); an agro-economic model of climate change impacts on agriculture (Adams et al. 1990, 1999; Parry et al. 2004); a statistical estimation model of yield/productivity changes under climate change (Schlenker and Roberts 2009; Welch et al. 2010); an ecosystem model of climate change impacts (Dudal 1980; Fischer et al. 2005). The four traditions have been reported, to a degree or another, to the policy-makers and the general public through the successive IPCC reports (Reilly et al. 1996; IPCC 2000, 2001, 2007, 2014). Each of these scientific traditions will be given a brief introduction in the following sections.

The tall task of configuring the GCF institution cannot be successfully completed without looking into a large array of already-existing internationally transferred funds and programs. Some of these global funds have been in existence for many decades, e.g., United Nations specialized funds, while others are just emerging to be recognized, e.g., the Bitcoin system (Bohme et al. 2015). The need for the examinations of other global funds arises from two evident connections. The first is that the GCF and the other global funds share many overlapping areas of concern. Many existing global funds have already been providing funds to the projects related to, *inter alia*, food security, sustainable development, weather/climate disasters, natural disasters, and human health.

The second connection is that the Green Climate Fund is also one of the existing global funds, more specifically, the United Nations funds and programs. The United Nations Framework Convention on Climate Change (UNFCCC) is one of the UN's specialized funds and the GCF is the principal financial instrument of the UNFCCC (GCF 2011; UNSCEB 2018).

Other than the United Nations specialized funds, emerging global funds and currencies have the potential to offer novel insights on the economic aspects of the transfers of global funds and money. A new financing model called crowdfunding arose from the advance of personal computers and smartphones. The GCF can be seen as one kind of the crowdfunding techniques because of a purely voluntary nature of the GCF pledges (Freeman and Nutting 2015).

The CO2 permit that has been implemented throughout the European Union (EU) through the vehicle of the Emission Trading Scheme (ETS) is another novel global fund, the price of which can offer valuable information to this book's task of evaluating a large number of project proposals made to the GCF Board whose outcomes are framed in terms of the avoided CO2-equivalent emissions (EC 2016).

1.2 Paris Agreements and the GCF

The GCF was conceived and born during the international negotiation process which resulted eventually in the Paris Agreements in December 2015 (See Table 1.1). Therefore, the global climate fund cannot be understood independently of the Paris Agreements. Indeed, it can be said that the GCF is the heart of the accomplishments of the Paris Agreements that were highly praised by many.

The Paris Agreements (PA) was reached with the understanding that it would be a voluntary agreement by participating parties, meaning that there is no legal force in it. All the things agreed upon in the Agreements were submitted by the parties of the UN climate conference voluntarily as intended contributions. Further, the PA is a non-binding agreement. There are no legal liabilities enforced for violating the intended national contributions included in the Agreements. Further, national implementations of the national components included in the PA are monitored by a national governmental agency, that is to say, not by an international committee of inspectors (UNFCCC 2015; Seo 2017a).

Because of the voluntary and non-binding nature of the PA, the Green Climate Fund, which is an essential part of the Paris Agreements, is probably the only tangible concrete element of the Agreements. Specifically, a global temperature increase would be quite small in the next decade or so, making it difficult to strictly enforce a 1.5 degree Celsius temperature threshold across the members, a key component of the PA. Also, national greenhouse gas (GHG) mitigation efforts, another key component of the PA, would be intractable and unmeasurable by scientists and inspectors in the framework of the Paris Agreements.

By contrasts, the Green Climate Fund is a tangible and highly visual policy instrument. To give you an idea, countries' pledges should be collected from the rich countries to form the GCF fund, from which each successful project proposal made by one of the developing countries is awarded the GCF funding. The distributions of the global climate fund have actually begun already. The first project to receive the GCF fund in November 2015 was the project submitted by Peru entitled "Building the Resilience of Wetlands in the Province of Datem del Marañón, Peru." The project received 9.1 million US dollars (GCF 2018e).

To comprehend the relationship between the GCF and the Paris Agreement, we need to trace back to the Copenhagen Conference in 2009, then further back to the Kyoto Protocol in 1997, and then even further back to the United Nations Framework Convention on Climate Change in 1992 (UNFCCC 1992, 1997, 2009). The Kyoto Protocol was the first and only international treaty so far on climate change that committed the parties of the UN conference to the reduction of the emissions of carbon dioxide and other greenhouse gases to the level, on aggregate, 5% below the 1990 level. It was a legally binding agreement signed at the level of an international treaty, but excluded developing countries from any responsibility in cutting greenhouse gases. The obligations to mitigate greenhouse gas emissions were forced to only developed countries, with China, India, and other developing countries excluded from obligations.

The exclusion of the developing countries from the mitigation responsibilities formalized in the Kyoto Protocol can be interpreted from the first principle of the United Nations Framework Convention on Climate Change established through the Rio Earth Summit in 1992. The first principle, that is, Article 3 of the convention, declared that the parties of the UN convention should protect the climate system in accordance with their "common but differentiated responsibilities and respective capabilities" and committed the developed country parties to take the lead in combating climate change challenges (UNFCCC 1992).

The Kyoto Protocol entered into force in 2005, without the ratification by the United States. The first five-year implementation period began in 2008 and ended in

2012, during which the European Union member countries relied on the EU's Emission Trading Scheme (ETS) as the primary policy instrument. However, its effectiveness in controlling the global emissions was questioned because of a low participation rate as measured by the fraction of global total emissions that was subjected to the Kyoto treaty (Nordhaus 2008).

In December 2009, national policy-makers convened in Copenhagen with high expectation to negotiate the second phase of the Kyoto Protocol. The ultimate goal was to negotiate the Kyoto treaty II by including all United Nations' member countries into the Protocol for GHG mitigation responsibilities, making it a truly global treaty (UNFCCC 2009).

The Copenhagen Conference was widely broadcasted by the media around the globe because nearly a year of building-up of expectations had taken place since the election of Barak Obama as the 44th President of the United States in November 2008. But, the conference ended in disarray, failing to achieve the ultimate goal. The disparities of views between the developing country parties and the developed country parties were too large, leaving the Copenhagen COP into shambles (Nordhaus 2010).

To rescue the global conference, Hilary Clinton as new Secretary of State of the United States arrived at Copenhagen at the last moments of the conference. She announced that the United States would generate the fund of US$100 billion per annum to support poor countries for adapting to climate change threats, which was soon inscribed into the Copenhagen Accord as the Copenhagen Green Climate Fund (UNFCCC 2009).

The road to Paris which had been scheduled to host the UN Conference at the end of 2015 soon followed. The goal of the Paris conference was to complete the unaccomplished task of the Copenhagen COP. That is, the ultimate goal was to achieve a legally binding international climate treaty in which all parties of the UN climate convention are subjected to be responsible for cutting GHG emissions. This process was launched in the so-called Durban Platform for Enhanced Action (UNFCCC 2011a).

This process gave birth eventually to the Paris Agreements in December 2015 during the 21st conference of the parties of the UN convention. It was successful in one aspect of the Durban Platform but a failure in another aspect. It was a success in that it committed all parties of the convention to signing the agreement. It was a failure in that the agreement was not legally binding and only consists of voluntary pledges by the parties.

Through the Paris Agreements, the truly victorious was the Green Climate Fund. Through the process that led up to the Paris conference, it had become evident that without the international aids through the GCF funding, developing countries would not commit to the legal obligation to cut their countries' carbon dioxide and other greenhouse gas emissions. In the end, the Paris agreements ended up with voluntary mitigation pledges that were specified in dependence upon the transfer of funding through the GCF to their countries.

To give several examples, the United States submitted its Intended Nationally Determined Contribution (INDC) to be 26–28% reduction of GHG emissions by

2025 from the baseline level of 2005. However, not all countries submitted the intended contribution, e.g., China and India, in the form of a percentage reduction by a specified time period (Seo 2017b).

For another example, Vietnam proposes to cut GHG emissions by 8% by 2030 from the business as usual scenario without funding from the GCF and other funds, and by 25% with the funding from the above sources. Other countries, e.g., Thailand, also committed to GHG emission pledges conditional on the realization of the GCF funding (Seo 2017b).

1.3 Voluntary Pledges

In a truly international fund like the GCF, setting a fund-raising goal would be a complicated and difficult negotiation process. It would be similarly challenging to agree upon a fund-raising method because of the large number of participating countries. When first announced, the GCF immediately took on the fund-raising form of voluntary pledges and contributions by rich countries.

The size of the GCF, which is a fund-raising goal of the GCF scheme, was first mentioned by Hilary Clinton, US$100 billion per annum, or upon her arrival at Copenhagen. That figure quickly entered into the Copenhagen Accord and has remained in all subsequent UNFCCC conference documents. But, there were also calls from COP negotiators to increase the fund size tenfold to US$1 trillion per annum, which was about 1.5% of the global GDP in 2015 (UNFCCC 2009, 2011a, b, 2015). A serious conceptual gap with regard to these targets is that there is no strong rationale to justify either US$ 100 billion or 1 trillion target. To put it differently, what should be the anchor for the GCF fund size?

As of April 2018, the GCF reported that it had received pledges of about US $10 billion up to that point from COP member countries (GCF 2018d). A closer examination of the history of pledges made to the GCF reveals that the pledges were made mostly during the period leading up to the Paris Conference in December 2015.

Further, the pledges were made mostly by five richest countries: the US, Japan, Germany, the UK, and France. About 80% of the GCF pledges was made by the five countries (Refer to Fig. 2.1, Chap. 2). Although developing country parties can contribute to the GCF, there has been little, if not none, contribution made up to this point. The prospect of a further increase in the GCF contributions was dimmed when President Trump announced the U.S. withdrawal from the Paris Agreements in June 1, 2017 as well as the U.S. contributions to the GCF (White House 2017). This was followed by the replacement of the Clean Power Plan (CPP), a cornerstone legislation designed for a successful implementation of the U.S. commitments to the Paris Agreements, with the Affordable Clean Energy (ACE) rule by the Trump Administration (USEPA 2014, 2018; Seo 2019).

One of the key questions for the success of the Green Climate Fund is therefore whether the current level of financial pledges and contributions at US$ 10 billion can be increased tenfold to US$ 100 billion, the initial promised size by climate negotiators. If not US$ 100 billion, is there any more rational or justifiable size?

Another key question is how the UN COPs can encourage other countries than the aforementioned five to give money to the GCF, including the developing countries. As of April 2018, almost all pledges to the GCF, 95% more specifically, came from about 15 developed countries. It is not the case that developing country parties, i.e., non Annex B parties, did not contribute at all. But, their contributions remained nominal (Refer to Table 2.1, Chap. 2). Should the poor countries be asked to contribute to the GCF? What should be the criterion for assigning the burden of financial contributions to each of the developed country parties?

1.4 Making Decisions on the GCF Allocations

An even harder question facing the GCF Board and the United Nations' Conference is to come up with a decision rule on allocating the GCF funds to a proposal from a pool of proposals made to the Board. The Board should determine which proposal receives the award, once determined, how much for what period of time.

Nonetheless, an initial allocation of the GCF to a family of proposed projects has already begun in November 2015, timed for the period of the Paris Conference. The first nine projects were awarded the GCF funding, whose geographical distribution showed a well-balanced spread across Africa, Asia, Latin America and the Caribbean, and South Pacific small island nations.

In the African continent, the GCF award went to the project submitted by Malawi, Senegal, Rwanda, Kenya, respectively. In Latin America and the Caribbean, the award went to the project from Peru and another from Mexico. In Asia, Bangladesh received the GCF initial funding. In Pacific islands, Maldives and Fiji each received the initial GCF funds.

In the nine recipient projects, the size of the GCF award varied. The smallest grant was US$ 8.2 million given to the project submitted by a team from Senegal entitled "Increasing Resilience of Ecosystems and Communities through Restoration of the Productive Bases of Salinized Lands." The project is classified by the GCF as an adaptation project. The largest grant was given to the project from a team from Fiji entitled "Urban Water Supply and Wastewater Management Project" which received US$ 405.1 million. The project is also classified by the GCF as an adaptation project (See Tables 2.4, 2.5, 2.6, 2.7, 2.8, 2.9 and 2.10 in Chap. 2).

As of May 2018, 82 projects were awarded the GCF grant, from which five projects were cancelled or withdrawn. The geographical distribution of the GCF funding by the size of GCF funding is as follows: Latin America and the Caribbean (30%), Africa (25%), Asia (20%), multi-continent (17%), Pacific Islands (7%), Eastern Europe (1%) (Refer to Fig. 2.2, Chap. 2). Therefore, the GCF grant recipients were exclusively the non-Annex B country parties of the UNFCCC, that is, developing and transitional economies.

Notable is a large fraction of the GCF funding allocated to the projects submitted by a team of multiple countries from multiple continents. Although there were only two multi-continent projects that received funding, they received the largest and the second largest grant. The project entitled "GCF-EBRD Sustainable Energy Financing Facilities" received the grant of US$ 1.4 billion, which was the largest award given by the GCF. It has participating countries in Africa, Asia, and Eastern Europe: Armenia, Egypt, Georgia, Jordan, Republic of Moldova, Mongolia, Morocco, Serbia, Tajikistan, and Tunisia.

The framework for the funding decisions by the GCF is laid out by a set of criteria named the "investment criteria" by the GCF policy board (GCF 2018b, c). The investment framework, which will be analyzed in detail in Chaps. 2 and 7, espouses multiple criteria including expected contribution, paradigm shift potential, sustainable development goals, national needs, and costs. The project document for an awarded project, however, specifies only the expected contribution by the project as the primary criterion for the funding decision. The expected contribution is expressed in terms of either the amount of CO2-equivalent emissions that can be avoided by the project or the number of beneficiaries from the implementation of the project.

To give several examples, the Peruvian project, one of the initial batch of recipients, entitled "Building the Resilience of Wetlands in the Province of Datem del Marañón, Peru" was specified to have the expected contribution of 2.6 million tons of CO2-equivalent emissions avoided. The Rwanda's energy project entitled "KawiSafi Ventures Fund in East Africa" was specified to have the expected contribution of 1.5 million tons of CO2-equivalent emissions reduction. The Mexico's project entitled "Energy Efficiency Green Bonds in Latin America and the Caribbean" was specified to have the expected contribution of 2.6 million tons of CO2-equivalent emissions reduction.

The investment criteria put forth by the GCF do not declare explicitly the ultimate goal that the full portfolio of the fund-allocated projects purport to achieve, for comparisons, in the way that the goal of the Paris Agreements is expressed (IPCC 2018). Further, it does not describe how each of the funded projects would contribute to that ultimate goal and how much.

From another angle, an investment decision by the GCF board is made by considering a basket of criteria mentioned above such as paradigm shift potential, national needs, cost efficiency, avoided CO2-equivalent emission, number of beneficiaries. The GCF governing instrument, however, does not specify the way how each of these criteria are weighed against each other. There are no weights assigned on these criteria.

The procedures and rationales involved in the long list of GCF grant decisions would offer rich resources, to say the least, to economists and policy analysts who endeavor to comprehend and design an international protocol or a set of principles for many of the globally transferred funds based on an international agreement or a novel technological possibility. This book certainly taps into this treasure trove.

1.5 Economics of the Global Green Climate Fund

Economic aspects of the Green Climate Fund comprise many intriguing elements and concepts which even go far beyond the economics of climate change and global warming, although these may not seem obvious at first to the observers (Nordhaus 1994, 2013, Wagner and Weitzman 2015). This book is well positioned to expose what have been treated as implicit processes or entirely amiss in the above-cited important works of climate change economics.

The economics of the GCF will be explained at length throughout this book and most directly in Chap. 7. We can begin with the most basic economics question: Why should the government or an international body such as the GCF Board provide the money for the supply of certain resources and actions in the society?

An elegant answer to this question was provided by Paul Samuelson through "The pure theory of public expenditure" in 1954 (Samuelson 1954, 1955). Samuelson argued that a certain good or service in the society is a public good, that is, a jointly consumed good by the public. For the public good, he showed that without a government provision, it will not be efficiently supplied in the market.

The global atmosphere is a public good in that it is shared by the people and beings on Earth, more specifically, the CO2 concentration in the atmosphere. An alteration of the global atmosphere, e.g., through CO2 emissions and climate change, is a public bad if the alteration would be harmful to the global community. If left alone, an individual or an individual country cannot provide the public good of the global atmosphere or the climate stability. A 'government' should provide it.

The climate change impact literature finds that emitting carbon dioxide and other greenhouse gases excessively will be harmful to the global community at large by altering the climate system, therefore, is a global public bad (Pearce et al. 1996; Mendelsohn and Neumann 1999; Tol 2009). A removal and reduction of excessive carbon emissions is therefore a global public good, which should be provided by the global public.

Given the absence of any global government, it can be justified in accordance with the Samuelson's pure theory that the Green Climate Fund Board acts as a global government on behalf of the global public, therefore is given the authority by the global public to allocate the GCF funds to the projects that propose to cut carbon dioxide and other GHGs or to mitigate harmful effects of climate change on potential victims.

Of the 93+ projects that have been awarded the GCF funding until today, about half of them is classified as a mitigation project and the other half as an adaptation project. The awarded mitigation projects propose a variety of activities that aim to build low-carbon or zero-carbon energy generation plants or production processes, for example, solar panels, hydroelectric power plants, wind turbines, to name but a few.

With this justification in mind, the GCF policy intervention adopts an alternative approach to the traditional economic policy approach for the provision of a public good in the current context providing reductions of carbon dioxide and other greenhouse gases, that is, carbon price-based policy instruments such as carbon tax or CO2 emissions permits (Baumol and Oates 1988; Nordhaus 2007; Ellerman et al. 2016). In the traditional economic approach, carbon penalty is charged for a unit of emissions of carbon dioxide equivalent emissions. The tax or penalty would discourage people and firms from emitting greenhouse gases without considering the cost of such behaviors, thereby would achieve an optimal mitigation at the societal level.

In the price-based policy instruments, the central economics question has been what the optimal level of the carbon dioxide price, put differently, the social cost of carbon dioxide, should be. Through the dynamic social welfare optimization framework that economists have developed, an optimal—efficient—carbon tax level is determined at each time period, which would get ramped up over the course of time (Nordhaus 1994). To give you an idea, a recent study by the foremost proponent who won the Nobel Prize in economics in 2018 proposed that a carbon tax (or penalty) should be charged at $25 per ton of carbon dioxide emissions today, which should increase to $50 per ton by the middle of the century, and $150 per ton of carbon dioxide emissions by the century's end (Nordhaus 2013).

Compared with the price-based policy, does the GCF offer as strong a policy intervention to addressing the accumulating emissions of carbon dioxide and other GHGs? Does the GCF present as clear a policy goal? What does the GCF allocations say about the price of carbon dioxide or the social cost of carbon dioxide? Does the GCF allocation induce the reduction of earth-heating pollutants from the sources that can achieve it with less costly options? A host of questions will be raised and addressed, including these mentioned here, in Chap. 7 with regard to the theory of public goods.

The second fascinating aspect of the economics of the GCF that the present author will elaborate in this book is concerned on the efficiency consequences in natural resource uses of a GCF grant to a particular project which promises to perform a certain action over all other possible actions. From the perspective of an individual resource manager, the GCF grant causes reallocations of her current resource uses. A critical economic question should be asked on this public intervention: Are such resource reallocations induced by the GCF grant beneficial to the resource manager and further to the society (Ricardo 1817; Hartwick and Olewiler 1997)?

The question itself may seem ordinary to many observers. However, when the question is posed in connection with future changes in the climate system, it becomes a fascinating subject to examine. This is because any such public intervention would be certain to cause welfare loss to the grant-receiving resource manager if the public agency that is responsible for the grant distribution could not predict the future climate conditions precisely (Seo 2006, 2016c). A public grant given today taking the current climate system as given would turn out to be damaging, that is, not helping, under the future climate system.

To clarify this point, let's consider a few examples. Let's take a hypothetical GCF funding to a crop-growing region in Sub-Sahara where the average crop yield is below those achieved in more productive crop-growing regions in the continent.

Let's say that the funding is allocated to a maize (corn) farmer. The GCF Board may decide to allocate the international fund to the maize farmer on the ground that the farmer is under-performing because of a harsher climate condition at the present time. The grant would allow the farmer to purchase needed inputs to improve the yield of the crop.

Let's further assume that the change in the climate system should unfold in a way that the crop-growing region becomes more arid, causing croplands to give way to grasslands. In this scenario, the GCF grant would encourage the maize farmer to stick to the crop farming even though the change in the climate system make the maize farming less productive year by year (Seo 2010a, 2014a, 2016a).

The question of efficiency is whether the international funding would promote efficient land and resource uses by the local recipients under changed climate conditions or not. Put differently, will the GCF grant will be able to change the recipient's behaviors in a way that the most profitable uses of the resources and lands are adopted by the grant recipient (Ricardo 1817; von Thunen 1826)?

The concept of efficiency in resource uses, which can be traced back nearly two centuries to Ricardo, von Thunen, and Pareto, allows us a conceptual tool for evaluating the GCF-funded projects (Pareto 1906). Applying the efficiency criterion, some of the GCF-funded projects may be evaluated definitively to be inefficient, taking into considerations the evidence and knowledge that the scientific literature has offered up to this point. Similarly, other projects may be evaluated to be an efficient public intervention, taking into accounts empirical evidence provided by the climate change literature.

Pursuing the efficiency evaluation to the limit, we would be able to conclude whether a GCF grant would induce perverse incentives of grant recipients and whether it would cause a mal-adaptation of them (Seo 2014b, 2015b). A mal-adaptation, which is one of the most important concepts in climate change economics, is a behavioral change in resource uses that increases, that is, not decreases, the vulnerability and eventual damage of an individual grant recipient under the changed climate system.

Quite easily, a mal-adaptation can be promoted, albeit inadvertently, by the GCF funding if the funding body does not fully understand the incentives of the recipients. Also, a mal-adaptation can be promoted unintentionally if the GCF Board does not have the full knowledge of how the future climate would unfold. A mal-adaptation is a phenomenon that is extensively studied by the corpus of economics including an insurance, a contract theory, a principal-agent problem, and mechanism design (Hart and Holmstrom 1987).

An illustrative historical example of a mal-adaptation, often cited in the literature, is the US population shift to hurricane-prone coastal counties. The US citizens have moved massively to the coastal counties in the country after World War II, which may have increased the economic damages and human fatalities from the hurricanes over the time period (Bakkensen and Mendelsohn 2016; Brannon and Blask 2017). If the US federal government or the UN climate convention were to give grants/ subsidy to U.S. coastal residents who fell victim to hurricane strikes, it would only force a mal-adaptation of the residents to the future ocean warming and more frequent/intense hurricanes in the future by giving them an incentive to stay in the hurricane-prone coastal zones (Seo 2015c, 2017c; Seo and Bakkensen 2017).

The third aspect of the economics of the Green Climate Fund that this book explicates is the concept of a public adaptation. A public adaptation can be defined with reference to a private adaption in a similar way that a public good is defined as a private good. A private adaptation is an efficient adaptation behavior at the private level. A public adaptation is an efficient adaptation behavior to the future climate change at the public level. Therefore, a public adaptation behavior is not efficient if it is taken at the private level (Mendelsohn 2000; Hanemann 2000; Seo 2011).

A public adaptation can refer to a community action at different scales of the community that purports to adapt efficiently to a changed climate system. That is, it can be a community action at the scale of a village, of a geographically-defined group, of a municipality, of a province, or of a country.

Is there an 'efficient' public adaptation? What are some of the concrete examples of a public adaptation to climate change? Let's consider a coastal city or a coastal farming zone faced with rising sea levels attributable to global warming and globally warming oceans (Nicholls et al. 2007). An ever-higher sea level would also mean an increasing probability of a high storm surge of the sea and subsequent inundations in the community caused by more intense tropical cyclones.

For an individual house owner worried about such risks, it would be inane to build a wall that surrounds her house. Such a wall cannot stop inundations from either a sea level rise or a sea surge-caused inundation. An effective wall is one that should be built across the entire coastal line of concern. In the economist jargon, the effective wall involves a weakest-link production technology. Nonetheless, the individual does not have an incentive to build the entire coastal wall for the community. An efficient provision of the coastal wall in this case can be accomplished through the 'public sector', more precisely, the community of coastal residents, which would be given the authority to collect the fund from the members and build the wall.

In the contexts of the Green Climate Fund, the concept of a public adaptation is pertinent and also a practical one that can be utilized by the policy committee, which will become clearer through this book. However, there are a few obstacles that must be overcome in applying the concept to the GCF's funding decisions. For a starter, in an efficient public adaptation such as the coastal wall example in the above, the fund should be collected from the members that are beneficiaries from the public project. In the current GCF funding framework, this is not the case (GCF 2018b, e).

Another complication is to determine whether a public adaptation is provided at the efficient amount to the public. To be more specific, it may turn out to be the case of an over-supply or an under-supply of the adaptation measure provided by the project (Seo 2011). Since the fund for the public adaptation project is generated from elsewhere in the GCF-funded projects, the project taker may have an incentive to supply as much as possible, that is, not the efficient amount. However, this is only one of the host of incentives for a public adaption project to be inefficient, which is discussed at length in Chap. 7.

1.6 Evaluating the GCF's Allocations

Having followed the historical development of the GCF, we realize that the identity of the Green Climate Fund has never been unequivocally defined. People seem to be interpreting it in different ways. A number of competing arguments for the identity of the GCF could be made, including the economic arguments described in the preceding section. Its ultimate goal has not been defined. It is understood by many involved in its range of activities to be an aid fund to developing countries.

Further, there has been no set of rules agreed upon by the parties of the UN conference with regard to the allocations of the global fund (Seo 2015b). In fact, neither is there an agreed-upon principle with regard to how much money each country party should contribute to the GCF.

The wide-ranging allocation projects of the GCF, with a large number of the fund-receiving projects kicked off already, would without doubt call for a range of post-evaluations about the outcomes of each of the funded projects as well as the allocation decisions by the GCF Board. The rationale and necessity for the evaluations of the GCF projects will most certainly intensify in the coming years as political environments get more acquainted with the realities of the GCF redistributions.

As evidenced by the lack of the literature on this topic, the research challenges of evaluating policy interventions by the GCF will turn out to be high, to say the least. In the efforts to answer to the challenges, political interests are as diverse, deep, and heterogeneous as can be. Further, it will turn out that research interests are as divergent, protective, and distinct as can be across the academic disciples.

With the goal of laying out the conceptual framework for approaching the difficult task, this book will present a family of analytical and conceptual tools that can be employed for evaluating the selected projects by the GCF. The analytical tools are developed along the lines of the three economic aspects of the GCF outlined in the previous section: the economics of public goods and expenditure, the efficiency in resource reallocations, and the economics of public adaptation.

How can these basic economic concepts be applied to the research work of GCF project evaluations? Let's first consider the economics of a global public good and expenditure. The present author will begin by asking how much the GCF funded projects can contribute to the mitigation of greenhouse gas emissions over the twenty-first century which is the very global public good that the global policy-making body has intervened to provide. A closely related inquiry is whether the entire bundle of the GCF-awarded projects will be sufficient to avert global warming or stop a run-away climate catastrophe (Wagner and Weitzman 2015)?

Another more difficult question to address, in relation to the theory of public goods, is whether an individual mitigation project that was awarded would achieve the specified goal (in the project proposal document), whatever it is, at a cost-effective manner. The point of assessment here is that it would be deemed undesirable if the GCF funding leads to an overly costly way of achieving a certain mitigation goal, even if that mitigation goal would be met.

This evaluation approach is in the same line with the existing public sector project evaluations of the environmental programs and regulations, the results of which are expressed in terms of a benefit-cost ratio, that is, the total benefit over the total cost of a program (Hahn and Dudley 2007; USEPA 1999).

The second analytical tool can be explained in terms of the Ricardo-Pareto efficiency in proposed changes in an individual's behaviors and resource uses. It may be said that the efficiency concern has been the cornerstone of the economics of climate change adaptations for the past two decades and more (Seo 2016a, b). Put differently, climate economists are deeply involved in the ways that people and society could adjust its resources' uses efficiently in response to and anticipation of future climates (Mendelsohn 2000).

A policy-relevant consequence of efficient adaptations to climate change is the reduction in the vulnerability of a concerned economic agent as well as eventual economic damages. With an efficient manner of changes in the present behaviors or systems, the agent can take advantage of the changed conditions, thereby avoid a large loss of value and income that is expected upon the existing system without such behavioral changes.

The same analysis can be extended to the GCF funding programs. When it comes to the GCF grants and loans, will the GCF aids promote or discourage efficient changes of behaviors by the economic agent? Let's take a corn subsidy, for example, to a Sahelian country. A corn farmer insurance supported by the government subsidy, similar to the catastrophic coverage crop insurance in the United States, would encourage the today's corn farmers to remain as a corn farm, even if the crop is expected to suffer sizably owing to future climate changes, because any weather-related loss in corn production will be covered by the government-supported crop insurance (Sumner and Zulauf 2012). A GCF grant to a corn producer in a Sahelian country is likely to end up with similarly unintended consequences.

An analysis of efficiency and distributional impacts of a public intervention program, a tax or a subsidy, is deeply rooted in the economics of policy analyses (Friedman 1962). Readers may refer to Mas-Colell et al. (1995) for the general economics literature on this topic and Hartwick and Olewiler (1997) for the environmental and natural resource economics literature on policy interventions.

An analysis of a crop insurance subsidy is only one of many heavily-researched policy areas in this literature (Halcrow 1949). A government subsidized flood insurance in the United States is another example, which is called the already-mentioned National Flood Insurance Program (NFIP) (Knowles and Kunreuther 2014; Brannon and Blask 2017). Such an analysis is abundantly found with regard to environmental regulations such as, to name but a few, sulfar dioxide (SO2) allowance trading, CAFE (Corporate Average Fuel Economy) standards for automobiles (Burtraw and Szambelan 2009; McConnell 2013).

On the other hand, the literature of economics of adaptation to climate change offers a rich basket of empirical data and useful analyses on this regard. The literature abundantly documented potential adaptation strategies and measures which can be employed at specific geographical locations in specific climate change scenarios (Seo 2006, 2016c, 2017a). These climate adaptation studies can, coupled with the aforementioned policy analysis traditions, offer a powerful means for evaluating an individual GCF project in terms of efficiency consequences.

Some of the most-researched adaptation strategies include selection of climate-resilient animal species for the livestock management (Seo and Mendelsohn 2008), adoption of a diversified crop-livestock farming (Seo 2010a, b), an integrated system of crops-livestock-forests (Seo 2010c, 2016b), an irrigated crop production (Kurukulasuriya et al. 2011; Seo 2011), high yield crop varieties (Evenson and Gollin 2003; World Bank 2009), changes of genomes in insects and pests (Aksoy et al. 2014), indoor air-conditioning and heating (Barreca et al. 2016), coastal walls against sea level rise (Yohe and Schlesinger 1998; Ng and Mendelsohn 2006), tropical cyclone trajectory projection techniques (Seo 2015b), cyclone shelters (Seo 2017c), typhoon resilient structures and preparedness (Bakkensen and Mendelsohn 2016, Seo 2018), solar energy generations and adoptions (MIT 2015; Graziano and Gillingham 2015), hybrid and electric vehicles (NRC 2013), Light Emitting Diode (LED) lighting revolutions (Akasaki et al. 2014), carbon capture-storage-reuse (Lackner et al. 2012), nuclear fusion energy (ITER 2015), hydroelectricity (IPCC 2011), solar reflector outside Earth (NRC 2015).

The third avenue for an empirical evaluation of the GCF-funded projects lies in the economics of public adaptations. At the core of the theory is the argument that a public adaptation measure cannot be provided efficiently by a private agent. A slight extension of the argument would imply that a particular public adaptation project can in no way be provided efficiently by 'any' public entity. The funding recipient who will carry out a particular public adaptation project should be a public agency or community at an appropriate level that is most suited to execute the project on behalf of the individual members of the community. More concretely, a GCF grant given to a national level agency/group for the purpose of carrying out a village level adaptation project would be inefficient.

Similarly, a GCF grant for a particular private adaptation project should be given, if ever, only to the individual who will implement the private project. According to this theory, a private adaptation strategy cannot be efficiently carried out if a public agency has to implement it on behalf of the individual. The central point here is that a public adaptation project should be allocated a GCF grant through the most pertinent public agency, that is, not any public agency.

In the third avenue of evaluation, we will pay close attention to the nature of each GCF-granted project and determine if the grant recipient is in match with the project itself and can be an ideal project undertaker. A clear mismatch, supported by the empirical climate change literatures, will indicate that there will be welfare loss resulting from the particular decision by the GCF policy board.

Relying on the theory of public adaptations, another test of the GCF-funded projects is possible by way of a deliberate examination of the size of a grant given to a particular public adaptation project. The test asks whether the GCF grant is an optimal amount of funding that can ensure an optimal size of the public adaptation measure in concern. At the end, we may be able to judge that the grant is either too small or too large, given the nature of the public project (Seo 2011). For example, a grant to a public irrigation project in a South American country may turn out to be too large so that the irrigation scheme is stretched even to the places where irrigation provides little added value.

1.7 Five Scientific Traditions of Global Warming

Having outlined the three avenues of evaluation of the GCF projects, we are well-positioned to introduce the five scientific traditions of climate change that will provide empirical supports and evidence for each project evaluation. This book indeed offers an innovative, perhaps unprecedented, way to reinterpret as well as utilize the empirical scientific literatures of climate change in the context of the GCF funding decisions.

The five traditions are as follows, as summarized in Table 1.2: a microbehavioral economic model of behavioral adaptations, an agro-economic model of climate change impacts on agriculture, a batch of statistical models of yield/productivity changes under global warming, a family of ecosystem models of climate change impacts, and a dynamic integrated climate and economy model.

The five traditions are not chosen at random. The first four traditions in Table 1.2 cover the entire range of climate change impact studies as well as behavioral adaptation studies on agricultural and natural resources (Seo 2015a). Further, these modeling traditions were developed and can be easily put into the contexts of developing countries, which makes them particularly relevant to an analysis of the GCF funding projects which were awarded to developing countries only. The fifth modeling tradition is highly regarded across the climate research community as the inevitable policy analysis framework, especially, of mitigation proposals, and is the first Nobel Prize awarded to a climate change researcher (Nobel Committee 2018).

The compilation of the literatures in Table 1.2 does not include the climate change impact and adaptation literatures on ocean-based ecosystems such as the effects of sea level rises on coastal communities, the effects of changes in hurricanes, and the

Table 1.2 Five scientific traditions of climate change

Scientific traditions	Academic disciplines	Representative studies
A microbehavioral economic model of behavioral adaptations	Microeconomics/ Economics	Seo (2016c)
An Agro-Economic Model of climate change impacts on agriculture	Agronomy	Adams et al. (1999)
Statistical models of yield changes	Statistics/ Econometrics	Schlenker and Roberts (2009)
Ecosystem models of climate change impacts	Ecology	Fischer et al. (2005)
A Dynamic Integrated Climate and Economy Model	Macroeconomics/ Economics	Nordhaus (1994)

impacts on fisheries (Bakkensen and Mendelsohn 2016; Seo 2015c, 2017c). Although this book will not devote an individual chapter to describe the ocean-based impacts of climate change, it will refer to the relevant studies that turn out to be important in evaluating some of the GCF-funded projects.

That there have been few economic analyses of the Green Climate Fund since the adoption of the Paris Agreements in December 2015 was already mentioned. The present author provided an early commentary just before the convening of the Paris Conference in December 2015 and subsequently more broadly on the Paris Agreements itself (Seo 2015b, 2017a, b). By contrasts, there have been a flood of analyses and media commentaries on the insufficiency of the national emission reduction commitments agreed in Paris (UNEP 2017; Klein et al. 2017; IPCC 2018; Washington Post 2018).

This book provides the first analytical evaluation of the decisions of the GCF and funded projects, but it does not end there. Through the evaluations of the GCF's decisions, the book will offer novel insights on the Paris Agreements' national voluntary commitments on emissions reductions and the PA's future, which will be done in a distinct way from the previous analyses.

Now, let me briefly introduce each of the five scientific traditions to the readers. Each of these traditions will be elaborated extensively in Chaps. 3, 4, 5, 6, and 7 of this book, with each chapter devoted exclusively to one of the five traditions.

First, we consider a microbehavioral model of adaptation behaviors to climate change. This literature refers to a family of economic models that is developed to quantify adaptation behaviors of an individual economic agent at the micro level in response to local changes in the climate system (Seo 2006, 2010a, 2016a, b; Seo and Mendelsohn 2008). This economic modeling is described in depth in the book entitled "Microbehavioral Econometric Methods" published by the present author in 2016 (Seo 2016c).

In the microbehavioral modeling, a researcher examines the portfolio of natural resource assets owned by a rural farm who has adapted to the present climate system historically. By associating an array of distinct portfolios found across different farm households with a range of different local climate regimes, s/he can estimate changes in the farm's portfolios that can be ascribed to changes in climate conditions. The associated differences are then fed into the microbehavioral model to predict the farm's future portfolio of resource assets after the farm's efficient adaptations to the future climate system are fully taken into account. The microbehavioral model relies on the future climate system forecast by one of the family of climate models made available by climate scientists.

In the same way, changes in the value of the farm's productions that can be attributed to changes in climate regimes can be predicted by the microbehavioral economic model. Let's take for example a natural resource enterprise, say, a crops-only enterprise in South America. For the specific enterprise, the researcher associates the values of farm productions found across the range of farm households with the range of local climate regimes. The associated differences are then fed into the microbehavioral model to predict the value of the farm's productions in the future climate system which is predicted by one of the climate change models. This procedure is repeated for each farm in the research sample that chose the aforementioned natural resource enterprise, that is, the crops-only enterprise.

There are more than one natural resource enterprises in the rural areas of any continent or any country. In South America, the following seven natural resource enterprises are prevalent: a crops-only, a livestock-only, a forests-only, a crops-livestock, a crops-forests, a livestock-forests, a crops-livestock-forests enterprise (Seo 2012, 2016b). The first three enterprises are a specialized system while the latter four are a diversified system. In Africa, dominant natural resource enterprises are a crops-only, a crops-livestock, a crops-forests, a crops-livestock-forests (Seo 2010c).

As will be clarified in Chap. 3, quantitative results and predictions from the microbehavioral models offer essential information, based upon which the evaluations of the GCF decisions and funding projects can be conducted, because the GCF grants are directed to a specific natural resource system located in a specific climate regime, for example, a maize farming system in the semi-arid climate zones of North Africa (Seo 2014b, 2015b; GCF 2018e).

The second scientific literature that this book refers to is the agro-economic modeling (AEM) literature (Adams et al. 1990, 1999). The agro-economic modeling is the first climate change impact economic assessment model originated from the agricultural sector of all the subsequent impact literatures (Pearce et al. 1996). The agro-economic model relies on the agronomic experiment on the crop of interest in a controlled setting. The experiment measures the change in the yield of the selected crop for investigation in two carbon dioxide-temperature conditions: one with the present level of carbon dioxide concentration and the other with a doubling of the present carbon dioxide concentration. The experiment can be either an agronomic crop yield simulator or a Free Air CO2 Enrichment (FACE) experiment (Tubiello and Ewert 2002; Ainsworth and Long 2005).

For the experiment to be without bias, it must be conducted after either controlling or randomizing all other factors that affect the yield of the crop, which includes, *inter alia*, soil types and quality, solar radiation, wind direction, slope of the plots, water availability, input uses, skills of farmers (Keuhl 1999). The degree to which these covariates are controlled or randomized varies across the experimental models, so is the ease—or cost—with which these covariates can be controlled.

The experiment is therefore always site-specific, but the experiment can be conducted on multiple sites. The site-specific results from the experiments are then extrapolated to the national economy such as the US agriculture. This extrapolation is another most tricky part of the agro-economic modeling which cannot be done without assuming, among other things, a certain future scenario of land uses across the country which is most often assumed to be exactly same as the present-day land uses.

The final output variables of the agro-economic modeling are the set of predictions of changes in national-level yields of the most important grains of the country of study. In the United States, they are the four dominant crops: wheat, maize, soybeans, and cotton. These predictions are coupled with price change predictions to forecast changes in producer surplus and consumer surplus. Agro-economic modelers often venture further to calculating other policy variables such as the number of people in poverty, hunger, or mal-nutrition (Rosenzweig and Parry 1994; Parry et al. 2004).

Readers will soon realize, upon examining the fund-allocated GCF projects, to be explained in Chap. 2, that many of the project proposals refer to the agro-economic modelling studies or the two types of crop yield experiments. The AEM studies produce quantitative predictions of climate change-induced changes in crops' yields, producer surplus, hunger, mal-nutrition, all of which are key climate policy variables, especially concerning agriculture, that can be fed into the GCF proposals.

The third scientific tradition this book elucidates is a statistical modeling of yield changes (SMY) of a selected crop or changes of another measure of productivity. Unlike the AEM literature which relies on experimental data, this tradition makes use of the actual yield data of the crop of importance observed annually and seasonally from the farmers. In order to replicate the controls and randomizations of the crop experiment, the SMY researchers employ a range of intensive data analysis tools such as non-parametric statistics, panel data analysis, simulation methods, spatial statistics, and Bayesian statistics (Schlenker and Roberts 2009; Welch et al. 2010).

Another unique feature is that the SMY tradition takes for analysis national statistics on agriculture. For the U.S. studies, national statistics compiled from the series of the U.S. Agricultural Census conducted and compiled every 5 year, available continuously since the 1970s, are taken for analysis by the researchers (Deschenes and Greenstone 2007; Schlenker and Roberts 2009). The statistics are compiled at the U.S. county level, the 2nd administrative division of the United States government. There are more than 2300 counties in the US. The semi-decadal agricultural census is designed and conducted by the National Agricultural Statistics Service (NASS) of the US Department of Agriculture. The NASS maintains the yield data of crops (NASS 2018).

The SMY researchers are primarily concerned about the impacts of climate change on the most important crops in the nation. For example, the U.S. studies concentrated on the four major crops: maize, wheat, soybeans, and cotton. For the study of South Asian agriculture, researchers focused on the most prominent grain of the region, that is, rice (Auffhammer et al. 2006; Welch et al. 2010).

The fourth scientific literature this book elaborates is an ecosystem modeling of climate change impacts (EM). The EM literature's focal point is ecosystems, or alternatively ecological zones, that are present in the globe today but are expected to undergo significant changes owing to future climatic changes. An ecosystem or an ecological zone is an encompassing concept *vis a vis* a crop variety or an animal species since it is defined by a multitude of crops and animals that share ecological similarities.

A prime example of the early ecosystem modeling is the classification of agro-ecological zones (AEZ), which was the first ecosystem change modeling in which climate is the primary factor, of the African continent (FAO 1978; Dudal 1980). The AEZ was the classification of ecological zones developed for the purpose of understanding agricultural productivity differences across the geographical regions of Africa.

For a particular region, there can be more than one AEZ classifications. That is, a review of the AEZ classification systems reveals that there were multiple ways to classify the AEZs depending upon the emphasis of the modeler. It also shows that an

AEZ classification of one region, e.g., African continent, is different from that of another region, e.g., South Asia.

In Sub-Saharan Africa, for example, an AEZ classification may split the continent into sixteen AEZs. Based on climate, soil characteristics, water availability, the continent can be divided into six AEZ zones: a desert zone, an arid zone, a semi-arid zone, an arid savannah zone, a moist savannah zone, and a humid forest zone. These six AEZs may be further divided into a lowland, a mid-latitude, and a highland zone (Seo et al. 2009).

In the literature of climate science, a prominent research field has long been the modeling of changes in the ecological zones that would be caused by future climate changes, which would also mean changes in the distribution of the ecological zones across the geographical area of concern, say, Sub-Saharan Africa (Schlesinger 1991). Given a rather simple method to classify the current ecological zones, as in the AEZ classification, an EM researcher can readily predict the changes in the existing AEZ classification zones after assuming a particular climate change scenario.

It is also possible, also occasionally done, to map the changes in ecological zones using the observational data from Earth-orbiting satellites. EM researchers can examine the many decades-long observational satellite data on the African continent to report, among other things, that the Sahara Desert or the Namib desert has expanded over the satellite-monitoring period or shrunk (Matthews 1983). This change may be attributed to the changes in the climate system that has occurred during the satellite era.

A more difficult scientific undertaking by the EM tradition is to answer whether a particular change in the ecological zones is good or bad for, say, the agricultural sector or global food supply. EM researchers took this to task to developing an analytical method by linking measures of agricultural productivities to classifications of ecological zones. Based on the linkages thus made, they went on to predict changes in agricultural productivities by inferring from the changes in ecological zones which are in turn induced by climate changes (Fischer et al. 2005).

The EM impact modeling exposes many onerous issues that are hard to overcome by the EM modelers (Seo 2014a). For one thing, it is full of pitfalls to associate a certain crop's productivity with an ecological zone, one of which is that, across the single ecological zone, the productivity of the crop is varied greatly from one farm to another. For another, the crop of choice in one ecological zone is not the same as the crop of choice in another ecological zone, which makes the EM impact analysis to err by comparing apples to oranges.

Finally, the fifth scientific literature this book refers to is a global dynamic welfare optimization model developed for a climate change policy analysis. Pioneered and elucidated elegantly by William Nordhaus, this literature defines CO2-equivalent emissions as negative externality to the global economy and searches for the right price of the externality through a global welfare optimization framework over many centuries (Nordhaus 1991, 1992). It took decades to develop a quantitative model which was named the Dynamic Integrated Climate and Economy (DICE) model, and further the regionally-specific model of the DICE called the RICE (Nordhaus 1994; Nordhaus and Yang 1996).

The DICE is an integrated assessment model (IAM) that integrates the entire processes from smokestack/tailpipe emissions of GHGs, changes in the climate system, economic impacts, and back to policy decisions on smokestack/tailpipe emissions. Other similar integrated assessment models such as the MERGE (A Model for Evaluating the Regional and Global Effects of GHG reduction policies), FUND (The Climate Framework for Uncertainty, Negotiation and Distribution), and PAGE (A Policy Analysis of the Greenhouse Effect) have followed suit (Manne et al. 1995; Stern 2009).

In the DICE model, the complex relationships between the global economic system and the climate system as well as the complexity of each system are simplified into about two dozen mathematical equations. The global economy and the climate system are linked through the carbon cycle models developed by climate scientists on the one hand and the climate change impact models developed by climate economists on the other.

As a byproduct of a host of economic activities, carbon dioxide and other greenhouse gases enter the DICE model, whose ultimate consequences are anthropogenic alterations of the global climate system which causes a loss in the global economic production. Therefore, the global community would be better off by imposing a penalty on carbon dioxide-equivalent emissions to force polluters to reduce the level of emissions. The global community that has struggled so much to address the problem of climate change needs only to adopt the carbon tax or penalty given out by the DICE model. This powerful and so lucid policy proposal has anchored policy debates on global warming for the past three decades, among both economists and policy negotiators.

Regarding the tasks undertaken by this book, the DICE model and other IAMs offer a useful benchmark for assessing the GCF funding decisions. As will be made clear in Chap. 7, an individual GCF funding project places a value on a ton of carbon dioxide emissions, that is, a price of carbon dioxide, because it is awarded for the potential contribution expressed by the expected ton of carbon dioxide emissions avoided (GCF 2018b). Does it have anything to do with the carbon price from the IAMs?

From another perspective, we can evaluate the effectiveness of the GCF funding programs through the DICE models. For example, we can add up the suggested carbon dioxide equivalent emissions avoided by the GCF-funded projects and enter the quantity into the DICE model. This exercise would tell us the degree of temperature change that would be incurred by the total of the GCF-funded projects. This exercise would also tell us the net cost of the GCF funding programs, that is, the total cost minus the total benefit from the GCF portfolio of projects and programs.

1.8 Global Funds, Programs, Currencies

The Green Climate Fund is only one of the many global funds. The most well-established global fund is the United Nations' funds. The United Nations' activities are supported by the UN regular budget and the UN peacekeeping budget. In addition to the two UN funds, there are more than three dozen UN specialized funds, programs, and special agencies (UNSCEB 2018). The GCF can be considered one of the UN's specialized funds, whose size is expected to far exceed other UN funds.

The funding programs of the GCF or the investment decisions of the GCF may be framed after the framework in the existing UN specialized funds. This book will examine the United Nations funds rather at length to clarify the funding and contribution rules of some of the well-recognized UN funds, some of which may turn out to be useful to the analysis of the GCF projects.

The UN system's related organizations and specialized funds can conflict with the GCF programs in the areas where their interests overlap. A GCF funding recipient can also be a recipient of another United Nations funds on a similar project. Such UN specialized funds include the United Nations Development Programme (UNDP), United Nations Environment Programme (UNEP), Food and Agriculture Organization (FAO), World Bank, United Nations International Strategy for Disaster Reduction (UNISDR), International Fund for Agricultural Development (IFAD), World Meteorological Organization (WMO), International Civil Aviation Organization (ICAO), International Maritime Organization (IMO), and others.

Similar to the GCF contributions, these funds are maintained by the UN member countries' contributions. Beside the national governments, donors also include industry sectors as well as private donors in many of these funds, for example, the UNDP, World Food Programme (WFP), and the United Nations Children's Emergency Relief Fund (UNICEF). The size of annual budget for the UNDP in 2018 was US$ 5.6 billion; The size of annual expenses for the UNICEF was US$5.4 billion in 2018; The size of annual contributions to the WFP was US$ 6.8 billion in 2017 (UNICEF 2018; UNDP 2018; WFP 2018).

Another salient phenomenon as per a global fund is that, with the rise of the Internet and, especially, smartphone phone technologies in the twenty-first century, a novel global scale fund is becoming ever easier to being set up, in which the currency/money that is generated or collected globally through an Internet platform for a certain purpose is transferred globally. Such funds include a long list of crowdfunding-generated funds and a cryptocurrency such as the Bitcoin.

For the expansion of a global money, a host of social networking services available in the Internet provides an excellent platform because the social networking sites are linked globally, beyond national boundaries. A Facebook page can be linked to any page in the world, barring national censorship by a totalitarian regime.

The crowdfunding refers to a method of financing a project by gathering small contributions through an online platform from the crowd, that is, potential donors. This method of financing began in 2003 as an online platform for helping finance the production of a jazz album. It was quickly adopted as a financing method for other areas and commercial areas beyond music and art projects (Freedman and Nutting 2015). The Kickstarter, one of the most successful crowdfunding platforms, started in 2009, have had over a hundred thousand successful funding projects, with 15 million donors and US$ 3.6 billion total funding (Statistica 2018).

GoFundMe is the largest crowdfunding platform which raised more than US$ 5.0 billion since 2010, as of April 2018 (See Table 1.3). It is also unique from other

Table 1.3 Global funds and currencies

Funds	Examples	Fund size
Green Climate Fund		About US$ 100 billion per annum (planned)
United Nations specialized funds and programs	United Nations Development Programme (UNDP), World Food Programme (WFP), etc.	US$ 5.6 billion for UNDP per annum; US$ 6.8 billion for WFP
Crowdfunding	GoFundMe	Raised over US$ 5 billion since 2010
Cryptocurrency	Bitcoin	About US$ 100 billion to 1.2 trillion by end 2018
Environmental funds	CO2 permit trading, SO2 allowance trading	About 2 billion emissions permits traded at about 15–25 euros per permit

crowdfunding platforms in that it is neither incentive-based nor commercially-motivated. In other words, donors of the GoFundMe project do not receive a reward. Nor are the GoFundMe projects for commercial profits. Most GoFundMe projects are for a particular cause, e.g., saving an endangered species, or for a particular event, e.g., a graduation event, or for a particular person, e.g., a politician. The GoFundMe platform is more international than other crowdfunding platforms (GoFundMe 2018).

The Green Climate Fund is a form of crowdfunding. Donations to the GCF come from the contributions from the parties of the UN climate convention, industrial sectors, and private individuals, which are then distributed to various projects proposed by poor developing countries and selected by the GCF Board.

There is a major difference between the Green Climate Fund and the other crowdfunding platforms. In the crowdfunding platform, allocations of the raised project fund into specific projects are determined by backers, i.e., donors of the project. On the other hand, there is no formal mechanism that determines the allocations of the global money raised for the GCF into developing country parties or project proposals. The allocation is determined by the ad hoc decisions of the Green Climate Fund Board which is composed of 24 members, among which 12 members are developed country parties and the other 12 members are developing country parties.

An even more prominent global money has emerged recently, also through the technological revolution of the internet and smartphone technologies. The Bitcoin was the first of its kind of currencies, called cryptocurrency, that followed soon afterwards (Bohme et al. 2015). The market value of the Bitcoin may top US$ 1.2 trillion by the end of 2018 (See Table 1.3).

The cryptocurrency, i.e., the bitcoin, was created purely from an Internet communication protocol in 2009 uploaded to the internet by Nakamoto which can be a person, or an agency, or something else (Nakamoto 2008). There is no central bank that mints the money. Further, there is no government in the world that backs the currency. It is created by an internet protocol and then traded purely on the basis of the trust on the internet transactions called the block chain which is designed to be verified by anyone in the world. Therefore, the currency is truly global in its concept.

There are more than a dozen Bitcoin exchanges available, such as Bitfinex and Coinbase, in which a bitcoin can be exchanged for a government-backed currency such as the US dollars and euros. The bitcoins can be exchanged for any major international currencies in these exchanges (Bitcoinity 2018). According to exchange rate data, the dollar value of a unit of bitcoin was almost zero in July 2010. By February 1, 2017, the exchange rate surpassed US$1000 per bitcoin. Through the rest of 2017, the price of a bitcoin climbed up steeply to reach a peak at US$16,000 by December 1, 2017. The price then fell precipitously. By April, 2018, the exchange rate fell to about $7000 per bitcoin.

Where does the value of a bitcoin come from? It comes from the block chain technology first developed and uploaded by Satoshi Nakamoto to the Internet. A block is a record of transactions of bitcoins. The entire history from the beginning of a bitcoin transaction is made available to anyone in the Bitcoin system by the block chain, that is, the chain of blocks. A bitcoin is created by the Bitcoin system and awarded to the person who compiles a new block and links it to the existing block chain. The block-chain makers are the record-keepers of bitcoin transactions, like the traditional banking service, for which they are awarded the bitcoins.

The price of a bitcoin is determined by the strength of confidence of the crowd in the bitcoin as a currency. The bitcoin gained value in the initial years because there were people who believed in the bitcoin currency and traded goods and services for it. The first traders of the bitcoin were those who cannot make payments using the traditional currencies, most prominently, narcotics dealers, e.g., marihuana and prescription drugs, and gambling sites. Despite the FBI seizure of the Silk Road, the first online market place that accepted the Bitcoin transactions exclusively, more retailers started to accept bitcoins as a form of payments, whose examples include Overstock.com (online retailer), Expedia (travel), and TigerDirect (electronics) (Bohme et al. 2015). There are emerging signs that institutional investors started to invest in the cryptocurrency (Liu and Tsybinski 2018).

An incredible aspect of the Bitcoin is that its value arises from the social networking or connectivity through the Internet. The stronger the social networking, the higher the value of the bitcoin would become. Similarly, the more global the social connectivity is, the higher the value of the bitcoin gets.

The Bitcoin system and the GCF fund are both a global fund, but different in many aspects. The Bitcoin has a market price while the GCF does not. The Bitcoin system does not have a central authority that operates the system while the GCF does have one. The Bitcoin sets no limit in the volume of bitcoins while the GCF does have the limit in the size of total funding.

Of the internationally traded funds, there is also a green fund that was developed explicitly to tackle an environmental problem similarly as the GCF was developed to tackle the climate change problem. The European Union's Emissions Trading System (EU ETS) is a prime example of such environmental funds (EC 2016). The total value of the emissions allowances is approximately 40 billion euros, that is, about 2 billion allowances times 15–25 euros per allowance (See Fig. 8.5, Chap. 8).

Although the two funds seem little related at a first glance, a brief reasoning will reveal that they share many similarities and, in fact, the future of the GCF system cannot be discussed satisfactorily without the relationship with the CO2 emissions permit trading system clarified. For a starter, the ETS creates the market price of a ton of carbon dioxide emissions whose history can be traced back to 2008 when the first implementation phase of Kyoto Protocol began. By contrasts, the GCF project funding decision creates an implicit price of a ton of carbon dioxide emissions because the total money given is associated with the tons of carbon dioxide equivalent emissions (GCF 2018e).

Are the two prices, that is, the explicit price and the implicit price of CO2-equivalent emissions are approximately same? Theoretically, which should be explained later in this book, the two prices should be same for some reasons and different for other reasons. Extending this line of inquiry, this book will explore the possibility of associating the GCF funding to the market price of the EU ETS allowance.

1.9 Road Map of the Book

I began the book in this chapter with an overview of the inquiries and undertakings of this book. As you walk through the book, you will come across and ponder about the following characters and questions. The next chapter, Chap. 2, offers a central place for the gatherings of all the components of the Green Climate Fund system. Convened are a history of the GCF establishment, its standing in the Paris Agreement, institutional organizations, pledges and contributions, and allocations of the grants and loans to selected projects.

From Chap. 3 to Chap. 6, the above-described four scientific traditions are discussed in each chapter: A microbehavioral economic model of behavioral adaptations to climate change in Chap. 3, an agro-economic modeling tradition in Chap. 4, a statistical modeling of yield and productivity changes in Chap. 5, an ecosystem modeling of climate change impacts in Chap. 6.

Based on the scientific traditions, in Chap. 7, a series of evaluations of the GCF's allocations of funds into the host of selected projects are conducted. The evaluations are conducted based on the three theoretical criteria: a theory of public expenditure, efficiency consequences in GCF-induced re-allocations of natural resources, and a theory of group adaptations.

In Chap. 8, the evaluations of the GCF system and decisions are extended to other global funds and programs. Of particular emphasis is the overlapping of the GCF programs with the other United Nations specialized funds and programs. A particularly intriguing possibility of linking the GCF fund allocations to one of the other global currencies will be explored.

References

Adams R, Rosenzweig C, Peart RM, Ritchie JT, McCarl BA, Glyer JD, Curry RB, Jones JW, Boote KJ, Allen LH (1990) Global climate change and US agriculture. Nature 345:219–224
Adams R, McCarl BA, Segerson K, Rosenzweig C, Bryant K, Dixon BL, Conner R, Evenson RE, Ojima D (1999) The economic effects of climate change on US agriculture. In: Mendelsohn R, Neumann J (eds) The impact of climate change on the United States economy. Cambridge University Press, Cambridge
Ainsworth EA, Long SP (2005) What have we learned from 15 years of free-air CO2 enrichment (FACE)? A meta-analytic review of the responses of photosynthesis, canopy properties and plant production to rising CO2. New Phytol 165:351–371
Akasaki I, Amano H, Nakamura S (2014) Blue LEDs e filling the world with new light. Nobel Prize Lecture. The Nobel Foundation, Stockholm. Available at: http://www.nobelprize.org/nobel_prizes/physics/laureates/2014/popular-physicsprize2014.pdf
Aksoy S, Attardo G et al (2014) Genome sequence of the tsetse fly (Glossina morsitans): vector of African Trypanosomiasis. Science 344(6182):380–386
Auffhammer M, Ramaathan V, Vincent JR (2006) Integrated model shows that atmospheric Brown clouds and greenhouse gases have reduced rice harvests in India. Proc Natl Acad Sci U S A 103:19668–19672
Bakkensen LA, Mendelsohn R (2016) Risk and adaptation: evidence from global hurricane damages and fatalities. J Assoc Environ Resour Econ 3:555–587
Barreca A, Clay K, Deschenes O, Greenstone M, Shapiro JS (2016) Adapting to climate change: the remarkable decline in the US temperature-mortality relationship over the twentieth century. J Polit Econ 124:105–159
Baumol WJ, Oates OA (1988) The theory of environmental policy, 2nd edn. Cambridge University Press, Cambridge
Bitcoinity.org (2018) Markets. Available at https://bitcoinity.org/markets/list. Accessed May 8 2018
Bohme R, Christin N, Edelman B, Moore T (2015) Bitcoin: economics, technology, and governance. J Econ Perspect 29:213–238
Brannon I, Blask A (2017) Reforming the national flood insurance program: toward private insurance. Policy Anal 817:1–20
Burtraw D, Szambelan SJ (2009) U.S. emissions trading markets for SO2 and NOx. Resources for the future Discussion Paper 09-40. Resources for the Future, Washington, DC
Carbon Brief (2018) Eyes on ministers to intervene as UN climate talks get mired in old battles. Published on May 10, 2018. Accessed from https://www.carbonbrief.org/daily-brief/eyes-on-ministers-to-intervene-un-climate-talks
Columbia Broadcasting System (CBS) (2016) Obama calls Paris agreement “a turning point for our planet”. Published on Oct 15, 2016. Accessed from https://www.cbsnews.com/news/obama-calls-paris-agreement-a-turning-point-for-our planet/
Deschenes O, Greenstone M (2007) The economic impacts of climate change: evidence from agricultural output and random fluctuations in weather. Am Econ Rev 97:354–385
Dudal R (1980) Soil-related constraints to agricultural development in the tropics. International Rice Research Institute, Los Banos
Economic Times of India (2015) Green Climate Fund of $100 billion/year not enough to tackle climate change: India. Published on July 14, 2015. Accessed from https://economictimes.indiatimes.com/news/politics-and-nation/green-climate-fund-of-100-billion/year-not-enough-to-tackle-climate-change-india/articleshow/48070235.cms
Ellerman AD, Marcantonini C, Zaklan A (2016) The European Union Emissions Trading System: ten years and counting. Rev Environ Econ Policy 10:89–107
European Commission (EC) (2016) EU ETS handbook. European Commission, Brussels
Evenson R, Gollin D (2003) Assessing the impact of the green revolution 1960–2000. Science 300:758–762

Fischer G, Shah M, Tubiello FN, van Velhuizen H (2005) Socio-economic and climate change impacts on agriculture: an integrated assessment, 1990–2080. Philos Trans R Soc B 360:2067–2083
Food and Agriculture Organization (FAO) (1978) Report on agro-ecological zones vol. 1: methodology and results for Africa. FAO, Rome
Freeman DM, Nutting MR (2015) A brief history of crowdfunding including rewards, Donation, Debt, and Equity Platforms in the USA
Friedman M (1962) Price theory. Transaction Publishers, New Brunswick
Global Environment Facility (GEF) (2017) Comparative advantage, adequacy of funding/financing, health of the expanded GEF partnership and governance structure. GEF, UNDP, New York
GoFundMe (2018) How GoFundMe works. The GoFundMe, Red Wood City. Accessed from https://support.gofundme.com/hc/en-us/articles/115011182268-How-GoFundMe-Works-
Graziano M, Gillingham K (2015) Spatial patterns of solar photovoltaic system adoption: the influence of neighbors and the built environment. J Econ Geogr 15:815–839
Green Climate Fund (GCF) (2011) Governing instrument for the Green Climate Fund. GCF, Songdo City
Green Climate Fund (GCF) (2018a) About the fund. GCF, Songdo City
Green Climate Fund (GCF) (2018b) Annex III: investment framework. GCF, Songdo City
Green Climate Fund (GCF) (2018c) Annex IX: results management framework. GCF, Songdo City
Green Climate Fund (GCF) (2018d) Status of pledges and contributions made to the Green Climate Fund, status date: 29 January 2018. GCF, Songdo City
Green Climate Fund (GCF) (2018e) Projects + programmes. GCF, Songdo City. Accessed from https://www.greenclimate.fund/what-we-do/projects-programmes
Hahn RW, Dudley PM (2007) How well does the U.S. government do benefit-cost analysis? Rev Environ Econ Policy 1:192–211
Hanemann WM (2000) Adaptation and its management. Clim Chang 45:571–581
Hart O, Holmstrom B (1987) The theory of contracts. In: Bewley T (ed) Advances in economics and econometrics. Cambridge University Press, Cambridge
Hartwick JM, Olewiler ND (1997) The economics of natural resource use, 2nd edn. Pearson, New York
Halcrow HG (1949) Actuarial structures for crop insurance. J Farm Econ 31:418–443
Intergovernmental Panel on Climate Change (IPCC) (2000) Special report on land use, land-use change, and forestry. Cambridge University Press, Cambridge
Intergovernmental Panel on Climate Change (IPCC) (2001) Climate change 2001: impacts, adaptation, and vulnerability. A Report of the Working Group II of the IPCC. Cambridge University Press, Cambridge
Intergovernmental Panel on Climate Change (IPCC) (2007) Climate change 2007: impacts, adaptation, and vulnerability. Contribution of the Working Group II to the Fourth Assessment Report of the IPCC. Cambridge University Press, Cambridge
Intergovernmental Panel on Climate Change (IPCC) (2011) Special report on renewable energy sources and climate change mitigation. Cambridge University Press, Cambridge
Intergovernmental Panel on Climate Change (IPCC) (2014) Climate change 2014: impacts, adaptation, and vulnerability. Contribution of the Working Group II to the Fourth Assessment Report of the IPCC. Cambridge University Press, Cambridge
Intergovernmental Panel on Climate Change (IPCC) (2018) Special report on global warming of 1.5 °C. Cambridge University Press, Cambridge
International Thermonuclear Experimental Reactor (ITER) (2015) ITER: the world's largest Tokamak. Available at: https://www.iter.org/mach
Keuhl RO (1999) Design of experiments: statistical principles of research design and analysis, 2nd edn. Duxbury Press, Pacific Grove
Klein D, Carazo MP, Doelle M, Bulmer J, Higam A (2017) The Paris agreement on climate change: analysis and commentary. Oxford University Press, Oxford

Knowles SG, Kunreuther HC (2014) Troubled waters: the national flood insurance program in historical perspective. J Policy Hist 26:325–353
Kurukulasuriya P, Kala N, Mendelsohn R (2011) Adaptation and climate change impacts: a structural Ricardian model of irrigation and farm income in Africa. Clim Change Econ 2:149–174
Lackner KS, Brennana S, Matter JM, Park AA, Wright A, Zwaan BV (2012) The urgency of the development of CO2 capture from ambient air. Proc Natl Acad Sci U S A 109(33):13156–13162
Liu Y, Tsybinski A (2018) Risks and returns of cryptocurrency. NBER Working Paper No. 24877. NBER, Boston
Manne AS, Mendelsohn R, Richels RG (1995) MERGE: a model for evaluating regional and global effects of GHG reduction policies. Energy Policy 23:17–34
Mas-Colell A, Whinston MD, Green JR (1995) Microeconomic theory. Oxford University Press, Oxford
Massachusetts Institute of Technology (MIT) (2015) The future of solar energy: an interdisciplinary MIT study. MIT, Cambridge, MA
Matthews E (1983) Global vegetation and land use: new high-resolution data bases for climate studies. J Clim Appl Meteorol 22:474–487. Available from: http://data.giss.nasa.gov/landuse/vegeem.html
McConnell V (2013) The new CAFE standards: are they enough on their own? Resources For the Future Discussion Paper 13–14. RFF, Washington, DC
Mendelsohn R (2000) Efficient adaptation to climate change. Clim Chang 45:583–600
Mendelsohn R, Neumann J (1999) The impact of climate change on the United States economy. Cambridge University Press, Cambridge
Nakamoto S (2008) Bitcoin: a peer-to-peer electronic cash system. Available at https://bitcoin.org/bitcoin.pdf. Accessed May 5 2018
National Agricultural Statistical Service (NASS) (2018) Census of agriculture. US Department of Agriculture, Washington, DC. http://www.agcensus.usda.gov/index.php
National Public Radio (NPR) (2016) Nations clinch landmark pact to fight climate change. Published on October 15, 2016. Accessed from https://www.npr.org/sections/thetwo-way/2016/10/15/498056486/nations-clinch-landmark-pact-to-fight-climate-change
National Research Council (NRC) (2013) Transitions to alternative vehicles and fuels. The National Academies Press, Washington, DC
National Research Council (NRC) (2015) Climate intervention: reflecting sunlight to cool earth. Committee on Geoengineering climate: technical evaluation and discussion of impacts. The National Academies Press, Washington, DC
New York Times (NYT) (2015) Trust and money at core of crucial Paris talks on climate change. Published on December 6, 2015. Accessed from https://www.nytimes.com/2015/12/07/world/europe/trust-and-money-at-core-of-crucial-paris-talks-on-climate-change.html
Ng N-S, Mendelsohn R (2006) The economic impact of sea-level rise on nonmarket lands in Singapore. Ambio 35:289–296
Nicholls RJ, Wong PP, Burkett VR, Codignotto JO, Hay JE, McLean RF, Ragoonaden S, Woodroffe CD (2007) Coastal systems and low-lying areas. In: Parry ML, Canziani OF, Palutikof JP, van der Linden PJ, Hanson CE (eds) Climate change 2007: impacts, adaptation and vulnerability. The fourth assessment report of the intergovernmental panel on climate change. Cambridge University Press, Cambridge
Nobel Committee (2018) The Sveriges Riksbank prize in economic sciences in memory of Alfred Nobel 2018. The Royal Swedish Academy of Sciences, Stockholm. https://www.nobelprize.org/prizes/economic-sciences/2018/summary
Nordhaus W (1991) To slow or not to slow: the economics of the greenhouse effects. Econ J 101:920–937
Nordhaus W (1992) An optimal transition path for controlling greenhouse gases. Science 258:1315–1319
Nordhaus W (1994) Managing the global commons. MIT Press, Cambridge, MA

Nordhaus W (2007) To tax or not to tax: alternative approaches to slowing global warming. Rev Environ Econ Policy 1(1):26–44
Nordhaus WD (2008) A question of balance—weighing the options on global warming policies. Yale University Press, New Haven
Nordhaus W (2010) Economic aspects of global warming in a post-Copenhagen environment. Proc Natl Acad Sci U S A 107(26):11721–11726
Nordhaus W (2013) The climate casino: risk, uncertainty, and economics for a warming world. Yale University Press, New Haven
Nordhaus W, Yang Z (1996) A regional dynamic general-equilibrium model of alternative climate change strategies. Am Econ Rev 86:741–765
Pareto V (1906) Manual for political economy. Oxford University Press, Oxford. [Edited by Montesano A, Zanni A, Bruni L, Chipman JS, McLure M in (2014)]
Parry ML, Rosenzweig CP, Iglesias A, Livermore M, Fischer G (2004) Effects of climate change on global food production under SRES emissions and socioeconomic scenarios. Glob Environ Chang 14:53–67
Pearce D, Cline WR, Achanta A, Fankhauser S, Pachauri R, Tol R, Vellinga P (1996) The social costs of climate change: greenhouse damage and benefits of control. In: Bruce J, Lee H, Haites E (eds) Climate change 1995: economic and social dimensions of climate change. Cambridge University Press, Cambridge
Politico (2018) Hot dispute over money threatens climate deal. Published on May 10, 2018. Accessed from https://www.politico.eu/article/climate-change-goals-stalled-by-money-dispute-paris-agreement-katowice-summit/
Reilly J, Baethgen W, Chege F, Van de Geijn S, Enda L, Iglesias A, Kenny G, Patterson D, Rogasik J, Rotter R, Rosenzweig C, Sombroek W, Westbrook J (1996) Agriculture in a changing climate: impacts and adaptations. In: Watson R, Zinyowera M, Moss R, Dokken D (eds) Climate change 1995: impacts, adaptations, and mitigation of climate change. Intergovernmental Panel on Climate Change (IPCC). Cambridge University Press, Cambridge
Ricardo D (1817) On the principles of political economy and taxation. John Murray, London
Roberts T, Weikmans R (2015) Unfinished agenda of the Paris climate talks: finance to the global south. Brookings Institution, Washington, DC
Rosenzweig C, Parry M (1994) Potential impact of climate change on world food supply. Nature 367:133–138
Samuelson P (1954) The pure theory of public expenditure. Rev Econ Stat 36:387–389
Samuelson P (1955) Diagrammatic exposition of a theory of public expenditure. Rev Econ Stat 37:350–356
Schlenker W, Roberts M (2009) Nonlinear temperature effects indicate severe damages to crop yields under climate change. Proc Natl Acad Sci U S A 106(37):15594–15598
Schlesinger WH (1991) Biogeochemistry: an analysis of global change. Academic, San Diego
Seo SN (2006) Modeling farmer responses to climate change: climate change impacts and adaptations in livestock management in Africa. PhD dissertation, Yale University, New Haven
Seo SN (2010a) A microeconometric analysis of adapting portfolios to climate change: adoption of agricultural systems in Latin America. Appl Econ Perspect Policy 32:489–514
Seo SN (2010b) Is an integrated farm more resilient against climate change?: a micro-econometric analysis of portfolio diversification in African agriculture? Food Policy 35:32–40
Seo SN (2010c) Managing forests, livestock, and crops under global warming: a micro-econometric analysis of land use changes in Africa. Aust J Agric Resour Econ 54(2):239–258
Seo SN (2011) An analysis of public adaptation to climate change using agricultural water schemes in South America. Ecol Econ 70:825–834
Seo SN (2012) Adaptation behaviors across ecosystems under global warming: a spatial microeconometric model of the rural economy in South America. Pap Reg Sci 91:849–871
Seo SN (2014a) Evaluation of Agro-Ecological Zone methods for the study of climate change with micro farming decisions in sub-Saharan Africa. Eur J Agron 52:157–165

Seo SN (2014b) Adapting sensibly when global warming turns the field brown or blue: a comment on the 2014 IPCC report. Econ Aff 34:399–401
Seo SN (2015a) Micro-behavioral economics of global warming: modeling adaptation strategies in agricultural and natural resource enterprises. Springer, Cham
Seo SN (2015b) Helping low-latitude, poor countries with climate change. Regulation 38(4):6–8. Winter 2015–2016
Seo SN (2015c) Fatalities of neglect: adapt to more intense hurricanes? Int J Climatol 35:3505–3514
Seo SN (2016a) Modeling farmer adaptations to climate change in South America: a micro-behavioral economic perspective. Environ Ecol Stat 23:1–21
Seo SN (2016b) The micro-behavioral framework for estimating total damage of global warming on natural resource enterprises with full adaptations. J Agric Biol Environ Stat 21:328–347
Seo SN (2016c) Microbehavioral econometric methods: theories, models, and applications for the study of environmental and natural resources. Academic/Elsevier, Amsterdam
Seo SN (2017a) The Behavioral economics of climate change: adaptation behaviors, global public goods, breakthrough technologies, and policy-making. Academic, Amsterdam
Seo SN (2017b) Beyond the Paris agreement: climate change policy negotiations and future directions. Reg Sci Policy Pract 9:121–140
Seo SN (2017c) Measuring policy benefits of the cyclone shelter program in the North Indian Ocean: protection from intense winds or high storm surges? Clim Change Econ 8(4):1–18
Seo SN (2018) Two tales of super-typhoons and super-wealth in Northwest Pacific: will global-warming-fueled cyclones ravage East and Southeast Asia? J Extreme Events 5:1850012. https://doi.org/10.1142/S2345737618500124
Seo SN (2019) Economic questions on global warming during the trump years. J Public Aff. https://doi.org/10.1002/pa.1914
Seo SN, Mendelsohn R (2008) Measuring impacts and adaptations to climate change: a structural Ricardian model of African livestock management. Agric Econ 38:151–165
Seo SN, Mendelsohn R, Dinar A, Hassan R, Kurukulasuriya P (2009) A Ricardian analysis of the distribution of climate change impacts on agriculture across Agro-Ecological Zones in Africa. Environ Resour Econ 43:313–332
Seo SN, Bakkensen LA (2017) Is tropical cyclone surge, not intensity, what kills so many people in South Asia? Weather Clim Soc 9:71–81
Statistica (2018) Number of successfully funded projects on Kickstarter.com as of April 2018, by project category. Statistica, Hamburg. Available at https://www.statista.com/statistics/222460/number-of-successfully-pledged-kickstarter-projects-by-category/
Stern N (2009) The economics of climate change: the Stern review. Cambridge University Press, Cambridge
Sumner DA, Zulauf C (2012) Economic & environmental effects of agricultural insurance programs. The Council on Food, Agricultural & Resource Economics (C-FARE), Washington, DC
Tol RSJ (2009) The economic effects of climate change. J Econ Perspect 23:29–51
Tubiello FN, Ewert F (2002) Simulating the effects of elevated CO2 on crops: approaches and applications for climate change. Eur J Agron 18:57–74
United Nations (2017) Assessment of Member States' advances to the Working Capital Fund for the biennium 2018–2019 and contributions to the United Nations regular budget for 2018. Secretariat, United Nations, New York
United Nations (2018) United Nations peacekeeping: how we are funded. Accessed from https://peacekeeping.un.org/en/how-we-are-funded
United Nations Children's Fund (UNICEF) (2018) UNICEF integrated budget 2018–2021. The UNICEF, New York
United Nations Development Program (UNDP) (2018) Our funding: top contributors. UNDP, New York. Accessed from http://www.undp.org/content/undp/en/home/funding/top-contributors.html

United Nations Environment Programme (2017) The emissions gap report 2017: a UN environment synthesis report. UNEP, Nairobi
United Nations Framework Convention on Climate Change (UNFCCC) (1992) United Nations framework convention on climate change. UNFCCC, New York
United Nations Framework Convention on Climate Change (UNFCCC) (1997) Kyoto protocol to the United Nations framework convention on climate change. UNFCCC, New York
United Nations Framework Convention on Climate Change (UNFCCC) (2009) Copenhagen accord. UNFCCC, New York
United Nations Framework Convention on Climate Change (UNFCCC) (2010) Cancun agreements. UNFCCC, New York
United Nations Framework Convention on Climate Change (UNFCCC) (2011a) The Durban platform for enhanced action. UNFCCC, New York
United Nations Framework Convention on Climate Change (UNFCCC) (2011b) Report of the transitional committee for the design of Green Climate Fund. UNFCCC, New York
United Nations Framework Convention on Climate Change (UNFCCC) (2012) Decisions adopted by the conference of the parties on its eighteenth session, held in Doha from 26 November to 8 December 2012. UNFCCC, New York
United Nations Framework Convention on Climate Change (UNFCCC) (2015) The Paris agreement. Conference of the Parties (COP) 21, UNFCCC, New York
United Nations Framework Convention on Climate Change (UNFCCC) (2018) Decisions adopted at the Climate Change Conference in Katowice, Poland, 2–14 December 2018. Conference of the Parties (COP) 24, UNFCCC, New York
United Nations System Chief Executive Board (UNSCEB) (2018) UN system. UN, New York. Accessed from http://www.unsceb.org/directory
United States Environmental Protection Agency (US EPA) (1999) The benefits and costs of the clean air act: 1990–2010. EPA report to Congress. EPA 410-R-99-001. Washington, DC
United States Environmental Protection Agency (US EPA) (2014) Carbon pollution emission guidelines for existing stationary sources: electric utility generating units. US EPA, Washington, DC
United States Environmental Protection Agency (US EPA) (2018) Proposal: affordable clean energy (ACE) rule. US EPA, Washington, DC. Accessed from https://www.epa.gov/stationary-sources-air-pollution/proposal-affordable-clean-energy-ace-rule
von Thunen JH (1826) The isolated state (CM Wartenberg, trans.). Pergamon Press, Oxford/New York [1966]
Wagner G, Weitzman ML (2015) Climate shock: the economic consequences of a hotter planet. Princeton University Press, Princeton
Washington Post (2018) Countries made only modest climate-change promises in Paris. They're falling short anyway. Published on February 19, 2018. WP, Washington, DC
Welch JR, Vincent JR, Auffhammer M, Moya PF, Dobermann A, Dawe D (2010) Rice yields in tropical/subtropical Asia exhibit large but opposing sensitivities to minimum and maximum temperatures. Proc Natl Acad Sci U S A 107:14562–14567
White House (2017) Statement by President Trump on the Paris Climate Accord. White House, Washington, DC
World Bank (2009) Awakening Africa's sleeping giant: prospects for commercial agriculture in the Guinea savannah zone and beyond. World Bank and FAO, Washington, DC
World Food Programme (WFP) (2018) Contributions to WFP in 2017. The WFP, Rome. Accessed from https://www.wfp.org/funding/year/2017
Yohe GW, Schlesinger ME (1998) Sea level change: the expected economic cost of protection or abandonment in the United States. Clim Chang 38:337–342

Chapter 2
The Green Climate Fund: History, Institution, Pledges, Investment Criteria

Abstract This chapter provides an encompassing description of the Green Climate Fund. A historical development of the GCF is described from its first conception, to the establishment, to the stage of funding allocations, which has been closely tied to the development of the United Nations conferences up to the Paris Agreement. The pledges and contributions to the GCF made by major rich countries are analyzed. The institutional structure and legal status of the GCF are explained. The first batch of funding allocations began in November 2015, timed for the Paris Conference. This chapter analyzes the 93+ projects that received the funding since then. Allocations of the fund are analyzed by geo-political areas as well as the results areas defined by the GCF Board. This chapter takes a close look at the investment criteria which are relied upon in making funding decisions, which include various concepts such as mitigation impact, adaptation impact, paradigm shift, sustainable development, needs of the recipients, country ownership, and efficiency and effectiveness.

Keywords Green Climate Fund · Paris Agreement · Pledges · Governing instrument · Investment criteria

2.1 Introduction

The Green Climate Fund (GCF) is a global fund created by the parties of the United Nations convention on climate change whose mission is to assist underdeveloped countries to 'cope with future climatic changes.' The GCF must serve the agreement or treaty by the parties of the UN conferences, which is, as of 2018, the Paris Agreement achieved in November 2015 and entered into force the following year (GCF 2011; UNFCCC 2015).

The GCF is a distinct global environmental—green— fund in many aspects. First of all, the Fund's scale is truly global. It has 194 member-countries which are the parties of the United Nations Framework Convention on Climate Change (UNFCCC). Therefore, it is not a private fund, nor is it a public fund. It is a global

S. N. Seo, *The Economics of Global Allocations of the Green Climate Fund*,
https://doi.org/10.1007/978-3-030-18274-8_2

public fund (GCF 2018a). As explained in the previous chapter, the declared size of the Fund is also far larger than other global-scale environmental funds.

Second, the Fund was set up and maintained by voluntary pledges and contributions dominantly by developed country parties of the UNFCCC. Only a handful of richest countries contributed nearly all of the GCF fund, as of December 2018, including the United States, Germany, France, Japan, United Kingdom (GCF 2018d). A large majority of the GCF member countries is a recipient of grants and loans from the GCF (GCF 2018e). Since there is no legal responsibility for developed countries to contribute to the Fund, its continuity is not guaranteed and on a fragile ground.

Third, the Fund is earmarked to climate change adaptation programs and also mitigation programs that are and will be undertaken by underdeveloped countries (GCF 2011). A specific program is developed by an entity in developing country-parties with the help of the GCF Secretariat and Board which should then decide to award or not award a grant to the project from a large number of proposed projects. The Fund, as it is set up legally, cannot be used for other purposes than climate change mitigation and adaptation actions. The GCF is the only stand-alone financial instrument that supports the UNFCCC agreements (GCF 2011). Other funds are multi-purpose environmental funds that support also non-climate policy programs.

Fourth, the Fund's allocation decisions are made by the GCF Board based on the investment guidelines developed by the GCF. The GCF Board is composed of 24 board members equally selected from 12 developing country-parties and 12 developed country-parties. The investment criteria are the key to understanding the Fund's allocation decisions to individual projects. The criteria are manifold, but include many non-climate change policy goals: expected mitigation of carbon dioxide-equivalent emissions, expected number of beneficiaries, paradigm shift potential, sustainable development goals, the needs of a recipient, country ownership, efficiency, effectiveness (GCF 2018b, c).

The GCF was first conceived at the Copenhagen Conference in 2009 as an adaptation fund that helps poor countries to adapt to climate changes (UNFCCC 2009; Nordhaus 2010). Its importance has risen drastically since then through the series of the UNFCCC conferences. According to the GCF website, "When the Paris Agreement was reached in 2015, the Green Climate Fund was given an important role in serving the agreement and supporting the goal of keeping climate change well below 2 degrees Celsius (GCF 2018a)."

From a different angle, it can be said that the promise of the GCF was the most important factor in the highly-acclaimed achievement of the Paris Agreement in 2015 (UNFCCC 2015; Seo 2015; Roberts and Weikmans 2015). Many developing country parties of the Paris Agreement signed the Paris Agreement on the condition that international aids from the GCF will flow to their countries to help them achieve the mitigation goals submitted voluntarily through the negotiation scheme called in the Agreement as the Intended Nationally Determined Contribution (INDC) (Seo 2017a, b).

In fact, concerned observers would go further to state that the GCF is the most important instrument of the current UNFCCC climate agreements (UNEP 2017;

IPCC 2018; Seo 2017a). This is because the only binding legal treaty on climate change, the Kyoto Protocol, became defunct at the end of the first phase in 2012 and will be replaced by the Paris Agreement from 2020. By then, the binding mitigation commitments under the Kyoto Protocol will become non-binding legally since the agreements at the Paris are non-binding (UNFCCC 1997, 2011a). It would leave the GCF as the only hope for future negotiations or mitigation commitments of individual country parties.

Great challenges that the GCF poses to the international community were already explained in Chap. 1. This book provides a rare, but much awaited, analysis of the GCF through the framework of the economics of global allocations of the global fund. Considering the size of the fund which was proclaimed to be US$ 100 billion annually in the initial years from 2020 compared to other global funds such as the UN's peacekeeping budget which is about US$ 8 billion per annum (UN 2018), the analyses and outputs from this book will surely provide invaluable insights to the global community into rewards and perils of the global money transfers.

2.2 A Historical Development of the GCF

At the 2009 Copenhagen Climate Conference, the Conference Of the Parties (COP) 15, the global climate communities and negotiators ran into a major disaster. The road to the COP 15 was full of high expectations, favorable political alignments, intense natural disasters in preceding years, which boded well for a major breakthrough in Copenhagen. The international negotiators were confident that they would come out of Copenhagen with a new international climate treaty that will replace the Kyoto Protocol which was considered incomplete and ineffective.

As summarized in Table 1.1 of Chap. 1, a series of events, that is, the Rio Earth Summit in 1992, the Kyoto Protocol in 1997, the Kyoto Protocol entering into force in 2005, and the Bali Action Plan in 2007, paved the way to the Copenhagen meeting (UNFCCC 1997, 2005, 2007; Manne and Richels 1999; Nordhaus and Boyer 1999).

The Copenhagen Conference ended in shambles, the process of which was broadcast vividly worldwide. The primary obstacle was the disagreement between developing country parties and developed country parties of the convention (NYT 2009a). Poor countries argued that rich countries are responsible for the global warming trend of the climate system because of the historical cumulative emission in the atmosphere and must bear the burden of climate change mitigations (Nordhaus 2010).

From the disarray of the conference arrived Hilary Clinton, then the first Secretary of the State of the Obama Administration. The year 2009 was the first year of President Obama's 8 years in office. Mrs. Clinton unveiled a major commitment to climate negotiations and pledged US$ 100 billion per annum to help developing country parties to 'cope with climate change' or 'adapt to climate change', which would later become referred to as the Copenhagen Green Climate Fund in the Copenhagen Accord (NYT 2009b; UNFCCC 2009).

The Copenhagen Accord, the outcome of the COP 15, describes the fund as follows:

> In the context of meaningful mitigation actions and transparency on implementation, developed countries commit to a goal of mobilizing jointly USD 100 billion dollars a year by 2020 to address the needs of developing countries.
>
> We decide that the Copenhagen Green Climate Fund shall be established as an operating entity of the financial mechanism of the Convention to support projects, programme, policies and other activities in developing countries related to mitigation including REDD-plus, adaptation, capacity building, technology development and transfer.

The following year saw the formal establishment of the Green Climate Fund through the COP 16 held in Cancun, Mexico in 2010. At the Cancun Conference, the Transitional Committee for the GCF was created and the interim trustee of the GCF was designated (UNFCCC 2010):

> Decides to establish a Green Climate Fund, to be designated as an operating entity of the financial mechanism of the Convention. . . . . .; Also decides that the Fund shall be governed by a Board of 24 members, comprising an equal number of members from developing and developed country Parties. . . . . . .; Invites the World Bank to serve as the interim trustee for the Green Climate Fund. . . . . .; Also decides that the Green Climate Fund shall be designed by a Transitional Committee in accordance with the terms of reference contained in appendix III to this decision. . . . . .

In 2011, the Durban Conference in South Africa approves the governing instrument for the Green Climate Fund developed by the transitional committee (UNFCCC 2011b). The governing instrument gives the GCF the mandate to make an ambitious contribution to the ultimate objective of the UNFCCC (GCF 2011; UNFCCC 2011b):

> Given the urgency and seriousness of climate change, the purpose of the Fund is to make a significant and ambitious contribution to the global efforts towards attaining the goals set by the international community to combat climate change.

For its effective functioning, the Fund will have juridical personality and necessary legal capacity. The governing instrument also defines the decision-making rule of the GCF (GCF 2011): (1) "The Fund will be governed and supervised by a Board that will have full responsibility for funding decisions"; (2) "Decisions of the Board will be taken by consensus of the Board members."

Following the governing instrument, the GCF's governing board held its first meetings in 2012. The governing board is composed of 24 members, half of which are selected from developing country parties and the other half of which are selected from developed country parties of the convention. The board is headed by two co-chairs: one from the developing country parties and the other from the developed country parties (GCF 2018a).

As of May 2018, the developing country parties of the GCF board are from Nicaragua (Chair), Bhutan, Egypt, Saudi Arabia, Chile, Samoa, Democratic Republic of Congo, Georgia, India, South Africa, China, Uruguay. The developed country parties of the board are from Sweden (Chair), Japan, Norway, Finland, Canada, United Kingdom, United States, France, Germany, Spain, Australia, the Netherlands.

At the Doha COP in 2012, Songdo city in the Republic of Korea was selected, out of six competing cities, as the host city of the GCF Secretariat by the GCF board's consensus decision, which was endorsed at the Doha Conference (UNFCCC 2012).

The GCF Secretariat was established in Songdo city in December 2013. The Secretariat is headed by the Executive Director, first appointed in 2013 and second in 2017, and senior management officials. As of April 2018, the Secretariat had 172 staff members who come from 55+ countries, in addition to a large number of consultants. The Executive Director abruptly stepped down in September 2018, owing in large part to a failure to raise the fund.

In 2014, the GCF began its initial resource mobilization period that would last to 2018. According to the website, it raised about US$ 10 billion equivalent in the first year, about 80% of which came from the five countries: The US (US$ 3 billion), Japan (US$ 1.5 billion), United Kingdom (US$ 1.2 billion), France (US$ 1 billion), and Germany (US$ 1 billion) (GCF 2018d).

In 2015, the first batch of allocation decisions of the GCF fund was made in November 2015, timed for the Paris Conference. Eight projects were awarded with a grant: Two projects from Latin America and the Caribbean (Peru, Mexico), one project from Asia Pacific (Bangladesh), two projects from Alliance of Small Island States (AOSIS) (Maldives, Fiji), and three projects from Africa (Malawi, Senegal, Rwanda).

In December 2015, the historic Paris Agreement was signed at the COP 21 by 195 countries. The following year marked the first full year of operations of the GCF. Additional 29 projects were approved by the end of 2016, some of which were withdrawn later.

The COP 24 held at Katowice, Poland at the end of 2018 requests developed country parties to communicate biennially the projected levels of financial resources that are available to developing country parties and further decides to convene the high-level ministerial dialogue every 2 years after 2021 to update the information. The Katowice decisions also allow developed country parties to count both grants and loans as financial resources, one of the most contentious issues in the conference (UNFCCC 2018).

In June 2017, Trump announced the U.S. will withdraw from the Paris Agreement and a formal letter of withdrawal was sent in July 2017 to the United Nations (White House 2017; Seo 2019). Citing an unfair deal to the U.S. against China and India, President Trump also left open a possibility for re-negotiation for a better deal for the United States.

As of April 2018, a total of 81 GCF funding projects were approved. The first project was approved in February 2018 in accordance with the Simplified Approval Process (SAP) for small-scale projects (SAP). The first recipient was Namibia's improved rangeland management practices. In October 2018, additional 19 projects were approved. As of December 2018, 93 projects in total received money from the GCF and the size of the approved funding amounted to US$ 4.6 billion (GCF 2018f).

As of the end of 2018, the total pledge announced and signed remained at US$ 10 billion, of which about 60% was disbursed to various projects. Almost all disbursements were in the form of a grant and only a small fraction of the funding allocations was in the form of a loan (GCF 2018d).

2.3 An Analysis of GCF Pledges

The initial resource mobilization for the establishment of the GCF began in 2014. As of January 29, 2018, the total of GCF pledges announced and signed is about US$ 10 billion, counting both grant pledges and loan pledges. In Fig. 2.1, top country-level pledges are drawn for the pledges that were at least US$ 100 million. The dollar amount in the figure shows total pledges made as a grant, excluding loan pledges.

There were four countries that pledged more than US$ 1 billion in grants, five countries if both grants and loans are counted: The US, Japan, the UK, Germany, and France. There were 14 countries that pledged more than US$ 100 million: In addition to the five countries, Sweden, Norway, Italy, Australia, Spain, Canada, the Netherlands, Republic of Korea, and Switzerland. The share of contribution by the 14 countries is about 98% of the total GCF fund, taking into account both grants and loans. The share of contribution by the five top contributors is about 78%.

In Table 2.1, the full record of the pledges made as of the above-mentioned date is presented. In addition to the 14 countries, there are 30 other countries contributed to the GCF fund. Notably, many developing countries also contributed to the fund, but their contributions appear quite nominal, that is, 0.1 million USD pledged by many developing countries.

Of the pledges that were made in the form of either a grant or a loan, the grant pledges accounted for about 95% of the total GCF fund while the loan pledges about 5% or about US$ 480 million contributed by France and Canada. The loan pledge is given to the GCF in return for a certain rate of interest payment.

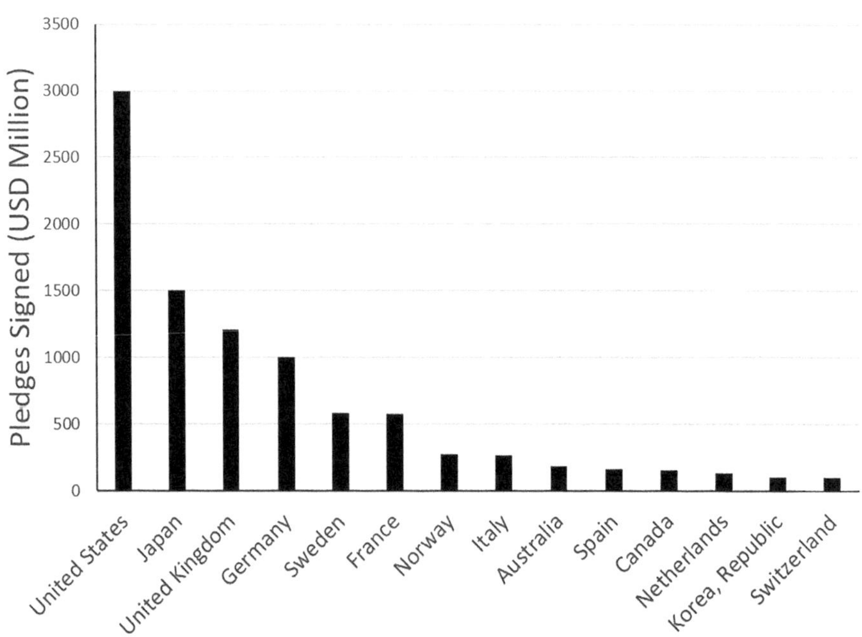

Fig. 2.1 Top GCF pledges (grant only). (As of January 29, 2018)

Table 2.1 Pledges announced, signed, disbursed. (As of January 29, 2018)

	Pledges announced in US$ (m.)	Pledges signed in US$ (m.)	% Pledges disbursed cash and deposited promissory notes
Announced, Signed			
Australia	187.2	187.2	83%
Austria	34.8	34.8	91%
Belgium	66.9	66.9	100%
Belgium Brussels	4.8	4.8	100%
Flanders	13	13	100%
Wallonia 1	1.5	1.5	100%
Wallonia 2	9.4	9.4	100%
Bulgaria	0.1	0.1	100%
Canada, grant	155.1	155.1	100%
Loan	101.6	101.6	0%
Cushion	20.3	20.3	0%
Chile	0.3	0.3	100%
Colombia	0.3	0.3	100%
Cyprus	0.5	0.5	0%
Czech	5.3	5.3	100%
Demark	71.8	71.8	100%
Estonia	1.3	1.3	100%
Finland	46.4	46.4	100%
France, grant	577.9	577.9	63%
Loan	381.3	381.3	100%
Cushion	76.3	76.3	100%
Paris	1.3	1.3	70%
Germany	1003.30	1003.30	75%
Hungary	4.3	4.3	100%
Iceland	0.7	0.7	100%
Indonesia	0.3	0.3	67%
Ireland	5.4	5.4	100%
Italy	267.5	267.5	75%
Japan	1500.00	1500.00	75%
Latvia	0.5	0.5	100%
Liechtenstein	0.1	0.1	100%
Lithuania	0.1	0.1	100%
Luxembourg	33.4	33.4	100%
Malta	0.3	0.3	150%
Mexico	10	10	100%
Mongolia	0	0	0%
Monaco	1.3	1.3	100%
Netherlands	133.8	133.8	34%
New Zealand	2.6	2.6	100%
Norway	270.8	270.8	77%

(continued)

Table 2.1 (continued)

	Pledges announced in US$ (m.)	Pledges signed in US$ (m.)	% Pledges disbursed cash and deposited promissory notes
Panama	1	1	100%
Poland	0.1	0.1	100%
Portugal	2.7	2.7	100%
Korea, Republic	100	100	47%
Romania	0.1	0.1	100%
Spain	160.5	160.5	14%
Sweden	581.2	581.2	100%
Switzerland	100	100	100%
United Kingdom	1211.00	1211.00	73%
United States	3000.00	3000.00	33%
	10,148.40	10,148.40	
Announced, not Signed			
Colombia	5.7		
Finland	60.6		
Iceland	0.3		
Italy	66.9		
Luxembourg	13.4		
Peru	6		
Vietnam	1		
	153.9		

Note: The data are from GCF (2018d)

Shown at the bottom panel of Table 2.1, there are also announced-but-not-yet-signed pledges, but the total amount of which is only US$ 153 million, so it does not add much to the total size of the GCF fund.

2.4 An Analysis of GCF Funding Decisions

An allocation of the US$ 10 billion GCF began in November 2015 at which time the first batch of projects were approved. As of March 2018, 77 GCF projects were approved and the total spending approved for these projects over the courses of project implementation periods is about US$ 12 billion.

All approved projects were chosen from developing countries in Africa, Latin America and the Caribbean, Asia Pacific, and Eastern Europe. As shown in Fig. 2.2, 30% of the total funding went to Latin American and the Caribbean countries, 25% to African countries, 20% to Asia Pacific, 7% to Pacific Islands, and 1% to Eastern European countries. There were several projects that engage multiple continents, which received 17% of the funding.

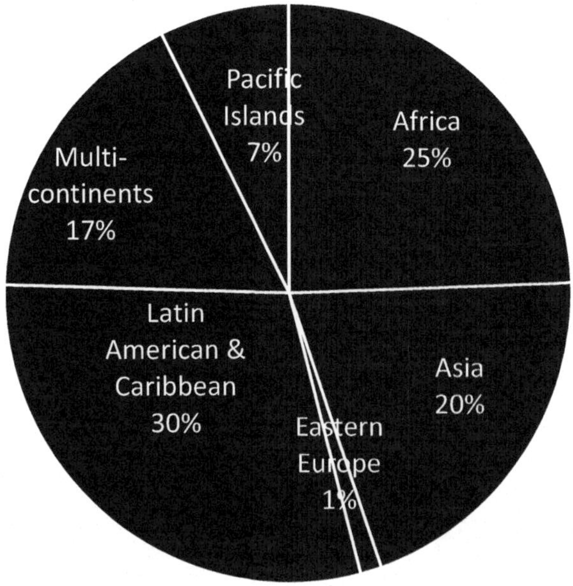

Fig. 2.2 GCF allocations by geopolitical region

The disbursements of the GCF funding can be analyzed by the policy 'results areas' defined by the GCF Board. As shown in Fig. 2.3, the largest share of the fund allocations went to the results area of 'energy generation and access', which received almost US$ 7 billion. About US$ 1.8 billion was approved for the projects in the results area of 'buildings, cities, industries, and appliances' and another US$ 1.8 billion for the projects on the results area of 'health, food security, and water security.'

Other policy results areas that received the GCF funding are as follows: the results area of transport (US$ 0.6 million), the results area of livelihoods of people and community (US$ 0.68 billion), the results area of forests and land use (US$ 0.4 billion), the results area of infrastructure and built environment (US$ 0.2 billion).

The GCF grants and loans on each GCF-defined policy results area can be further broken down by the continent to which the fund is distributed. As shown in Table 2.2, when it comes to the projects on the results area of the energy generation and access, the awards were by and large evenly split into Africa, Latin America and the Caribbean, and Asia.

In the results area of the health, food, and water security, a majority of the funding (50%) was given to the African continent. A large fraction of the GCF funding (26%) within this results area was also allocated to the projects from Asia Pacific islands.

Of the projects in the results area of the buildings, cities, industries, and appliances, a vast majority of the funding (87%) was allocated to Latin America and the Caribbean countries. The continent also received all of the GCF funding allocated to the results area of the infrastructure and built environment.

In the results area of the transport, Asia received by far the largest share of the GCF funding (89%). The rest was distributed to Pacific Islands.

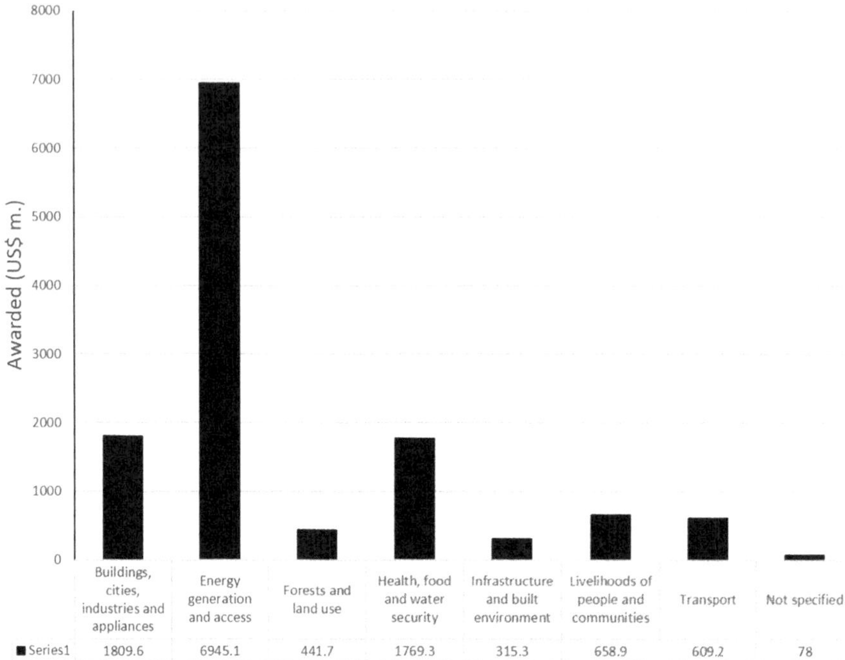

Fig. 2.3 GCF allocations by policy results area

For the results area of the forests and land use, the largest share of the GCF grants (57%) is given to Latin America and the Caribbean countries. The second largest share (32%) is given to the Asian projects.

For the projects in the results area of the livelihoods of people and community, the GCF funding was split to Africa (40%), Asia (35%), and Pacific Islands (22%).

Besides the results areas, the GCF Board classifies the awarded projects by the sustainable development goals which were suggested by another United Nations conference document, that is, the millennium sustainable development goals (SDG) defined by the UN Millennium Declaration (UN 2000).

An examination of the GCF projects' documents reveals that the GCF grants and loans were awarded to the following seven sustainable development goals: zero hunger; clean water and sanitation; affordable and clean energy; industry, innovation and infrastructure; sustainable cities and communities; climate action; life on land.

The disbursements of the GCF dollars can be analyzed by the UN sustainable development goals (UN 2000). It should be emphasized that although the problems of climate change and global warming are mostly not the same as the sustainable development problems, the GCF Board asks applicants to specify the area of UN sustainable development goals that the project addresses. It is also notable that there were 21 funded projects that did not specify the UNSDG area.

As presented in Table 2.3, of the total 77 GCF-funded projects only one is classified to be a climate change-related project, that is, the area of climate action.

Table 2.2 GCF funding allocations by results area and geopolitical region

Policy results areas	Geopolitical regions	Funding size (US$ m.)
Buildings, cities, industries and appliances	Africa	
	Asia	88
	Eastern Europe	152.4
	Latin America & Caribbean	1569.2
	Pacific Islands	
	Total	1809.6
Energy generation and access	Africa	1894.1
	Asia	1218.1
	Eastern Europe	
	Latin America & Caribbean	1407.9
	Pacific Islands	260
	Multi-continents	2165
	Total	6945.1
Forests and land use	Africa	49.2
	Asia	141.4
	Eastern Europe	
	Latin America & Caribbean	251.1
	Pacific Islands	
	Total	441.7
Health, food and water security	Africa	884
	Asia	261.4
	Eastern Europe	
	Latin America & Caribbean	168.8
	Pacific Islands	455.1
	Total	1769.3
Infrastructure and built environment	Africa	
	Asia	
	Eastern Europe	
	Latin America & Caribbean	315.3
	Pacific Islands	
	Total	315.3
Livelihoods of people and communities	Africa	261.1
	Asia	229.1
	Eastern Europe	
	Latin America & Caribbean	20
	Pacific Islands	148.7
	Total	658.9

(continued)

Table 2.2 (continued)

Policy results areas	Geopolitical regions	Funding size (US$ m.)
Transport	Africa	
	Asia	544
	Eastern Europe	
	Latin America & Caribbean	
	Pacific Islands	65.2
	Total	609.2

Table 2.3 Allocations of the fund by UN sustainable development goal

Sustainable development goals	Funding size (US$ m.)	Number of projects funded
#2 Zero hunger	224	3
#6 Clean water and sanitation	1447	10
#7 Affordable and clean energy	6442.5	19
#9 Industry, innovation and infrastructure	365.9	7
#11 Sustainable cities and communities	433.9	9
#13 Climate action	9.1	1
#15 Life on land	608.2	7
Not specified	3096.5	21
	12,627.1	77

Only 9 million dollars are assigned to this sole climate project out of the total about 13 billion US$, less than one tenth of one percent of the GCF funding allocations.

The largest share of the GCF funding was allocated to the UNSDG categorized as the goal of affordable and clean energy. About 6.4 billion US$ (about 50%) was allocated to this SDG which is supported by 19 approved projects.

The next UNSDG area that received the second-most funding was the goal of clean water and sanitation which received 1.4 billion US$ funding divided into ten approved projects. It is followed by the UNSDG of sustainable cities and communities (9 funded projects), that of industry, innovation, and infrastructure (7 funded projects), that of life on land (7 funded projects), and that of zero hunger (3 funded projects).

2.5 An Analysis of Individual GCF Awards by Results Area

We can go deeper in the analysis of the allocations of the GCF fund to the level of individual GCF grants. Individual grants and loans can be analyzed by the GCF-defined results area. Let's take a look at each group of projects one by one.

In Table 2.4, individual projects in the results area of the energy generation and access are presented by the date of approval. For each project, the table shows recipient countries, project title, expected contributions, and grant/loan size.

As mentioned above, the awarded energy-generation-access projects were chosen almost evenly from the three geopolitical regions: Africa, Asia, and Latin America and the Caribbean countries. A limited funding was also given to the Eastern Europe and Pacific Islands.

The energy-generation-access projects in Table 2.4 can be categorized into two types. The first type is a direct grant to a renewable energy generation and access project through solar, hydro-electric, or wind energy. The second type is the support for establishing a renewable energy fund from which loans can be given in its own discretion to selected different renewable energy projects.

The first type of the energy-generation-access projects includes the following: Climate Action and Solar Energy Development Programme in the Tarapacá Region in Chile (Chile); Sustainable Energy Facility for the Eastern Caribbean (Caribbean countries); Sustainable Landscapes in Eastern Madagascar (Madagascar); Universal Green Energy Access Programme (African countries); Catalyzing Private Investment in Sustainable Energy in Argentina (Argentina); Accelerating the Transformational Shift to a Low-Carbon Economy in the Republic of Mauritius (Mauritius); Tajikistan's Scaling Up Hydropower Sector Climate Resilience (Tajikistan); Tina River Hydropower Development Project (Solomon Islands); Renewable Energy Program #1—Solar (Mongolia); Global Clean Cooking Program—Bangladesh (Bangladesh); Zambia Renewable Energy Financing Framework (Zambia); Line of Credit for Solar Rooftop Segment for Commercial, Industrial and Residential Housing sectors (India); Strengthening Climate Resilience of Rural Communities in Northern Rwanda (Rwanda).

The second type of the energy-generation-access projects includes the following: KawiSafi Ventures Fund in East Africa (Rwanda, Kenya); GCF-EBRD (European Bank for Reconstruction and Development) Sustainable Energy Financing Facilities (multi-continents); SCF Capital Solutions (South Africa); Geeref Next (Caribbean).

Of this results area, the largest funding was given to the GCF-EBRD Sustainable Energy Financing Facilities which is a multi-continent project across Asia, Africa, Latin America, and Eastern Europe, with the grant of US$ 1.4 billion. The project's expected contribution is 27.5 million tons of CO2-equivalent reduction. The second largest funding was given to the Egypt Renewable Energy Financing Framework project, with US$ 1.0 billion grant. The expected contribution was proposed to be 18.9 million tons of CO2-equivalent reduction.

The smallest grant was given to the Pacific Islands Renewable Energy Investment Program (Cook Islands, Marshall Islands, etc), with the award of US$ 26 million. The expected outcome of the project is estimated to be 3 million tons of CO2-equivalent reduction.

The next set of the funded projects we examine is from the results area of the buildings, cities, industries, and appliances, summarized in Table 2.5. The table also includes the projects on the results area of the infrastructure and built environment.

Table 2.4 GCF-funded projects on energy generation and access

Recipient	Project title	Expected contribution	Funding size	Approval date
GCF results area: energy generation and access				
Rwanda, Kenya	KawiSafi Ventures Fund in East Africa	1.5 m tons	$110.0 m	2015-11
Chile	Climate Action and Solar Energy Development Programme in the Tarapacá Region in Chile	3.7 m tons	$265.0 m	2016-06
Dominica, Grenada etc	Sustainable Energy Facility for the Eastern Caribbean	9.4 m tons	$190.5 m	2016-10
Armenia, Egypt, Georgia, etc	GCF-EBRD Sustainable Energy Financing Facilities	27.5 m tons	$1.4 b	2016-10
Madagascar	Sustainable Landscapes in Eastern Madagascar	10.0 m tons	$69.8 m	2016-10
Benin, Kenya, etc	Universal Green Energy Access Programme	50.6 m tons	$301.6 m	2016-10
Mongolia	Business loan programme for GHG emissions reduction	1.2 m tons	$60.0 m	2016-12
Argentina	Catalyzing private investment in sustainable energy in Argentina – Part 1	15.3 m tons	$653.0 m	2016-12
Mauritius	Accelerating the Transformational Shift to a Low-Carbon Economy in the Republic of Mauritius	4.3 m tons	$191.4 m	2016-12
Cook Islands, Marshall Islands, etc	Pacific Islands Renewable Energy Investment Program	3.0 m tons	$26.0 m	2016-12
South Africa	SCF Capital Solutions	33.0 m tons	$34.1 m	2016-12, W*
Bahamas, Barbados, etc	Geeref Next	769.0 m tons	$765.0 m	2017-04
Egypt	Egypt Renewable Energy Financing Framework	18.9 m tons	$1.0 b	2017-04
Tajikistan	Tajikistan: Scaling Up Hydropower Sector Climate Resilience	7.5 m tons	$133.0 m	2017-04
Solomon Islands	Tina River Hydropower Development Project	2.5 m tons	$234.0 m	2017-04
Mongolia	Renewable Energy Program #1 – Solar	306.7 k tons	$17.6 m	2017-10
Kazakhstan	GCF-EBRD Kazakhstan Renewables Framework	12.9 m tons	$557.0 m	2017-10
Bhutan	Bhutan for Life	35.1 m tons	$118.3 m	2017-10
Barbados	Water Sector Resilience Nexus for Sustainability in Barbados (WSRN S-Barbados)	220.2 k tons	$45.2 m	2018-02
Paraguay	Poverty, Reforestation, Energy and Climate Change Project (PROEZA)	7.9 m tons	$90.3 m	2018-02

(continued)

Table 2.4 (continued)

Recipient	Project title	Expected contribution	Funding size	Approval date
Argentina	Promoting risk mitigation instruments and finance for renewable energy and energy efficiency investments	9.1 m tons	$163.9 m	2018-02
Bangladesh	Global Clean Cooking Program—Bangladesh	2.9 m tons	$82.2 m	2018-02
Zambia	Zambia Renewable Energy Financing Framework	4.0 m tons	$154.0 m	2018-02
India	Line of Credit for Solar rooftop segment for Commercial, Industrial and Residential Housing sectors	5.2 m tons	$250.0 m	2018-02
Rwanda	Strengthening climate resilience of rural communities in Northern Rwanda (SCRNRP)	273.7 k people	$33.2 m	2018-03

Note: *b* billion, *m* million, *k* thousand, *W*∗ Withdrawn

Table 2.5 GCF-funded projects on buildings, cities, industries, and appliances

Recipient	Project title	Expected contribution	Funding size	Approval date
GCF results area: buildings, cities, industries, and appliances				
Mexico	Energy Efficiency Green Bonds in Latin America and the Caribbean	2.6 m tons	$184.5 m	2015-11
El Salvador	Energy savings insurance for private energy efficiency investments by Small and Medium-Sized Enterprises	562.0 k tons	$41.7 m	2016-06
Armenia	De-risking and scaling-up investment in energy efficient building retrofits in Armenia	1.4 m tons	$29.8 m	2016-06
Bosnia and Herzegovina	Scaling-up Investment in Low-Carbon Public Buildings	2.0 m tons	$122.6 m	2017-10
Paraguay	Promoting private sector investments in energy efficiency in the industrial sector in Paraguay	4.0 m tons	$43.0 m	2018-02
Brazil	Financial Instruments for Brazil Energy Efficient Cities (FinBRAZEEC)	17.4 m tons	$1.3 b	2018-02
Vietnam	Scaling Up Energy Efficiency for Industrial Enterprises in Vietnam	120.0 m tons	$88.0 m	2018-03
GCF Results Area: Infrastructure and Built Environment				
Argentina	Implementation Project of the Integral Management Plan of the Lujan River Basin	2.8 m people	$315.3 m	2017-10

Note: *b* billion, *m* million, *k* thousand

Table 2.6 GCF-funded projects on forests and land use

Recipient	Project title	Expected contribution	Funding size	Approval date
GCF results area: forests and land use				
Ecuador	Priming Financial and Land-Use Planning Instruments to Reduce Emissions from Deforestation	15.0 m tons	$84.0 m	2016-10
Morocco	Development of Argan orchards in Degraded Environment – DARED	604.2 k tons	$49.2 m	2016-10
Guatemala, Mexico	Low-Emission Climate Resilient Agriculture Risk Sharing Facility for MSMEs	9.2 m tons	$158.0 m	2017-10
Cambodia	Climate-Friendly Agribusiness Value Chains Sector Project	240.0 k tons	$141.4 m	2018-02
Peru	Building the Resilience of Wetlands in the Province of Datem del Marañón, Peru	2.6 m tons	$9.1 m	2015-11

Note: *m* million, *k* thousand

The projects in these areas purport either to improve energy efficiency in buildings, cities, and industries or to build low-carbon buildings.

The GCF-funded projects are as follows: Energy Savings Insurance for Private Energy Efficiency Investments by Small and Medium-Sized Enterprises (El Salvador); De-risking and Scaling-up Investment in Energy Efficient Building Retrofits in Armenia (Armenia); Scaling-up Investment in Low-Carbon Public Buildings (Bosnia and Herzegovina); Promoting Private Sector Investments in Energy Efficiency in the Industrial Sector in Paraguay (Paraguay); Financial Instruments for Brazil Energy Efficient Cities (Brazil); Scaling Up Energy Efficiency for Industrial Enterprises in Vietnam (Vietnam).

In the results area, the largest funding was given to the project entitled "Financial Instruments for Brazil Energy Efficient Cities," with the grant size of US$ 1.3 billion. With the investment, the expected reduction of the CO2-equivalent emissions is estimated to be 17.5 million tons.

The GCF-funded projects in the results area of the forests and land use are summarized in Table 2.6. There are five funded projects which cover a range of related topics: deforestation, orchards, agriculture, agribusiness chains, and wetlands. Three of the projects in this area are located in Latin America.

Individual projects are as follows: Priming Financial and Land-Use Planning Instruments to Reduce Emissions from Deforestation (Ecuador); Development of Argan Orchards in Degraded Environment (Morocco); Low-Emission Climate Resilient Agriculture Risk Sharing Facility for Micro, Small, and Medium Enterprises (MSMEs) (Guatemala, Mexico); Climate-Friendly Agribusiness Value Chains Sector Project (Cambodia); Building the Resilience of Wetlands in the Province of Datem del Marañón, Peru (Peru).

In the results area, the largest funding was given to the project entitled "Low-Emission Climate Resilient Agriculture Risk Sharing Facility for MSMEs" submitted by the two Central American countries, with a price tag of US$ 158 million. The smallest funding was given to the Peru's wetlands projects, with a price tag of US$ 9.1 million.

Next, the GCF-funded projects in the results area of the health, water, and food security are listed in Table 2.7. This results area can be further decomposed into the following three priority areas: water security, food security, health security. Each of these sub-results areas is treated, more often than not, as a separate research field. When we come to evaluate individual GCF projects in Chap. 7, we will also treat each of these sub-areas separately.

The water security projects are: Support of Vulnerable Communities in Maldives to Manage Climate Change-Induced Water Shortages (Maldives); Fiji Urban Water Supply and Wastewater Management Project (Fiji); Saïss Water Conservation Project (Morocco); Scaling Up Climate Resilient Water Management Practices for Vulnerable Communities in La Mojana (Colombia); Responding to the Increasing Risk of Drought, Building Gender-responsive Resilience of the Most Vulnerable Communities (Ethiopia); Climate-Resilient Water Sector in Grenada (Grenada).

The food security projects are: Strengthening the Resilience of Smallholder Farmers in the Dry Zone to Climate Variability and Extreme Events through an Integrated Approach to Water Management (Sri Lanka); Climate Resilient Agriculture in Three of the Vulnerable Extreme Northern Crop-growing Regions (CRAVE) (Namibia); Empower to Adapt, Creating Climate-Change Resilient Livelihoods through Community-Based Natural Resource Management in Namibia (Namibia); Irrigation Development and Adaptation of Irrigated Agriculture to Climate Change in Semi-arid Morocco (Morocco); Ground Water Recharge and Solar Micro Irrigation to Ensure Food Security and Enhance Resilience in Vulnerable Tribal Areas of Odisha (India); Building the Climate Resilience of Food-insecure Smallholder Farmers through Integrated Management of Climate Risks (Senegal); Strengthening Climate Resilience of Agricultural Livelihoods in Agro-Ecological Regions I and II in Zambia (Zambia); Africa Hydromet Program—Strengthening Climate Resilience in Sub-Saharan Africa, Burkina Faso Country Project (Burkina Faso); Improving Rangeland and Ecosystem Management Practices of Smallholder Farmers under Conditions of Climate Change in the Republic of Namibia (Namibia); Acumen Resilient Agriculture Fund (ARAF) (Uganda, Ghana, Nigeria).

The health-related projects are: Climate Information Services for Resilient Development in Vanuatu (Vanuatu); Simiyu Climate Resilient Development Programme (Tanzania); Enhancing Adaptive Capacities of Coastal Communities, Especially Women, to Cope With Climate Change Induced Salinity (Bangladesh).

Readers may feel distracted by the large number of projects, countries, localities, and concepts that appear in the preceding three paragraphs. However, I can assure you that you will feel rather comfortable with each of these projects by Chap. 7 where each project is brought forth for evaluation.

Importantly, unlike the projects described so far in the other results areas, the expected contribution of the project in this result area is expressed by the number of beneficiaries. The India's groundwater recharge and solar micro irrigation project in Odisha was estimated by the project applicant to benefit 16 million people, for which US$ 166 million was awarded. The Acumen Resilient Agriculture Fund (ARAF) in Uganda, Ghana, Nigeria was estimated to benefit 10 million people and awarded US $ 51 million.

Table 2.7 GCF-funded projects on health, food, and water security

Recipient	Project title	Expected contribution	Funding size	Approval date
GCF results area: health, food and water security				
Maldives	Support of Vulnerable Communities in Maldives to Manage Climate Change-Induced Water Shortages	105 k people	$28.2 m	2015-11
Fiji	Fiji Urban Water Supply and Wastewater Management Project	290854 people	$405.1 m	2015-11
Gambia	Large-scale Ecosystem-based Adaptation in the Gambia River Basin: developing a climate resilient, natural resource based economy	57750 people	$25.5 m	2016-06
Mali	Africa Hydromet Program – Strengthening Climate Resilience in Sub-Saharan Africa: Mali Country Project	5.3 m people	$27.3 m	2016-06
Sri Lanka	Strengthening the resilience of smallholder farmers in the Dry Zone to climate variability and extreme events through an integrated approach to water management	2.0 m people	$52.1 m	2016-06
Namibia	Climate Resilient Agriculture in three of the Vulnerable Extreme northern crop-growing regions (CRAVE)	21.0 k people	$10.0 m	2016-10
Namibia	Empower to Adapt: Creating Climate-Change Resilient Livelihoods through Community-Based Natural Resource Management in Namibia	76.5 k people	$10.0 m	2016-10
Vanuatu	Climate Information Services for Resilient Development in Vanuatu	260.7 k people	$21.8 m	2016-12
United Republic of Tanzania	Simiyu Climate Resilient Development Programme	3.0 m people	$176.0 m	2017-04
Morocco	Irrigation development and adaptation of irrigated agriculture to climate change in semi-arid Morocco	15.5 k people	$93.3 m	2017-04
Morocco	Saïss Water Conservation Project	2.2 m people	$253.6 m	2017-04
India	Ground Water Recharge and Solar Micro Irrigation to Ensure Food Security and Enhance Resilience in Vulnerable Tribal Areas of Odisha	16.0 m people	$166.3 m	2017-04
Senegal	Building the Climate Resilience of Food-insecure Smallholder Farmers through Integrated Management of Climate Risks (the R4 Rural Resilience Initiative)	526.5 k people	$10.0 m	2017-10
Colombia	Scaling Up Climate Resilient Water Management Practices for Vulnerable Communities in La Mojana	405.6 k people	$117.2 m	2017-10

(continued)

Table 2.7 (continued)

Recipient	Project title	Expected contribution	Funding size	Approval date
Ethiopia	Responding to the Increasing Risk of Drought: Building Gender-responsive Resilience of the Most Vulnerable Communities	1.3 m people	$50.0 m	2017-10
Grenada	Climate-Resilient Water Sector in Grenada (G-CREWS)	107.3 k people	$51.6 m	2018-02
Tajikistan	Building climate resilience of vulnerable and food insecure communities through capacity strengthening and livelihood diversification in mountainous regions of Tajikistan	120.0 k people	$10.0 m	2018-02
Bangladesh	Enhancing adaptive capacities of coastal communities, especially women, to cope with climate change induced salinity	719.2 k people	$33.0 m	2018-02
Zambia	Strengthening climate resilience of agricultural livelihoods in Agro-Ecological Regions I and II in Zambia	3.9 m tons	$137.3 m	2018-02
Burkina Faso	Africa Hydromet Program—Strengthening Climate Resilience in Sub-Saharan Africa: Burkina Faso Country Project	7.0 m people	$25.0 m	2018-02
Namibia	Improving rangeland and ecosystem management practices of smallholder farmers under conditions of climate change in Sesfontein, Fransfontein, and Warmquelle areas of the Republic of Namibia	44.4 k people	$10.0 m	2018-02
Uganda, Ghana, Nigeria	Acumen Resilient Agriculture Fund (ARAF)	10.0 m people	$56.0 m	2018-03

Note: *m* million, *k* thousand

The projects funded for the results area of the livelihoods of people and communities are presented in Table 2.8. The funded projects appear to embrace broadly the activities that might increase the resilience of various economic sectors and communities. It is notable that a majority of the projects in this area are concentrated on either coastal risk management or flood risk management. Some are the projects from the South Pacific small island nations. Examples are the project of Climate-Resilient Infrastructure Mainstreaming in Bangladesh (Bangladesh); Climate Adaptation and Mitigation Program for the Aral Sea Basin (Tajikistan, Uzbekistan); Tuvalu Coastal Adaptation Project (Tuvalu); Integrated Urban Flood Management Project (Senegal); Integrated Flood Management to Enhance Climate Resilience of the Vaisigano River Catchment in Samoa (Samoa); Enhancing Climate Change Adaptation in the North Coast and Nile Delta Regions in Egypt (Egypt); Pacific Resilience Project Phase II for the Republic of the Marshall Islands (Marshall Islands).

Table 2.8 GCF-funded projects on livelihoods of people and communities

Recipient	Project title	Expected contribution	Funding size	Approval date
GCF results area: livelihoods of people and communities				
Malawi	Scaling Up of Modernized Climate Information and Early Warning Systems in Malawi	2.1 m people	\$16.3 m	2015-11
Senegal	Increasing Resilience of Ecosystems and Communities through Restoration of the Productive Bases of Salinized Lands	129.8 k people	\$8.2 m	2015-11
Bangladesh	Climate-Resilient Infrastructure Mainstreaming in Bangladesh	10.5 m people	\$80.0 m	2015-11
Tajikistan, Uzbekistan	Climate Adaptation and Mitigation Program for the Aral Sea Basin (CAMP4ASB)	205.0 k people	\$68.8 m	2016-06
Tuvalu	Tuvalu Coastal Adaptation Project	6.6 k people	\$38.9 m	2016-06
Senegal	Senegal Integrated Urban Flood Management Project	2.2 m people	\$87.1 m	2016-10
Uganda	Building Resilient Communities, Wetlands Ecosystems and Associated Catchments in Uganda	4.8 m people	\$44.3 m	2016-12
Samoa	Integrated Flood Management to Enhance Climate Resilience of the Vaisigano River Catchment in Samoa	37.0 k people	\$65.7 m	2016-12
Egypt	Enhancing Climate Change Adaptation in the North Coast and Nile Delta Regions in Egypt	17.7 m people	\$105.2 m	2017-10
Marshall Islands	Pacific Resilience Project Phase II for the Republic of the Marshall Islands	39.8 k people	\$44.1 m	2018-02
Georgia	Scaling-up Multi-Hazard Early Warning System and the Use of Climate Information in Georgia	3.7 m people	\$70.3 m	2018-02
Tajikistan	Institutional Development of the State Agency for Hydrometeorology of Tajikistan	8.7 m people	\$10.0 m	2018-02
Antigua and Barbuda, Dominica, Grenada	Integrated physical adaptation and community resilience through an enhanced direct access pilot in the public, private, and civil society sectors of three Eastern Caribbean small island developing states	82.2 k people	\$20.0 m	2018-03

Note: *m* million, *k* thousand

Other projects funded through this results area are the project of Institutional Development of the State Agency for Hydrometeorology of Tajikistan (Tajikistan); Building Resilient Communities, Wetlands Ecosystems and Associated Catchments in Uganda (Uganda); Scaling-up Multi-Hazard Early Warning System and the Use of Climate Information in Georgia (Georgia).

Table 2.9 GCF-funded projects on transport

Recipient	Project title	Expected contribution	Funding size	Approval date
GCF Results Area: Transport				
Nauru	Sustainable and Climate Resilient Connectivity for Nauru	535.4 k tons	$65.2 m	2017-10
Mongolia	Ulaanbaatar Green Affordable Housing and Resilient Urban Renewal Project (AHURP)	7.9 m tons	$544.0 m	2018-02

The largest expected contribution is predicted to come from the Egypt's climate adaptation project in the Nile Delta regions with 17.7 million beneficiaries. The second largest project in terms of the number of beneficiaries is the Bangladesh's infrastructure project with the expected 10.5 million beneficiaries. The size of the grant was US$ 105 million for the Nile Delta project and US$ 80 million for the Bangladesh project.

Bangladesh has historically suffered the largest number of human deaths from tropical cyclones of any countries in the world, with Tropical Cyclone Bhola in 1970 alone resulting in half a million human deaths (Paul 2009; Seo and Bakkensen 2017). As such, many other international aids, besides the GCF grants shown in Table 2.8 and other tables in this chapter, have been directed to coastal risk management in the Bangladesh Delta (World Bank 2008; Seo 2017c).

The next results area is transport, shown in Table 2.9. There are two projects awarded for the transport area: The Sustainable and Climate Resilient Connectivity for Nauru (Nauru); Ulaanbaatar Green Affordable Housing and Resilient Urban Renewal Project (Mongolia). The Mongolian project was given US$ 544 million. It was estimated to cut 7.9 million tons of CO2-equivalent emissions.

As of April 2018, the two GCF-funded projects were not assigned to a particular results area (Table 2.10). The first is the Vietnam's coastal adaptation project entitled Improving the Resilience of Vulnerable Coastal Communities to Climate Change Related Impacts in Viet Nam. The second was the Pakistan's glacier program entitled Scaling-up of Glacial Lake Outburst Flood (GLOF) risk reduction in Northern Pakistan.

At the time of writing this book through the end of fall 2018, the projects summarized in Tables 2.4, 2.5, 2.6, 2.7, 2.8, 2.9 and 2.10 were the entirety of the projects on which the GCF Board approved its funding. In the fall of 2018 amid the resignation of the Director of the GCF Secretariat, additional projects were approved most recently, which are summarized in Table 2.11, at the 21st Board meeting of the GCF (GCF 2018f).

The recipient countries of these most recent awards are Indonesia, India, Pakistan, Guatemala, El Salvador, Tonga, Kiribati, Niger Basin, Burkina Faso, Comoros Islands, Democratic Republic of Congo, Kyrgyz Republic, Bahrain. There are several multi-country projects.

The results areas of these most recently funded projects were not specified at the time of writing. However, these projects are similar in the proposed activities to

Table 2.10 GCF-funded projects, no results area

Recipient	Project title	Expected contribution	Funding size	Approval date
Vietnam	Improving the resilience of vulnerable coastal communities to climate change related impacts in Viet Nam	1.9 m tons	$40.5 m	2016-06
Pakistan	Scaling-up of Glacial Lake Outburst Flood (GLOF) risk reduction in Northern Pakistan	29.2 m people	$37.5 m	2016-10

Note: *m* million

those approved before the 21st Board meeting on October 21, 2018 shown from Tables 2.4, 2.5, 2.6, 2.7, 2.8, 2.9, 2.10 and 2.11 while the recipients of these most recent grants are different geographical regions.

2.6 GCF Investment Criteria

How we should evaluate the GCF Board's decisions to fund the selected projects in Tables 2.4, 2.5, 2.6, 2.7, 2.8, 2.9, 2.10 and 2.11 is the central question of this book. The GCF's investment decisions are made by the set of investment criteria that the international organization has formulated. Therefore, we will succeed in assessing the GCF-funded projects only after an incisive examination of the investment criteria.

The GCF's investment criteria are formulated with multiple layers from the top down. At the top, there are a set of investment criteria, under which are a family of sub-criteria, under which are a set of assessment indicators. The six top-level criteria are, in the order established by the Board, impact potential, paradigm shift potential, sustainable development potential, needs of the recipient, country ownership, and effectiveness and efficiency. The impact potential criterion is further divided into mitigation impact potential and adaptation impact potential (GCF 2018b, c).

The GCF criteria are summarized in Tables 2.12 and 2.13. A cursory glance at the tables reveals two salient features of the investment framework put forth by the GCF. First, there is no single encompassing indictor that can tell us decisively which project is more or less important. The single investment indicator may be constructed by assigning an individual weight to each of the six criteria.

The second salient feature is that the sustainable development concept is again inserted as one of the key criteria and assessment indicators in the investment decisions of the GCF Board. We already observed in Table 2.3 that the United Nations' sustainable development goals must be attached to each of the GCF proposal documents. Despite the apparently pivotal weight given to the concept of sustainable development in the GCF project proposals and investment decisions, it is quite striking that there is no serious analysis conducted on the fact that the problem of climate change is very different from that of sustainable development.

Table 2.11 Approvals at the 21st board meeting of the GCF (October 21, 2018)

ID	Project description	Funding size
FP083	Indonesia Geothermal Resource Risk Mitigation Project with World Bank	USD 100 million
FP084	Enhancing climate resilience of India's coastal communities with UNDP	USD 43.4 million
FP085	Green BRT Karachi in Pakistan with ADB	USD 49 million
FP086	Green Cities Facility in Albania, Armenia, Georgia, Jordan, Moldova, Mongolia, Serbia, The Former Yugoslav Republic of Macedonia, and Tunisia with EBRD	USD 101.4 million
FP087	Building livelihood resilience to climate change in the upper basins of Guatemala's highlands with IUCN	USD 22 million
FP089	Upscaling climate resilience measures in the dry corridor agroecosystems of El Salvador (RECLIMA) with FAO	USD 35.8 million
FP090	Tonga Renewable Energy Project under the Pacific Islands Renewable Energy Investment Program with ADB	USD 29.9 million
FP091	South Tarawa Water Supply Project in Kiribati with ADB	USD 28.6 million
FP092	Programme for integrated development and adaptation to climate change in the Niger Basin (PIDACC/NB) in Benin, Burkina Faso, Cameroon, Chad, Cote d'Ivoire, Guinea, Mali, Niger (the), and Nigeria with AfDB	USD 67.8 million
FP093	Yeleen rural electrification project in Burkina Faso with AfDB	USD 28.3 million
FP094	Ensuring climate resilient water supplies in the Comoros Islands with UNDP	USD 41.9 million
FP095	Transforming financial systems for climate in Benin, Burkina Faso, Cameroon, Cote d'Ivoire, Ecuador, Egypt, Kenya, Madagascar, Mauritius, Morocco, Namibia, Nigeria, Senegal, South Africa, Tanzania, Togo, and Uganda with AFD	USD 279.7 million
FP096	DRC green mini-grid program in Democratic Republic of Congo with AfDB	USD 21 million
FP097	Productive investment initiative for adaptation to climate change (CAMBio II) in Costa Rica, Dominican Republic (the), El Salvador, Guatemala, Honduras, Nicaragua, and Panama with CABEI	USD 15.5 million
FP098	DBSA Climate finance facility in Eswatini, Lesotho, Namibia, South Africa with DBSA	USD 55.6 million
FP099	Climate Investor One in Burundi, Cameroon, Djibouti, Indonesia, Kenya, Madagascar, Malawi, Mongolia, Morocco, Nigeria, Uganda with FMO	USD 100 million
SAP002	Climate services and diversification of climate sensitive livelihoods to empower food insecure and vulnerable communities in the Kyrgyz Republic with WFP	USD 8.6 million
SAP003	Enhancing climate resilience of the water sector in Bahrain with UNEP	USD 2.16 million
SAP004	Energy efficient consumption loan programme in Mongolia with XacBank	USD 10 million

Note: GCF (2018f)

Table 2.12 Investment criteria: mitigation and adaptation impact

Criteria	Coverage area	Activity specific sub-criteria	Indicative assessment factors
Impact potential	Mitigation impact	Contribution to the shift to low-emission sustainable development pathways	(1) Expected tonnes of carbon dioxide equivalent to be reduced or avoided;
			(2) Degree to which activity avoids lock-in of long-lived, high-emission infrastructure;
			(3) Expected increase in the number of households with access to low emission energy
	Adaptation impact	Contribution to increased climate-resilient sustainable development	(1) Expected total number of direct and indirect beneficiaries;
			(2) Number of beneficiaries relative to total population, particularly the most vulnerable groups

Note: GCF (2018b)

Taking a closer look at Table 2.12, we find that the 'impact' criterion is divided into mitigation impact and adaptation impact. The sub-criterion for the mitigation impact criterion is specified to be the 'contribution to the shift to low-emission sustainable development pathways.' Again, with the sustainable development!

The indicators for the sub-criterion, called the indicative assessment factors by the GCF Board, include, among other things, (1) expected tons of carbon dioxide equivalent emissions to be reduced or avoided; (2) degree to which activity avoids lock-in of long-lived, high-emission infrastructure; (3) expected increase in the number of households with access to low emission energy.

With regard to the adaptation impact criterion, the sub-criterion is specified to be a 'contribution to increased climate-resilient sustainable development.' Again, with the sustainable development proclamation! The indicative assessment factors are, among other things, specified as follows: (1) expected total number of direct and indirect beneficiaries through either reduced vulnerability or increased resilience; (2) number of beneficiaries relative to total population, particularly the most vulnerable groups.

Again, the Board of the GCF does not seem to be aware that adaptation pathways to climate change will be largely different from sustainable development pathways. Across the GCF documents, approved projects, and the investment framework, there are many abuses of the term 'sustainable development' applied to wrong places, which makes precise interpretations of the criteria difficult.

A couple of more straightforward comments can be made about the adaptation impact criterion in Table 2.12. The indicative assessment indicators of the adaptation impact would overlook the situation that a GCF's adaptation project can harm a certain section of the recipient community. For example, a GCF's adaptation funding would relatively harm the sections of the community that are excluded from the funding.

Table 2.13 Investment criteria: others

Criteria	Coverage area	Indicative assessment factors/sub-criteria
Paradigm shift potential	Potential for scaling up and replication, and its overall contribution to global low-carbon development pathways being consistent with a temperature increase of less than 2 degrees Celsius (mitigation only)	(1) Opportunities for targeting innovative solutions, new market segments, developing or adopting new technologies, business models, modal shifts and/or processes;
		(2) Expected contributions to global low-carbon development pathways.
	Potential for knowledge and learning	Existence of a monitoring and evaluation plan and a plan for sharing lessons.
	Contribution to the creation of an enabling environment	Sustainability of outcomes and results beyond completion.
	Contribution to the regulatory framework and policies	Degree to which the project or programme advances the national/local regulatory or legal frameworks.
	Overall contribution to climate-resilient development pathways consistent with a country's climate change adaptation strategies and plans (adaptation only)	Scaling up the scope and impact of the intended project/programme without equally increasing the total costs of implementation.
Sustainable development potential	Environmental co-benefits	Degree to which the project or programme promotes positive environmental externalities such as air quality, soil quality, conservation, biodiversity, etc.
	Social co-benefits	Potential for externalities in the form of expected improvements, for women and men as relevant, in areas such as health and safety, access to education, improved regulation and/or cultural preservation
	Economic co-benefits	Potential for externalities in the form of expected improvements in areas such as expanded and enhanced job markets, job creation and poverty alleviation for women and men, increased and/or expanded involvement of local industries.
	Gender-sensitive development impact	Explanation of how the project activities will address the needs of women and men in order to correct prevailing inequalities in climate change vulnerability and risks.
Needs of the recipient	Vulnerability of the country (adaptation only)	Intensity of exposure to climate risks and the degree of vulnerability, including exposure to slow onset events.
	Vulnerable groups and gender aspects	Refer to the Note below.

(continued)

Table 2.13 (continued)

Criteria	Coverage area	Indicative assessment factors/sub-criteria
	Economic and social development level of the country	Refer to the Note below.
	Absence of alternative sources of financing	Refer to the Note below.
	Need for strengthening institutions and implementation capacity	Refer to the Note below.
Country ownership	Existence of a national climate strategy	Refer to the Note below.
	Coherence with existing policies	Refer to the Note below.
	Capacity of accredited entities or executing entities to deliver	Refer to the Note below.
	Engagement with civil society organizations and other relevant stakeholders	Refer to the Note below.
Efficiency and effectiveness	Cost-effectiveness and efficiency regarding financial and non-financial aspects	(1) Proposed financial structure is adequate and reasonable in order to achieve the proposal's objectives;
		(2) Demonstration that the proposed financial structure provides the least concessionality needed to make the proposal viable;
		(3) Demonstration that the Fund's support for the programme/project will not crowd out private and other public investment;
		(4) Estimated cost per tCO2eq as defined as total investment cost/ expected lifetime emission reductions, and relative to comparable opportunities.
	Amount of co-financing	Refer to the Note below.
	Programme/project financial viability and other financial indicators	Economic and financial rate of return with and without the Fund's support.
	Industry best practices	Explanations of how best available technologies and/or best practices, including those of indigenous peoples and local communities, are considered and applied

Note: When the indicative assessment indicators and the sub-criteria described in the GCF document do not add much details, the present author left the cell empty (GCF 2018b)

Another straightforward point is that the adaptation assessment indicators would fail to discern the situation in which a certain project would benefit a large number of people but the size of the overall benefit would be small. A project with a large overall benefit for a small group of people will be disadvantaged from the investment decisions. Put differently, the GCF investment criteria for adaptation projects have not yet established the assessment indicator which tells the Board the aggregate size of the benefits measured against the aggregate size of the harms which can occur through, e.g., increased income, increased yields, land and property values, increased interest payments.

Other than the impact potential, there are many other criteria that the GCF Board relies on in making a funding decision. The other investment criteria are summarized in Table 2.13. They are, in the order used by the GCF document, paradigm shift potential, sustainable development potential, needs of the recipient, country ownership, efficiency and effectiveness. At first glance, it is hard to grasp what each of these criteria means, how it is related to climate change policy, and in what ways it is practically used in decisions.

Let's first take a closer look at the paradigm shift potential, whose coverage areas are defined by the GCF by the following five components: (1) potential for scaling up and replication, and its overall contribution to global low-carbon development pathways being consistent with a temperature increase of less than 2 degrees Celsius; (2) potential for knowledge and learning; (3) contribution to the creation of an enabling environment; (4) contribution to the regulatory framework and policies; (5) overall contribution to climate-resilient development pathways consistent with a country's climate change adaptation strategies and plans (adaptation only).

To many readers of this book, these explanations may sound at odds with the idea of a paradigm shift. Further elaborations are attached by the GCF Board as indicative assessment factors or sub-criteria. The coverage area of 'potential for knowledge and learning' is further explained by the indicative assessment factor which is 'existence of a monitoring and evaluation plan and a plan for sharing lessons.' The coverage area of 'contribution to the creation of an enabling environment' is further elaborated by 'sustainability of outcomes and results beyond the project's completion as an indicative assessment factor.

These clarifications, however, do indicate that there is a critical problem of scaling-up each funding decision, but does not provide any solid reasoning on how the scaling-up issue can be addressed. In Chap. 7, this issue will turn up again as one of the important evaluation criteria of the GCF.

Next, let's consider the sustainable development potential criterion. The coverage areas of this criterion are defined by the GCF by the following four components: (1) environmental co-benefits; (2) social co-benefits; (3) economic co-benefits; (4) gender-sensitive development impact.

The environmental co-benefits are further elaborated by the GCF as the indicative assessment factor: 'Degree to which the project or programme promotes positive environmental externalities such as air quality, soil quality, conservation, biodiversity, etc.' The social co-benefits are elaborated by 'Potential for externalities in the form of expected improvements, for women and men as relevant, in areas such as health and safety, access to education, improved regulation and/or cultural preservation.' The economic co-benefits are further explained as 'Potential for externalities in the form of expected improvements in areas such as expanded and enhanced job markets, job creation and poverty alleviation for women and men, increased and/or expanded involvement of local industries.' The coverage area of gender-sensitive development impact is further elaborated as 'Explanation of how the project activities will address the needs of women and men in order to correct prevailing inequalities in climate change vulnerability and risks.'

Again, many readers may wonder how the various types of co-benefits are related in any way to a climate change policy goal or another. Most of these indicative factors, if not all, seem to have little to do with either mitigating the warming of the Planet or minimizing the impacts of climate change on humans and ecosystems.

Together with the above-described observation that all the GCF funding decisions are assigned to the goal of sustainable development, the investment criterion of sustainable development with its attached clarifications augment the perception that the GCF funding decisions are not directed to climate change policy goals.

The next two investment criteria seem to be related to each other, so I can put them together for explanation: needs of the recipient, country ownership. The coverage areas of the needs of the recipient criterion are defined as follows: (1) vulnerability of the country; (2) vulnerable groups and gender aspects; (3) economic and social development level of the country; (4) absence of alternative sources of financing; (5) need for strengthening institutions and implementation capacity.

The coverage areas of the country ownership criterion are defined as follows: (1) existence of a national climate strategy; (2) coherence with existing policies; (3) capacity of accredited entities or executing entities to deliver; (4) engagement with civil society organizations and other relevant stakeholders.

Again, the GCF's accounts of these criteria do not clearly convey what these criteria can achieve in terms of climate change policy goals. That is, these are not explicitly linked to either mitigation activities or adaptation activities. Nonetheless, these criteria and assessment factors can be interpreted as the attempts by the GCF Board to assist the poor vulnerable countries to participate in the global climate change agreements and protocols. The coverage areas seem to set the platform for justifications of a future climate-change foreign aid to developing countries.

These criteria, however, can be interpreted to establish a direct link of the GCF investments to the Article 3 of the UNFCCC which declared a principle of a common but differentiated responsibility (UNFCCC 1992).

For a funded project to be considered as being owned by a recipient country, the GCF project proposal should indicate either an accredited entity or an executing entity that is owned by the country. Since GCF recipient countries are in all cases underdeveloped countries, they do not often have the capacity to submit a GCF project proposal. It is often the case that there is neither an accredited entity in the proposal country. As such, a proposal submission is often done by an international entity such as the United Nations Development Programme, the World Bank, the United Nations Environment Programme, the Asian Development Bank (ADB), the Inter-American Development Bank (IADB), and the African Development Bank (AfDB). These international organizations are referred to as an accredited entity by the GCF.

Finally, the last investment criterion adopted by the GCF Board is efficiency and effectiveness. The coverage areas cover the following topics: (1) cost-effectiveness and efficiency regarding financial and non-financial aspects; (2) amount of co-financing; (3) programme/project financial viability and other financial indicators; (4) industry best practices.

Of the four coverage areas, it would be of much interest to the readers to know how the 'cost-effectiveness and efficiency' in the first coverage area are defined and measured. The four assessment indicators, as shown in Table 2.13, the first and second indicators are simply asking whether the GCF funding will be sufficient for a particular project.

The third assessment indicator is 'Demonstration that the Fund's support for the programme/project will not crowd out private and other public investment', which can be interpreted in more than one way. It can be interpreted that the GCF funding should not result in the termination of the funding from other agencies such as the UNDP or the World Bank.

The fourth assessment indicator is 'Estimated cost per tCO2eq as defined as total investment cost/expected lifetime emission reductions, and relative to comparable opportunities', which can be also interpreted in more than one way. However, the estimate of the cost per tCO2eq reduction is not included in any of the GCF proposals or project documents. Neither is there any discussion in any project document about a set of alternatives to the proposed project activity (GCF 2018b, c).

There are many other issues that need to be discussed with the regard to the 'efficiency and effectiveness' criteria. To do that, however, we must develop beforehand key economic theories and concepts with regard to efficiency, welfare optimization, and public goods (Ricardo 1817; von Thunen 1826; Pareto 1906; Samuelson 1954; Solow 1957; Nordhaus 1994), which is one of the central topics of Chap. 7. At this stage, the above discussion seems sufficient.

2.7 The Road to Evaluations of the GCF Projects

In this chapter devoted to the Green Climate Fund, I attempted to describe the entirety of the global climate fund, covering a historical timeline, associations with major UN climate conferences, governing instrument, pledges and contributions, institutional structure, allocations of the fund, and investment criteria. This chapter has set the stage for analyses and evaluations of the GCF projects in the upcoming chapters.

From this point on, the present author will elaborate the four scientific traditions of global warming introduced in Chap. 1 from Chaps. 3, 4, 5 and 6. Chapter 7 will follow up with the explanation of the fifth scientific literature, that is, the dynamic integrated climate and economy model which is an essential concept that cannot be left out during the evaluations of the GCF-funded projects conducted in Chap. 7 (Nordhaus 2008, 2013).

References

Green Climate Fund (GCF) (2011) Governing instrument for the Green Climate Fund. GCF, Songdo City
Green Climate Fund (GCF) (2018a) About the fund. GCF, Songdo City
Green Climate Fund (GCF) (2018b) Annex III: investment framework. GCF, Songdo City
Green Climate Fund (GCF) (2018c) Annex IX: results management framework. GCF, Songdo City
Green Climate Fund (GCF) (2018d) Status of pledges and contributions made to the Green Climate Fund, status date: 29 January 2018. GCF, Songdo City
Green Climate Fund (GCF) (2018e) Projects + programmes. GCF, Songdo City. Accessed from https://www.greenclimate.fund/what-we-do/projects-programmes
Green Climate Fund (GCF) (2018f) Green Climate Fund invests USD 1 billion for developing country climate action, launches first replenishment. Published on October 21, 2018. Accessed from https://www.greenclimate.fund/-/green-climate-fund-invests-usd-1-billion-for-developing-country-climate-action-launches-first-replenishment
Intergovernmental Panel on Climate Change (IPCC) (2018) Special report on global warming of 1.5 °C. Cambridge University Press, Cambridge
Manne AS, Richels RG (1999) The Kyoto Protocol: a cost-effective strategy for meeting environmental objectives? Energy J 20(Special Issue):1–23
New York Times (NYT) (2009a) An air of frustration for Europe at climate talks. Published on Dec 21, 2009. NYT, New York
New York Times (NYT) (2009b) Hillary Clinton pledges $100B for developing countries. Published on Dec 17, 2009. NYT, New York
Nordhaus W (1994) Managing the global commons. MIT Press, Massachusetts
Nordhaus WD (2008) A question of balance—weighing the options on global warming policies. Yale University Press, New Haven
Nordhaus W (2010) Economic aspects of global warming in a post-Copenhagen environment. Proc US Natl Acad Sci 107(26):11721–11726
Nordhaus W (2013) The climate casino: risk, uncertainty, and economics for a warming world. Yale University Press, New Haven
Nordhaus WD, Boyer JG (1999) Requiem for Kyoto: an economic analysis of the Kyoto Protocol. Energy J 20(Special Issue):93–130
Pareto V (1906) Manual for political economy. Oxford University Press, Oxford. [Edited by Montesano A, Zanni A, Bruni L, Chipman JS, Mc Lure M]
Paul BK (2009) Why relatively fewer people died? The case of Bangladesh's Cyclone Sidr. Nat Hazards 50:289–304
Ricardo D (1817) On the principles of political economy and taxation. John Murray, London
Roberts T, Weikmans R (2015) Unfinished agenda of the Paris climate talks: Finance to the global South. Brookings Institution, Washington, DC
Samuelson P (1954) The pure theory of public expenditure. Rev Econ Stat 36:387–389
Seo SN (2015) Helping low-latitude, poor countries with climate change. Regulation. Winter 2015–2016:6–8
Seo SN (2017a) The behavioral economics of climate change: adaptation behaviors, global public goods, breakthrough technologies, and policy-making. Academic Press, Amsterdam
Seo SN (2017b) Beyond the Paris Agreement: climate change policy negotiations and future directions. Reg Sci Policy Pract 9:121–140
Seo SN (2017c) Measuring policy benefits of the cyclone shelter program in North Indian Ocean: protection from high winds or high storm surges? Clim Chang Econ 8(4):1–18
Seo SN (2019) Economic questions on global warming during the Trump years. J Public Aff. https://doi.org/10.1002/pa.1914
Seo SN, Bakkensen LA (2017) Is tropical cyclone surge, not intensity, what kills so many people in South Asia? Weather Clim Soc 9:71–81
Solow RM (1957) A contribution to the theory of economic growth. Q J Econ 70:65–94

United Nations (2000) United Nations millennium declaration. UN, New York

United Nations (2018) United Nations peacekeeping: how we are funded. Accessed from https://peacekeeping.un.org/en/how-we-are-funded

United Nations Environment Programme (2017) The emissions gap report 2017: A UN environment synthesis report. UNEP, Nairobi

United Nations Framework Convention on Climate Change (UNFCCC) (1992) United Nations framework convention on climate change. UNFCCC, New York

United Nations Framework Convention on Climate Change (UNFCCC) (1997) Kyoto protocol to the United Nations dramework convention on climate change. UNFCCC, New York

United Nations Framework Convention on Climate Change (UNFCCC) (2005) Kyoto Protocol to enter into force 16 February 2005. UNFCCC, New York

United Nations Framework Convention on Climate Change (UNFCCC) (2007) Report of the conference of the parties on its thirteenth session, held in Bali from 3 to 15 December 2007. UNFCCC, New York

United Nations Framework Convention on Climate Change (UNFCCC) (2009) Copenhagen accord. UNFCCC, New York

United Nations Framework Convention on Climate Change (UNFCCC) (2010) Cancun agreements. UNFCCC, New York

United Nations Framework Convention on Climate Change (UNFCCC) (2011a) The Durban platform for enhanced action. UNFCCC, New York

United Nations Framework Convention on Climate Change (UNFCCC) (2011b) Report of the transitional committee for the design of Green Climate Fund. UNFCCC, New York

United Nations Framework Convention on Climate Change (UNFCCC) (2012) Decisions adopted by the conference of the parties on its eighteenth session, held in Doha from 26 November to 8 December 2012. UNFCCC, New York

United Nations Framework Convention on Climate Change (UNFCCC) (2015) The Paris Agreement. Conference of the parties (COP). UNFCCC, New York, p 21

United Nations Framework Convention on Climate Change (UNFCCC) (2018) Decisions adopted at the Climate Change Conference in Katowice, Poland, 2-14 December 2018. Conference Of the Parties (COP). UNFCCC, New York, p 24

von Thunen JH (1826) The Isolated State. Translated by Carla M. Wartenberg, Pergamon Press, Oxford, New York [1966]

White House (2017) Statement by President Trump on the Paris Climate Accord. White House, Washington, DC

World Bank (2008) Bangladesh — emergency 2007 cyclone recovery and restoration project. World Bank, Washington, DC. Accessed at http://documents.worldbank.org/curated/en/763581468013246012/Bangladesh-Emergency-2007-Cyclone-Recovery-and-Restoration-Project

Chapter 3
The Microbehavioral Economic Models of Adaptation Behaviors to Global Warming

Abstract From this chapter on, four scientific traditions that provide useful empirical information to both the decision-makers of the Green Climate Fund and the evaluators of the GCF decisions are explained, the first of which is the microbehavioral economic models of adaptation to global warming. The microbehavioral economics examines how an individual manager of agricultural and natural resources chooses an optimal portfolio of resources, taking the given climate factors into account, to maximize the profit earned from the portfolio over the long-term. This chapter explains major empirical findings from the microbehavioral studies conducted through the farm household surveys in Africa, Latin America, and South Asia. In response to or anticipation of climatic shifts, behavioral changes of individuals occur, among other things, by switching from one enterprise to another among the natural resource enterprises, or from one crop or animal species to another, or from a specialized portfolio to a diversified portfolio. The results offer valuable insights to the GCP policy-makers on which natural resource portfolio is most vulnerable as well as on how natural source mangers should adapt to future climatic changes.

Keywords Microbehavioral · Adaptation behaviors · Global warming · Natural resource portfolio · Land value

3.1 Introduction

The Green Climate Fund (GCF) is the sole stand-alone financial instrument that supports the implementation of the Paris Agreement signed in December 2015 by all the parties of the United Nations' climate convention (UNFCCC 2015). First proposed at the Copenhagen Conference in Denmark in 2009, the governing instrument of the GCF was adopted in 2011 at the Durban Conference (UNFCCC 2011; GCF 2011, 2018a, b, c).

The critical question has arisen and is outstanding still now, which is where and how much the funds should be allocated to selected projects from a large pool of

S. N. Seo, *The Economics of Global Allocations of the Green Climate Fund*,
https://doi.org/10.1007/978-3-030-18274-8_3

funding applications (Seo 2014b, 2015b). The climate fund was formed by voluntary pledges and contributions of the rich countries of the convention and the annual size of the GCF fund allocations is set to be US$ 100 billion by 2020. Despite the gravity of the allocation issue, climate scientists and negotiators have, by and large, set aside this critical question.

Many scientists and negotiators seem to be relying on the conviction that the GCF allocation should simply be a redistribution of wealth at a global scale. To them, there may be no need for a complicated calculation of whether a particular proposal should be awarded with a GCF grant and, if so, what size the grant should be. Many also may assert that the GCF funding is a moral responsibility of the rich countries to the poor countries since the rich countries have caused the global warming and the poor countries are and will be bearing the brunt of the damage from the warming Planet.

The problem with this simple-minded rationalization is that such an allocation rule will detach the GCF funding decisions from the problems of global warming. That is, there would be no relationship between the GCF grants and climate change policy goals. Fundamentally, the GCF fund allocations would become meaningless *vis a vis* climatic changes and global warming.

Any serious thinker of climate change challenges for Planet Earth will argue that the GCF funds should be carefully allocated in a way that helps address these challenges. What is the sensible, rational, and most fruitful way to spend the GCF dollars? The present author will tackle and answer this question and other related questions from this point onwards. The six remaining chapters of this book are devoted to this task, of which the next four chapters provide in-depth explanations of the four scientific traditions which are crucial to answering the question.

The four traditions are as follows. This chapter is devoted to the microbehavioral economic models of climate change focusing on adaptation behaviors of individuals (Seo 2010a, b, 2016c). The next chapter discusses the Agro-Economic Models (AEMs) of climate change which focus on agricultural impacts (Adams et al. 1990, 1999). Chapter 5 explains statistical models of yield and productivity changes caused by climate change (Schlenker and Roberts 2009; Welch et al. 2010). Chapter 6 is devoted to the review of ecosystem models of climate change (Dudal 1980; Fischer et al. 2005; FAO/IIASA 2005).

The microbehavioral models, to be presented in this chapter, were developed to reveal micro (individual) behaviors of adaptation of farm portfolios to climatic changes in a geographically explicit manner (G-MAP) (Seo 2010b, 2016a, b). The G-MAP model has been applied to study farming behaviors and natural resource decisions in low-latitude developing countries, specifically, Africa and Latin America. As such, empirical data and results of the G-MAP models provide ample insights into the primary question posed in this book, that is, for which adaptation activities the GCF funds should be allocated.

Further, the microbehavioral models were built to study micro behaviors relying on actual farm decisions. The data on farm decisions are obtained through a continental-level household survey in Africa or Latin America. This approach is in contrast to those of the other scientific traditions whose primary unit of analysis was either an agricultural product such as maize or a political unit such as the US county. This means that the microbehavioral models are far better positioned to unravel behavioral changes of resource managers and those who are affected by climatic changes, which is the essential knowledge needed for the allocation decisions of the GCF.

3.2 A Theory of the Microbehavioral Economics of Adaptation Behaviors

The microbehavioral economics in the literature of global warming can be defined as the field of research on how an individual or a group of individuals makes behavioral changes in response to or in anticipation of changes in the climate system taking into account non-climate production factors. More narrowly stated, the microbehavioral economics of global warming is the economic study of individuals' adaptation behaviors through changes in portfolios of assets and products in response to or in anticipation of climatic shifts (Seo 2015a, 2016d, 2017).

This tradition of research is founded on the long historical line of research on an individual's rational decision-making in economics. In particular, the theoretical foundation can be traced back to a rational expectation theory (Samuelson 1938; von Neumann and Morgenstern 1947; Nash 1951), a dynamic optimization theory in resource uses (Hotelling 1931; Koopmans 1951, 1965; Nordhaus 1973), a theory of land rent and spatial land uses (Ricardo 1817; von Thunen 1826), and behavioral economics of finance (Kahneman and Tversky 1979; Shiller 2003, 2005; Thaler 2015).

In addition, this tradition of climate research makes extensive uses of the corpus of microeconometric methods in order to empirically examine and predict behavioral changes of individuals (Train 2003). Particularly pertinent microeconometric concepts are binary and multinomial choice models (Bliss 1934a, b, Berkson 1944; McFadden 1974), discrete-continuous econometric models and selection bias (Heckman 1979; Lee 1983; Dubin and McFadden 1984), mixed models and simulation methods (Geweke 1989; Hajivassiliou and McFadden 1998; Keane 1994; McFadden and Train 2000), and spatial spillover effects (Moran 1950a, b; Anselin 1988; Beron and Vijberberg 2004).

The starting point of the microbehavioral economics of climate change is the following fundamental question with regard to climate science and policy decisions: Will an individual resource manager adapt to climate change? Although we take nowadays this question as a critical scientific question on climate change without resistance, environmental activists and climate scientists treated adaptation as a "dirty" word, especially during the first decade or so since the establishment of the United Nations Framework Convention on Climate Change (UNFCCC) in 1992 (UNFCCC 1992). It was a very hard question to ask at that time.

For the three decades since the first assessment reports from the Intergovernmental Panel on Climate Change (IPCC), climate researchers and activists have warned the general public of impending catastrophic consequences of climate change, based upon which they argued for a stringent climate change policy (IPCC 1990). For them, adaptation possibilities meant a smaller damage or even a benefit from global warming, hence would provide a rationale for a weaker climate change policy.

Nonetheless, once the literature of adaptation behaviors kicked off, it quickly expanded in an explosive way and has become unstoppable. Of all the reasons, this was because there already existed a large volume of works with regard to humans', animals', and plants' adaptation mechanisms in response to weather and climate variables. Some of this literature were formal in a scientific sense, but most of it were informal, experiential, and intuitive.

Let me begin by introducing some of this literature. Agricultural and development experts who were examining poverty and mal-nutrition in Sub-Saharan Africa identified that local climate and weather conditions play a key role in agricultural productions thereof. In Sub-Saharan Africa, agro-climatic conditions are adverse to crop farming in most parts of the continent where two-thirds of the rural population resides in arid, semi-arid, and desert zones (World Development Report 2008). Temperature, rainfall, and soil conditions in Sub-Sahara make many areas unsuitable for crop growth and keep the crop growing period limited for those areas that are suitable (FAO 1978; Dudal 1980). A weather and climate regime in Sub-Sahara is responsible for the frequent outbreaks of livestock diseases which kill often millions of cattle, but also for weeds, insects, and pests of crops (Ford and Katondo 1977; Ziska 2003). A multi-decadal shift in the rainfall regime in West Africa makes it extremely difficult for Sahelian farmers to grow crops successfully for many decades without relocations (Hulme et al. 2001; Shanahan et al. 2009).

In another continent, similar climate constraints and managerial adjustments are observed. South American farmers earn about 35% of their income from forest-related activities owing to the fact that the continent has the largest forest cover in the world, defined as a more than 50% cover. The forest cover accounts for 44% of the total land area in South America (WRI 2005; Vedeld et al. 2007). Argentina and Brazil are two of the world's largest exporters of beef cattle and the world's largest consumers (Steiger 2006; Magrin et al. 2007). The success of the livestock industry results from the availability of pasturelands in the continent which is again conditioned by climate regimes. More concretely, the size of pasturelands in Argentina is eight times that of the croplands; In Brazil, the pastureland size is four times larger than the cropland size in the country (Baethgen 1997). Further, climate and weather variables influence various aspects of farm animals' growth and productivity (Johnson 1965; Hahn 1981).

The dependence of these today's farming activities and agricultural productivities on the existent climate system means that the future activities and productivities will be altered if the present climate system were to be shifted. Agricultural and resource activities and productivities will be altered because a shift in the climate system would force changes in ecosystems and ecological zones which include crops, animals, trees, and plants (Schlesinger 1997; Ainsworth and Long 2005; Hahn et al. 2009).

What are these changes we can expect in agricultural resources, natural resources, ecosystems, or ecological zones? Experimental researchers, through laboratory and open field experiments, have made strides in addressing this question for over many decades. An increased CO_2 concentration in the atmosphere alters the productivities of various ecosystems by, *inter alia*, affecting their photosynthetic processes (Schlesinger 1997). A doubling in carbon dioxide concentration increases crop growth in the approximate range from 17% to 35% (Ainsworth and Long 2005; Tubiello et al. 2007). The productivity effects of the carbon dioxide doubling as well as temperature and precipitation changes would be varied across crops, grasslands, shrubs, and trees (Gitay et al. 2001; Ainsworth and Long 2005; Denman et al. 2007). Changes in climate and CO_2 level also lead to the changes in growth and distributions of weeds, insects, and plant diseases that are critical factors for crop and plant growth (Porter et al. 1991; Sutherst 1991; Ziska 2003).

For the animal systems, changes in CO_2 concentration, temperature, and precipitation patterns all influence the productivities of animals (Johnson 1965; Hahn 1999; Mader and Davis 2004; Mader et al. 2009). These changes affect heat exchanges between animals and the environment, which then directly affect weight growth, milk production, wool production, egg production, and even conception rates (Amundson et al. 2006; Hahn et al. 2009). Indirectly, forage quantity and quality can be altered by elevated CO_2 (Shaw et al. 2002). Changes in precipitation patterns associated with a warming world may also alter the frequency and severity of livestock diseases such as Nagana (Trypanosomiasis) carried by Tsetse flies in Africa, cattle tick in Australia, and blue tongue that affects sheep and goats in Europe (Ford and Katondo 1977; White et al. 2003; USAHA 2008; Fox et al. 2012).

Accepting these changes in biogeochemical processes and animal processes caused by future climate changes, what changes can we expect in behavioral decisions of those in agricultural and natural resource sectors? This is the fundamental question of the microbehavioral economics of global warming.

A microbehavioral model of adaptation behaviors starts with an individual farm. That is, the unit of analysis is a farm or an individual decision maker (Seo 2006). This is in stark contrast to the experimental models described in the above or the agro-economic models to be explained in the next chapter in which the basic unit of analysis is a crop or a plant (Adams et al. 1990; Tubiello and Ewert 2002). Put differently, an experiment is conducted on a physical unit, say, maize, not on a behavioral unit.

A microbehavioral model examines a large number of farms located across and sampled from the geographical area that is concerned by a modeler. The geographical area is usually an entire continent such as Africa or South America. From the concerned geographical area, a large number of farms are sampled to encompass a full variety of farm portfolios and climate conditions that exist in the area (Seo 2014a).

Therefore, the microbehavioral model of adaptation behaviors is different from a representative farm model in which a single representative farm is defined in the first stage and changes in decisions of the representative farm are simulated in the second stage through, e.g., a mathematical programming method or a computable general equilibrium model (Kaiser et al. 1993; Darwin et al. 1995).

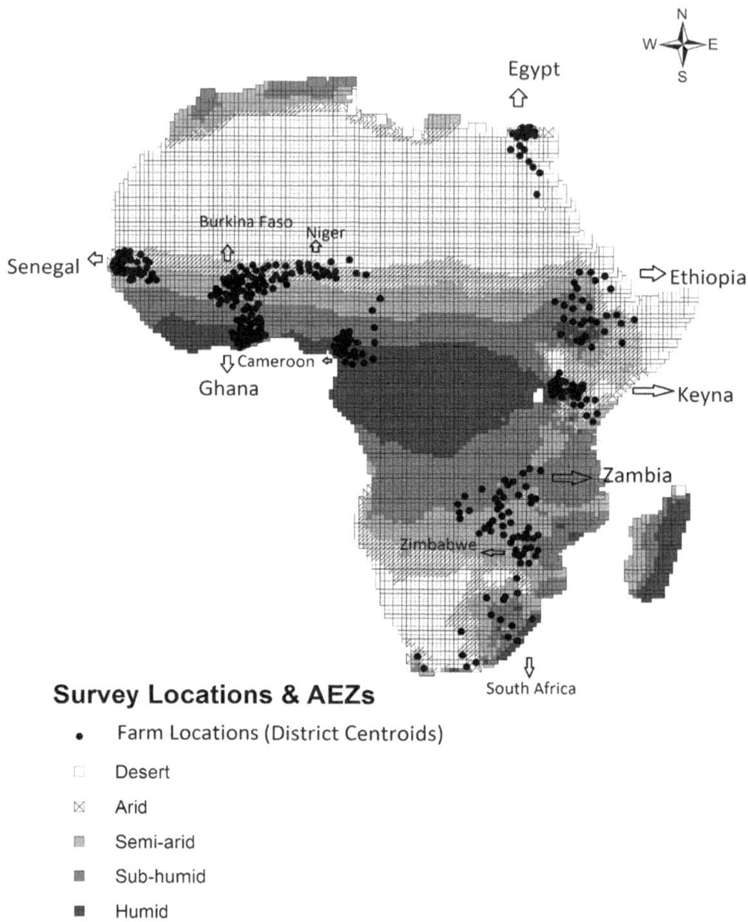

Fig. 3.1 Sampling of farms in Africa

How are farms, or rural households, selected for a microbehavioral model? In Fig. 3.1, the present author maps the locations of the household survey conducted as part of the World Bank project on climate change and African agriculture (Dinar et al. 2008; Seo et al. 2009). In total, the World Bank survey collected more than 11,000 completed survey responses. Farm locations on the map, marked as black dots, are centroids of the districts from which sampled farms are selected. A district is the second administrative division of a country's political system below a province or a state which is the first administrative division.

In the map, sampled farm locations (in district centroids) are overlaid on top of the five Agro-Ecological Zones (AEZs) as defined by the Food and Agriculture Organization (FAO) of the United Nations: deserts, arid, semi-arid, sub-humid, and humid zones (Dudal 1980; FAO 2005). The map shows that the filled-out surveys were collected in large numbers from each of the five AEZs.

In addition to representing all five agro-ecological zones of the continent, the map marks the 11 countries from which the completed surveys were gathered. The range of countries cover geopolitically-defined African regions: North Africa, West Africa, Central Africa, East Africa, and Southern Africa: South Africa, Zambia, and Zimbabwe for Southern Africa; Kenya and Ethiopia for East Africa; Cameroon for Central Africa; Ghana, Niger, Burkina Faso, and Senegal for West Africa; Egypt for North Africa.

The map also shows that the survey responses were obtained from all major geographies of the continent. That is, the range of geographies covers high mountain regions (in Kenya and Tanzania), low-land regions (in West Africa), grand rivers and lakes (in Lake Victoria and Niger Delta), landlocked regions (in Niger and Zimbabwe), and coastal regions.

In the microbehavioral model, a natural resource manager is assumed to maximize the long-term profit earned from managing the agricultural and natural resources that she owns, taking as given the external factors such as climate system, geographic, and geologic conditions. Conditional on the external factors, s/he would choose a natural resource portfolio from the full variety of possible portfolios and further make decisions on the array of inputs and outputs from the portfolio (Seo 2010a, 2016a, b).

If the current climate system were to be altered from the present state to a future state, as predicted by many climate scientists, the natural resource manager would also respond and adapt by switching from the current portfolio to an alternative portfolio as well as by altering the current combination of inputs and outputs of productions. These adaptation behaviors would result in the changes in the long-term profit earned by the resource manager.

Therefore, a microbehavioral modeler of climate change can reveal and make explicit, among other things, the following decision variables of great concern to global warming policy decisions. First, the modeler can reveal how choices of natural resource portfolios and choices of resource management practices are made currently and will be altered due to a shift in the climate system. Second, s/he can unravel how each of the agricultural and natural resource portfolios will suffer (or benefit) from a global warming scenario as measured by the changes in the long-term profit. Third, s/he can show how the agricultural and natural resource sector as a whole will fare under a variety of global warming scenarios in terms of its long-term profit (Seo 2014a, 2016d).

Henceforth, the microbehavioral model of adaptation behaviors to climate change is formally described as a natural resource portfolio selection model. Let's assume that a natural resource manager, or equivalently, a farmer (n) located in a particular climate regime chooses one portfolio from the family of available agricultural and natural resource portfolios (j) in order to maximize the expected long-term profit (π), given a range of immutable factors that affects the fam income. The problem can be stated succinctly in a mathematical form:

$$ArgMax_j\{\pi_{n1}, \pi_{n2}, \ldots, \pi_{nJ}\}. \tag{3.1}$$

In Eq. 3.1, what is the family of agricultural and natural resource portfolios that is available as the farmer's option? The number of possible natural resource portfolios is very large. As an illustration, let's consider the portfolios of activities undertaken by South American farmers (Seo and Mendelsohn 2008b; Seo 2012a). In the continent, a large number and variety of crops, animals, and forest products is managed by the rural managers. The major crops widely planted across the continent include the class of cereals such as wheat, maize, rice, millet, sorghum; the class of oil seeds such as soybeans, peanuts, canola, sunflowers; the class of vegetables such as potatoes, cassavas, tomatoes, onions, broccoli, garlic, carrots, lettuce, cabbage, spinach, and many others; and the class of specialty crops such as cotton, tobacco, mushrooms, sugarcane, and coffee (Mata et al. 2001).

In South America, livestock management outweighs crop farming in many aspects, which is a salient feature of South American agriculture. The dominant size of pasturelands over croplands was already mentioned in the above (Baethgen 1997). Also, South America leads the world's cattle export among the six continents in the globe (Steiger 2006). Along with different varieties of beef cattle, most frequently raised animals by South American farmers are dairy cattle, chickens, pigs, goats, and sheep (Seo et al. 2010).

Besides the crops and farm animals, forests and forest products are a vital component of the rural economy in South America. The continent is the world leader among the six continents in terms of the size of the area covered by dense forests (WRI 2005). The Amazonia covers 7.5 million km^2 and is the world's largest pluvial forest (Mata et al. 2001). Most common trees planted by rural farms include mango, pineapple, cashew, citrus, cacao, banana, palm, shea nut, apple, Kola, peach, almond, prune, apricot, avocado, cherry, hickory, eucalyptus, lemon, and Brazil nut (Seo 2012a).

From the large array of possible agricultural and natural resource portfolios a rural household can choose to manage, we can classify the following seven rural enterprises based on (1) whether an individual household manages crops or not, (2) whether it manages livestock or not, and (3) whether it manages forests or not:

$$\textit{Rural Economy} \sim \begin{cases} \textit{Specialized} \sim \begin{cases} \textit{Crops} - \textit{only} \\ \textit{Livestock} - \textit{only} \\ \textit{Forests} - \textit{only} \end{cases} \\ \textit{Diversified} \sim \begin{cases} \textit{Crops} - \textit{livestock} \\ \textit{Crops} - \textit{forests} \\ \textit{Livestock} - \textit{forests} \end{cases} \\ \textit{Fully Integrated} \sim \{\textit{Crops} - \textit{livestock} - \textit{forests} \end{cases} \tag{3.2}$$

The first three enterprises are a specialized enterprise. The middle three enterprises are a diversified enterprise: Each of these enterprises manages two of the three components. The last one is a fully integrated enterprise in that it contains all three components of the rural economy, that is, crops, livestock, and forests.

Having chosen one of the rural enterprises, the farmer will employ a host of technologies and practices in order to realize the maximum profit from the enterprise, which includes, *inter alia*, fertilizers, agricultural machineries, irrigation technologies, extension services, inter-cropping, trained workers.

Let the long-term profit that a rural farm (n) can earn from each of the seven enterprises defined in Eq. 3.2 be written as follows:

$$\pi_{n,j} \ where \ j = 1,2,\ldots,7. \tag{3.3}$$

Let's assume that the long-term profit of enterprise j be written as the sum of the observable component and the unobservable component, with the observable component being written as a linear function of the pertinent parameters (X) (McFadden 1974). For enterprise 1 and j:

$$\pi_{n1} = X_n\beta_1 + \mu_{n1}, \tag{3.4a}$$

$$\pi^*{}_{nj} = Z_n\gamma_j + \eta_{nj}, \ where \ j = 1,2,\ \ldots,J. \tag{3.4b}$$

The π denotes the observed profit of an enterprise which is only observable when the enterprise is actually chosen. The π^* denotes the latent profit, that is, the profit expected of an enterprise if the enterprise were to be chosen by the farm (n).

The vector Z is the set of explanatory variables pertinent to all the enterprises in the microbehavioral model and the vector X is the set of the explanatory variables that are pertinent only to the first alternative, that is, a crops-only enterprise.

Let's consider, for the sake of clarity of explanations and simplicity, the following three alternative model: $j = 1$ denotes a crops-only enterprise, $j = 2$ a crops-livestock enterprise, and $j = 3$ a livestock enterprise. In the model, the three choices are mutually exclusive and exhaustive (Seo 2010a, b). The three-enterprise model is easily extendable to a seven-enterprise model.

The vectors of explanatory variables, Z and X, include climate variables as an element. The sub-vector of climate variables includes both climate normals and climate risk normals (Seo 2012b). Seasonal climate variables are used to capture seasonal changes in the climate regime in spring, summer, fall, and winter as well as seasonal changes in farming activities (Mendelsohn et al. 1994). A summer season in the Southern Hemisphere corresponds to a winter season in the Northern Hemisphere (approximately December, January, and February) and a spring season in the Southern Hemisphere corresponds to a fall season in the Northern Hemisphere (approximately September, October, and November). The winter and fall seasons in the Southern Hemisphere are defined accordingly. In the monsoon climate regime, seasons can be distinguished in the microbehaviroal model by a monsoon season and a non-monsoon season (Seo 2016c).

The error term in the profit equation in Eq. 3.4a is assumed to be independently and identically distributed (iid) with the following mean and variance given the vectors of explanatory variables:

$$E(\mu_{n1}|X, Z) = 0, Var(\mu_{n1}|X, Z) = \sigma^2. \tag{3.5}$$

The probability to choose each of the natural resource enterprises is calculated from the latent profit equations in Eq. 3.4b and the choice decision rule in Eq. 3.1. An estimation of the choice equations calls for the specification of the unobservable term in Eq. 3.4b. Depending upon the assumptions that a researcher makes, the parameters of the choice equations are estimated parametrically or non-parametrically (Train 2003).

Assuming that $\eta_{nj}'s$ are independently and identically Gumbel distributed (McFadden 1974) and spatial neighborhood effects are accounted for by a re-sampling method (Anselin 1988; Beron and Vijberberg 2004; Seo 2011b, 2016d), the probability of the first natural resource enterprise being chosen by the farmer can be written succinctly as a Logit probability or a mixed Logit probability (McFadden and Train 2000; Train 2003):

$$P_{n1} = \frac{\exp(Z_n \gamma_1)}{\sum_{k=1}^{K} \exp(Z_n \gamma_k)}. \tag{3.6}$$

Having chosen enterprise 1, i.e., a crops-only enterprise, the farmer should make numerous decisions regarding a bundle of inputs and outputs to be employed as well as a basket of practices to maximize the long-term profit from managing the chosen enterprise. The conditional profit of the selected enterprise—the result of the series of her decisions—may be estimated directly using Eq. 3.4a and the vector of explanatory variables, *X*, with the same assumption on the error term as in Eq. 3.5.

However, because the first enterprise's profit is observed only for the farms that actually chose natural resource enterprise 1, a direct estimation of Eq. 3.4a will result in biased parameter estimates because of the selection decision that precedes the other management decisions (Heckman 1979). To obtain consistent parameter estimates of the profit function, selectivity or selection bias must then be corrected (McFadden 1999).

For a multinomial, or equivalently, polychotomous choice model, there are several selection bias correction methods that have been proposed since James Heckman's seminal paper. Of those, the Lee's generalized method, the Dahl's semi-parametric method, and the Dubin-Mcadden's method are most widely discussed and used (Lee 1983; Dubin and McFadden 1984; Dahl 2002). The Dubin-McFadden's method outperforms the other two methods because the other methods place severe restrictions on the correlation structure among choice alternatives (Schmertmann 1994; Bourguignon et al. 2004; Seo 2016d).

The microbehavioral models of global warming relied on the Dubin and McFadden method, the DM method henceforth, for the selection bias correction in a multinomial choice situation (Dubin and McFadden 1984). Allowing for a markedly more flexible correlation pattern than the other selection bias correction methods, the DM method assumes the following linearity condition, which states that the correlation coefficients (λ_{ij}) between the enterprise i profit and the enterprise j profit add up to one across j:

$$\sum_{j=1}^{J} \lambda_{ij} = 1 \text{ where } \lambda_{ij} = corr\left(\mu_i, \eta_j\right). \tag{3.7}$$

A second stage function is then estimated after incorporating the selectivity bias. The conditional long-term profit—land value in the empirical models—function for enterprise i is estimated as follows with the selection bias correction terms:

$$\pi_{ni} = X_n\varphi_i + \sigma_i \sum_{k\neq i}^{J} \lambda_{ik} \left[\frac{P_{nk} \ln P_{nk}}{1 - P_{nk}} + \ln P_{ni}\right] + \delta_{ni}. \tag{3.8}$$

In the above equation, δ is a white noise error term with its mean being zero. The conditional land value function for the crops-only enterprise, the crops-livestock enterprise, the livestock-only enterprise, and other enterprises can be estimated using Eq. 3.8 respectively.

The second term on the right-hand side of Eq. 3.8 is the set of selection bias correction terms. Note that the selection bias correction terms include selection probabilities of all enterprises in the model, that is, P_1, P_2, $P_{3,}$ ... ,P_7. This means that the estimation of the second-stage outcome equation in Eq. 3.8 takes into account selections of all seven alternatives in the model. This is not the case in the other selection bias correction methods (Bourguignon et al. 2004).

Why does the selection bias correction take the specific form shown in the second term on the right-hand side of Eq. 3.8? It naturally follows from the linearity assumption in Eq. 3.7. The derivation of the DM selection bias term in Eq. 3.8 was explained by the present author elsewhere (Seo 2016d), which relied on earlier works (Dubin and McFadden 1984; Bourguignon et al. 2004). Interested readers may refer to the above-cited book by the present author.

Since the microbehavioral model calls for a simultaneous estimation of a system of equations, that is, the set of the choice equations in Eq. 3.6 and the set of outcome equations in Eq. 3.8, the parameters in the model should be identified (Fisher 1966; Johnston and DiNardo 1997). In the applications of the microbehavioral economic model, the parameters are identified non-parametrically with an exclusion condition. More concretely, the identification strategy is to exclude from the outcome (land value) equations the identification variables that influences choices of the natural resource enterprises but not the land value equations (Seo and Mendelsohn 2008a; Seo 2010a).

Among the explanatory variables (X) in the microbehavioral models is a vector of climate variables, which is indeed a primary variable of interest to climate researchers. Climate variables can be either satellite-based climate data or ground weather station-based climate data. The satellite climate data are available from the late 1980s derived from the various instruments aboard the NASA (National Aeronautics and Space Administration) satellites and other satellites (Mendelsohn et al. 2007; NASA 2018; UAH 2018).

The weather station observational data are available and maintained by national weather/climate organizations around the world, from which a global climate data set is constructed for the time period from the beginning of the twentieth century and based on more than16,000 weather stations located around the world (New et al. 2002; Le Treut et al. 2007).

Other than the vector of climate variables, other control variables in the microbehavioral models are soils, topology, hydrology (water flows and runoff), market access (travel hours to major markets for exports, sales, or inputs), household characteristics (gender, education, number of family members, etc.), policy variables (extension service), and country dummies (Seo 2011b, 2016a). These data are obtained from various geographically-referenced data sources while the farm-related data are obtained directly from the household survey responses.

In the third stage of the microbehavioral model, the expected land value of the farm is calculated by making use of both the estimated choice probabilities of the enterprises in Eq. 3.6 and the conditional land values for the enterprises. The expected land value of the farm is the sum across the alternatives the estimated probability of each natural resource enterprise being chosen times the conditional land value of the enterprise once chosen. Put differently, it is the weighted average of the enterprise-specific land values by the choice probabilities of the enterprises. Formally, let E be the vector of climate variables. Then, the expected land value of the farm is derived as follows:

$$\Pi_n(E) = \sum_{j=1}^{J} P_{nj}(E|.) * \pi_{nj}(E|.) \tag{3.9}$$

Having the expected land value estimated through Eq. 3.9, we can now calculate the impact of climate change on the economic welfare of the farmer. Let's pick a climate change scenario from a host of climate scientists' predictions that are made available by climate organizations such as the Intergovernmental Panel on Climate Change (IPCC) (IPCC 2014). The scenario is such that the vector E takes on the value of E_0 at the present time and E_1 at the future time. The change in the economic welfare, $\Delta\Pi_n$, resulting from the actualization of the climate change scenario can be measured as the difference in the welfare measure after and before the climate change:

$$\begin{aligned} \Delta\Pi_n &= \Pi_n(E_1) - \Pi_n(E_0) \\ &= \sum_{j=1}^{J} P_{nj}(E_1) * \pi_{nj}(E_1) - \sum_{j=1}^{J} P_{nj}(E_0) * \pi_{nj}(E_0). \end{aligned} \tag{3.10}$$

The change in the expected land value in Eq. 3.10 captures both the changes in the choice probabilities of all possible enterprises and the changes in the conditional land values that would be generated by the chosen enterprises. The changes in the enterprise choice probabilities are calculated from Eq. 3.6. The changes in the conditional land values are calculated from Eq. 3.8.

The degree to which the estimate in Eq. 3.10 is uncertain is calculated by the confidence interval, e.g., 95% confidence interval. The bootstrapping method is used to calculate the standard error estimate based on a large number of subsamples drawn randomly from the original sample (Efron 1981; Seo and Mendelsohn 2008a). The standard errors for the changes in choice probabilities and for the changes in conditional land values can be calculated similarly by applying the bootstrapping method.

The microbehavioral economic model described thus far was developed to explain changes in behaviors of individuals in response to climatic changes. The model described up to this point provides a complete picture of the model's theory, structure, statistical components, and assumptions. An additional clarification is called for, however, with regard to what 'climate' and 'climate change' means in the microbehavioral model before we move on to the empirical sections. Put differently, there is the need for clarification with regard to which climate factors an individual decision-maker responds to, or in other words, makes decisions on.

The microbehavioral model, like other climate economics models, was initially developed to explain the impacts of changes in climate normals on individuals' behaviors and profits. That is, the goal of the microbehavioral models was not to quantify the impacts of weather changes in a specific year on farming behaviors and outcomes (Deschenes and Greenstone 2007; Schlenker and Roberts 2009; Seo 2013a). The distinctions between weather and climate normals are essentially important in the climate science literature (Le Treut et al. 2007). Weather may be hotter this year, colder next year, and so on. However, the fluctuations in yearly weather may or may not have something to do with the climate. A climate normal is defined to be the average of weather realizations for a sustained period of time, usually 30 years in the literature. For this reason, it is called a climate normal.

Most frequently used indicators of the climate system relied upon by the IPCC and climate scientists are temperature normals and precipitation normals. A temperature normal is a 30-year average of daily temperature while a precipitation normal is a 30-year average of daily precipitation. To capture different stages of crop and vegetation growth, the microbehavioral models have relied on seasonal climate normals: Spring, summer, fall, and winter. In the low-latitude tropical countries where four seasons are not distinct, summer and winter seasons are also used for the calibration of climate normals (Seo and Mendelsohn 2008a, b).

Over the course of development of climate science, scientists increasingly argued that climate risk may be of graver concern than the change in average climate, i.e., climate normals (Rahmstorf and Coumou 2011; Hansen et al. 2012). The crux of their arguments is that the variability of yearly weather may increase significantly, even though the average climate remains the same, due to greenhouse gas-enhanced warming. The microbehavioral researchers have followed suit to incorporate climate risks into the model (Seo 2012b, 2016c, 2018).

It should, however, be discerned that the climate risk is not the same as the weather risks (fluctuations) which past studies of African agriculture have strived to explicate (Udry 1995; Kazianga and Udry 2006). That is, an African village may suffer from occasional weather shocks such as a severe drought or a severe flooding but it can still be located in a low climate risk zone if such occurrences are not frequent or intense in the long-term timeframe. Applying the same reasoning, it should be noted that the climate risk is different from a drought severity index, of which there are quite a few (Dai et al. 2004).

A long-term variability (risk) of daily rainfall can be captured by the Coefficient of Variation in Precipitation (CVP) measured from many decades of observations, for example, for the 30-year period from 1961 to 1990 (Seo 2012b). The CVP is a measure of rainfall dispersion that does not depend on the size of average rainfall and can be defined as follows, with R_{jk} being monthly average precipitation in month j in year k (K = 30) and $\overline{R_j}$ being the 30-year average rainfall for month j:

$$\mathrm{CVP}_j = {}^{sd_j}/_{\bar{R}_j}, \text{with } sd_j = \sqrt{\frac{\sum_{k=1}^{K}\left(R_{jk} - \bar{R}_j\right)^2}{K-1}}. \tag{3.11}$$

In addition to the larger variation in the rainfall across the years over the 30-year period, the CVP measure can identify a regional rainfall pattern which exhibits a decadal or multi-decadal swing in yearly average rainfall caused by, for example, a change in an ocean current circulation. In Sub-Saharan Africa, the Atlantic Multi-decadal Oscillation (AMO) affects the rainfall regime of the continent, resulting in a multi-decadal swing (Janowiak 1988; Hulme et al. 2001). In the AMO affected regions of Sub-Sahara, the CVP value is higher than elsewhere (Seo 2012b).

Other ocean current events that cause a decadal or multidecadal fluctuation in a regional rainfall include the El Nino Southern Oscillation (ENSO) that affects the Pacific Rim countries and the Pacific Decadal Oscillation (PDO) that affects South Asia and the Pacific Rim countries (Curtis et al. 2001; Biondi et al. 2001).

Other than the CVP, the climate risk manifests as a larger temperature variability. For example, the climate system might exhibit a more frequent occurrence of extremely hot or cold days. Or, it might exhibit a more variable temperature pattern both within the year and across the years. Agricultural researchers often express their concern on an increase in daily maximum temperature or daily minimum temperature which affects crop growth (Easterling et al. 2000; Welch et al. 2010).

The temperature range can be measured by the Diurnal Temperature Range (DTR) (Seo 2012b). Average monthly DTRs for the 30-year period from 1961 to 1990 and beyond have been reported by the Climate Research Unit (CRU)'s global data sct, based on daily temperature data (New et al. 2002). Then, the Diurnal Temperature Range Variability (DTRV) can be defined to capture the average diurnal temperature variability over the 30-year period as another measure of climate risk.

Let $T_{k,\ m,\ j,\ max}$ be daily maximum temperature in day j, month m, and year k. Let $T_{k,\ m,\ j,\ min}$ be defined in the same way for daily minimum temperature. The DTRV for a particular month is defined as follows:

$$DTRV_m = \frac{\sum_{k=1}^{K}\sum_{j=1}^{J}\left(T_{k,m,j,max} - T_{k,m,j,min}\right)}{K^*J}. \tag{3.12}$$

A third measure of climate risk was constructed with regard to a monsoon climate regime (Seo 2016c). A monsoon climate regime is characterized as a regional climate regime in which an extremely heavy rainfall during the monsoon season is followed by an extremely low rainfall during the non-monsoon season. A monsoon regime is particularly strong in South Asian countries such as India, Bangladesh, Thailand.

At the present time, a monsoon regime severely constrains economic and agricultural activities of the regions under the grip of this regional climate system (Kala 2015; Seo 2016c). One of the challenging on-going climate science undertakings is to answer whether the monsoon risk will even get amplified owing to a warming world (Chung and Ramanathan 2006; Meehl and Hu 2006; May 2011).

The monsoon risk can be defined by a monsoon precipitation ratio normal (MPRN), which is averaged for a 30-year period. Let the monsoon season aggregate precipitation be PR_{ms} and the non-monsoon season aggregate precipitation be PR_{nm}. Then, the MPRN is defined:

$$MPRN_t = \frac{1}{30}\sum\nolimits_{k=t-30}^{k=t} \frac{PR_{ms,k}}{PR_{nm,k}}. \tag{3.13}$$

The higher the MPRN, the higher the monsoon climate risk for a region. The ratio of monsoon precipitation and non-monsoon precipitation can vary year by year. The MPR in Eq. 3.13 is the 30-year average of the ratio.

Another risk measure for a monsoon climate can be written based on the year-to-year variability of the MPR, i.e., the coefficient of variation in Monsoon Precipitation Ratio, which we can call the Monsoon Variability Index (MVI) (Seo 2019):

$$MVI_t = {}^{sd}(MPR_t)\big/_{MPRN_t}, \text{ with } MPR_t = \frac{PR_{ms,t}}{PR_{nm,t}}. \tag{3.14}$$

The MPRN is the average ratio while the MVI is the variability. Note that both risk measures are independent of scale. The MPRN is the monsoon season rainfall weighted by the non-monsoon season rainfall. The MVI is the standard deviation (sd) of MPR weighted by the MPRN, the long-term average of MPR.

The MVI or the coefficient of variation in monsoon precipitation ratio (CVMPR) is drawn for Indian States and Union Territories in Fig. 3.2. The MVI data are constructed by the present author based on the climate data provided by the Indian Institute of Tropical Meteorology (IITM) (IITM 2012; Seo 2016c).

The figure shows that the monsoon risk is very high in India. The MVI is close to and larger than 1 in all states and territories. That is, the yearly monsoon ratio varies by more than 100% of the MPRN, the 30-year average of the monsoon ratio. In seven States, the MVI exceeds 200%. In three States, the MVI exceeds 300%. In one State, the MVI exceeds 500%.

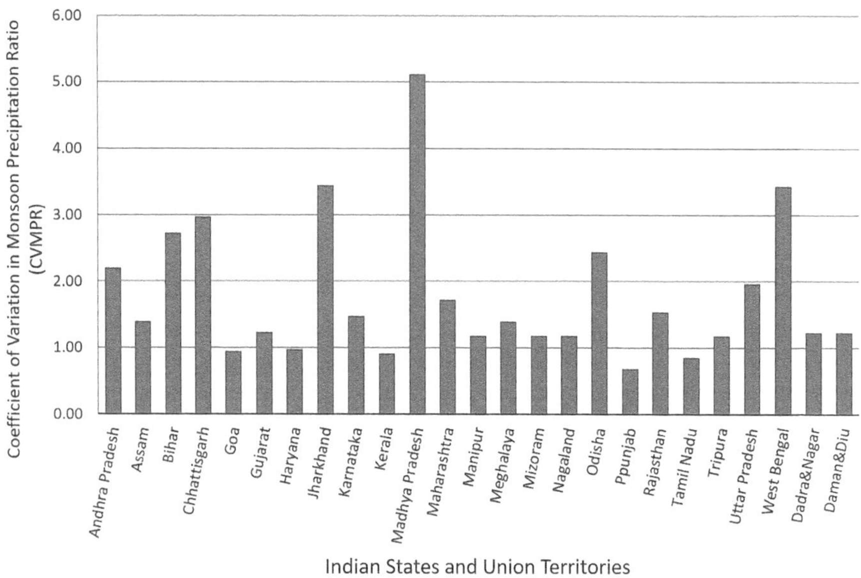

Fig. 3.2 Monsoon climate risk in India

For Thailand, another South Asian country, the trend in the Monsoon Precipitation Ratio Normal (MPRN) is shown in Fig. 3.3 using the Thailand weather data since 1900. The MPRN is defined as a previous 30-year average of monsoon precipitation ratio using Eq. 3.13, so the MPRN data are shown from 1930 to 2015. The yearly weather data based upon which the MPRN was constructed are obtained from the World Bank Climate Change Knowledge Portal which maintains a global climate data set supplied by the Climate Research Unit (CRU) at the University of East Anglia (Harris et al. 2014; World Bank 2017).

Figure 3.3 reveals that there is a reason to be concerned about a regional monsoon system under a globally warming world. First, it is noticeable that the MPRN is as large as 7.5 by the beginning of the twenty-first century. That is, the monsoon season precipitation is as large as 7.5 times the non-monsoon season precipitation. Second, the monsoon precipitation ratio during the past century has been increasing in a statistically significant way. The trend is an increase of 0.14 MPR per decade, which means an MPR increase of 1.4 per century.

3.3 Microbehavioral Data

The development of the microbehavioral economics of climate change was put off for more than a decade because of the lack of a micro-level economic dataset. International and national databases of agricultural activities were either product-based or politics-based. For example, the census data of agriculture provided by the United States Department of Agriculture are compiled at the level of political units

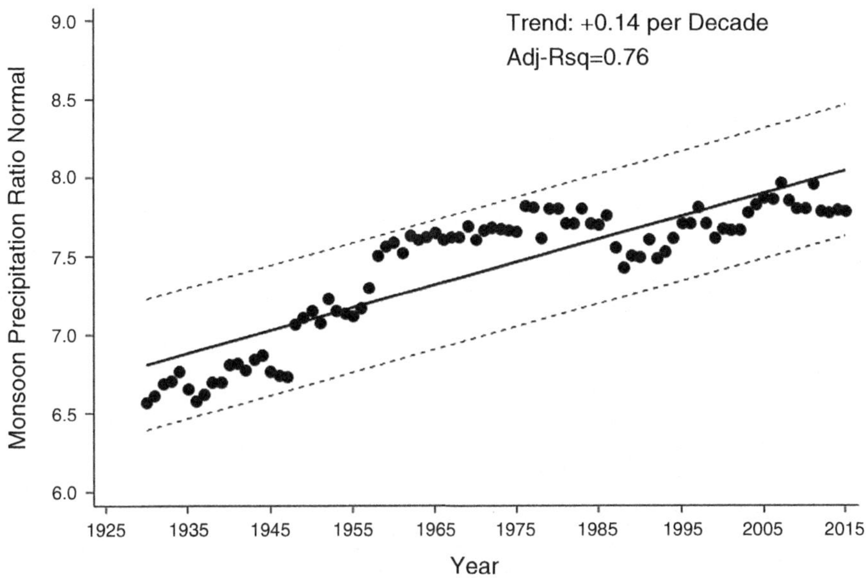

Fig. 3.3 Trend in monsoon precipitation ratio normal since 1900 in Thailand

such as the county (USDA 2012). For another, the FAOSTAT data provided by the Food and Agriculture Organization (FAO) are compiled for each agricultural product such as maize, wheat, rice, and others (FAO 2012).

During the 2nd decade of climate change economics, the micro behavioral data and economics came into the scene through the World Bank's projects on climate change and agriculture in Africa and Latin America. The African Project collected household surveys from more than 11,000 farms located in 11 African countries, as shown in Fig. 3.1 (Kurukulasuriya et al. 2006; Seo 2006; Dinar et al. 2008).

The World Bank's Latin American Project collected household surveys from more than 2500 households located in 7 countries (Seo and Mendelsohn 2008b; Seo 2010b). Of the seven, three countries were from the Andean region: Venezuela, Colombia, and Ecuador. Four countries were from the Southern Cone region of South America: Argentina, Uruguay, Brazil, and Chile.

The farm household surveys in both projects asked farm managers detailed questions on farming activities during the years of the surveys. The survey questions were organized into multiple sections: farm outputs, farm inputs, land characteristics, market access, household characteristics, perceptions of climate change, credit/subsidy, and identification variables of the respondent's farm.

The farm-level activities were then coupled with other geo-referenced data, the most important of which are climate data, climate change prediction data, soils, and geography data. These data are made available at the level of a grid cell whose size can be 1 degree by 1 degree or at a finer resolution of a 10 arc-minute cell (Seo 2013b).

Climate data are available at major climate research institutions in the world that keep track of the global weather and climate data which can go back to 1900 (New et al. 2002). The climate data can be constructed from either weather station measurements or satellite records. For the latter, the data are available from the middle of 1980s (Basist et al. 1998; UAH 2018).

Projections of the future climate system are made through supercomputer simulations referred to as the Atmospheric Oceanic General Circulation Model (AOGCM). At the time of writing this book, there are more than a dozen AOGCMs in the world that are relied upon by the Intergovernmental Panel on Climate Change (IPCC) for its scientific assessments (Taylor et al. 2012; IPCC 2014). In the AOGCMs, future climate predictions are made for multiple time periods, for example, the end of twenty-first century, the middle of twenty-first century, *etc*. The predictions are made for the entire Earth surface at the level of a grid cell whose size is often 1 degree by 1 degree, excluding some polar regions.

Of the climate models that are included in the Climate Model Inter-comparison Projects (CMIP), among the most frequently-used AOGCMs for the microbehavioral economic models are the Goddard Institute for Space Studies (GISS) model by the NASA, the Geophysical Fluid Dynamics Laboratory (GFDL) model by the NOAA, the Hadley center Climate Model (HadCM) by the UK Meteorology Office (UKMO), the Canadian Center for Climate Modeling and Analysis (CCCMA) model, the Commonwealth Scientific and Industrial Research Organization (CSIRO) model in Australia (Taylor et al. 2012; Gordon et al. 2000; Schmidt et al. 2005).

Soil data are available from the Food and Agriculture Organization (FAO)'s dominant soil map of the world and other national sources (Driessen et al. 2001; FAO 2003). For each 1-degree grid cell, the FAO map provides the dominant soil types and their percentages in the given cell. The FAO data also provide soil texture (sandy, clay) and soil slope (steep, flat) at the grid cell level.

A fine resolution elevation data set is available, *inter alia*, from the Global Multi-resolution Terrain Elevation Data 2010 (GMTED2010) data set which was constructed by the United States Geological Survey (USGS) and the National Geospatial-Intelligence Agency (NGA) (Danielson and Gesch 2011). The GMTED data were constructed by the USGS by merging a large number of source data sets whose resolution was in some sources as high as 1 arc-second.

A hydrology data set provides information on water availability for an area of interest, for example, a grid cell in a global data set or a political unit in a country. The amounts of seasonal waterflows and seasonal water runoffs for Africa are available from the University of Colorado hydrology model, among other sources (Strzepek and McCluskey 2006).

Lastly, another data set that was merged with the micro-level farming data is the market accessibility data. The accessibility of a local area to the nearest markets for purchases of farming inputs or sales of farm products can be measured by travel time (in hours) or a travel distance (in km) to a major city or a major port. This information was provided by the spatial data set constructed by the World Bank Project on Africa Infrastructure and Country Diagnostic (AICD) (World Bank 2009a).

3.4 Adaptation Behaviors

3.4.1 *Choice of Livestock Species*

The seminal microbehavioral model examined how African farmers choose farm animals in the climate conditions defined by the present climate system (Seo and Mendelsohn 2008a). The five animals are most frequently raised by African farmers: beef cattle, dairy cattle, goats, sheep, and chickens. Other animals less frequently favored include donkeys, horses, camels, pigs, beehives, dogs.

An application of the microbehavioral economic model described in the theory section in the above resulted in one of the most salient findings of climate change economics. The authors reported that the hotter the temperature normals become, the more often goats and sheep are adopted and the less often cattle and chickens are adopted. It was suspected that goats and sheep are more heat tolerant than the other animals.

The second important finding was that some animals are more favored by farmers in high precipitation zones. African farmers chose goats and chickens, especially goats, more often in higher rainfall areas while cattle and sheep are more preferred in drier zones. The result may indicate that goats are genetically more resilient to a high humidity condition, which is important to note because such a condition tends to lead to prevalence of livestock diseases such as trypanosomiasys (sleeping sickness) carried by tsetse flies which in some years kills millions of cattle in Sub-Sahara (Ford and Katondo 1977).

The result is also revealing in that agricultural or development experts had been concerned almost exclusively on the harmful effects of a drought or an arid condition. Under an arid condition, livestock management, especially of cattle and sheep, is preferred by African farmers and seems to become more profitable.

In Table 3.1, changes in adoption probabilities of the five primary livestock species by African farmers were calculated assuming two climate scenarios: The CCCma and the Parallel Climate Model (PCM). The CCCma is a hot and dry scenario and the PCM is a mildly hot and humid scenario. With the CCCma scenario, goats and sheep are predicted to be adopted more often by the year 2020.

Table 3.1 Changes in adoption probabilities of livestock species in Africa

		Beef cattle	Dairy cattle	Goats	Sheep	Chickens
Baseline probability		9.40%	27.3%	20.8%	21.8%	22.9%
% point changes under CCCma	CCCma: +1.6 °C, −3.7% in precipitation	−1.48%	−3.09%	+2.48%	+3.10%	−1.89%
% point changes under PCM	PCM: +0.6 °C, +12.5% in precipitation	−2.73%	−5.17%	+10.21%	−4.10%	−1.93%

On the other hand, with the PCM scenario, the choice of sheep is predicted to fall while that of goats is predicted to increase by a large percentage.

That some animals are more tolerant of or resilient to different climatic conditions, be they high temperature or high/low precipitation, has critical implications when it comes to farmers' responses to global warming (NRC 2015) as well as on the existing development aids to Sub-Saharan farmers which have tilted heavily to major staple crops (Byerlee and Eicher 1997; Kafle et al. 2016).

3.4.2 *Adoption of Agricultural Systems*

The model was soon followed by an agriculture-wide microbehavioral economic model (Seo 2010a, b). This follow-up model, applied to African agriculture and South American agriculture independently, revealed that crop farming and animal husbandry are indeed purposefully integrated as a strategy for a farmer to deal with harsher climatic conditions and changes.

The results from the microbehavioral model of agricultural systems are shown in Table 3.2 for Africa and in Table 3.3 for Latin America. In the model, a farmer is assumed to choose one from the three agricultural systems: A specialized crop system, a mixed crops-livestock system, and a specialized livestock system.

For Africa, an increase in temperature by 1-degree Celsius causes African farmers to switch away from specialized systems of agriculture to a mixed system. The adoption probability of the mixed crops-livestock system increases by 1.63% points. An increase in precipitation by 1% causes African farmers to switch away from the specialized livestock system to either a crops-only system or a crops-livestock system (Seo 2010b).

Table 3.2 Changes in adoption probabilities of agricultural systems in Africa

		Crops-only	Crops-livestock	Livestock-only
Baseline probability		34.23%	49.53%	16.24%
% point changes under temperature changes	Temperature change: +1 °C	−0.65%	+1.63%	−0.99%
% point changes under precipitation changes	Precipitation change: +1%	+0.05%	+0.05%	−0.10%

Table 3.3 Changes in adoption probabilities of agricultural systems in Latin America

		Crops-only	Crops-livestock	Livestock-only
Baseline probability		35.90%	42.00%	22.10%
% point changes under CCCma 2060	CCCma 2060	−4.14%	+2.10%	+2.04%
% point changes under PCM 2060	PCM 2060	+1.81%	+2.54%	−4.35%

For Latin America, there is a higher percentage of farmers who adopted a specialized livestock system, i.e., a livestock-only, as shown in Table 3.3 (Seo 2010a). About 22% of all farms in South America chose a specialized livestock system. This may come unsurprising to those who are acquainted with the iconic pasturelands in Latin America such as the lowland Pampas in Argentina, the highland Llanos in Colombia and Venezuela.

Under the CCCma 2060 scenario which predicts a hotter and drier climate, the probability of adoption of a specialized crop system is predicted to fall by 4.1% points. By contrast, the adoption probability of a mixed crops-livestock system is predicted to increase by 2.1% points and the probability of adoption of a specialized livestock system is predicted to increase by 2.0% points.

Under the PCM 2060 scenario which predicts a mildly hotter climate with higher precipitation, the probability of adoption of a livestock-only system is predicted to fall by 4.35% points. On the other hand, the choice probability of a crops-livestock system is predicted to increase by 2.5% points while that of a crops-only system to increase by 1.8% points.

3.4.3 Adaptation Behaviors to Climate Risk

The agricultural system model of the microbehavioral economics was further developed to explain how farmers rely on agricultural systems as a strategy to deal with climate risk (Seo 2012b). The climate risk indicators, the CVP and DTRV explained in the theory section above, were developed and matched with behavioral decisions using a spatial Logit analysis.

The results are summarized in Figs. 3.4 and 3.5. Figure 3.4 shows how adoption probability of each agricultural system varies across the precipitation risk, that is, the coefficient of variation in precipitation in Sub-Saharan Africa. It is notable that adoption of the mixed crops-livestock system increases as the precipitation risk increases. Adoption of the livestock-only system also increases as the CVP becomes larger, but it starts to fall when the CVP becomes too large.

On the contrary, the specialized crop system is chosen less frequently when the precipitation risk increases. This can be interpreted that when annual rainfall fluctuations are large, on a long-term average, farmers find it more difficult to manage crops successfully every year, thereby rely on alternative assets other than crops. Interestingly, adoption of the crops-only system ticks up at the highest CVP zones.

Sub-Saharan farmers are shown in Fig. 3.5 to strongly respond to temperature risk as measured by the diurnal temperature range variability. The higher the DTRV, the higher the adoption probability of a mixed crops-livestock system. The higher the DTRV, the lower the adoption probability of a crops-only system.

This may indicate that a higher diurnal temperature variability increases the probability of a crop failure, which causes farmers rely on an alternative asset such as livestock. The DTRV can increase by a higher daily maximum or a lower daily minimum temperature (Easterling et al. 2000; Welch et al. 2010). Another possible

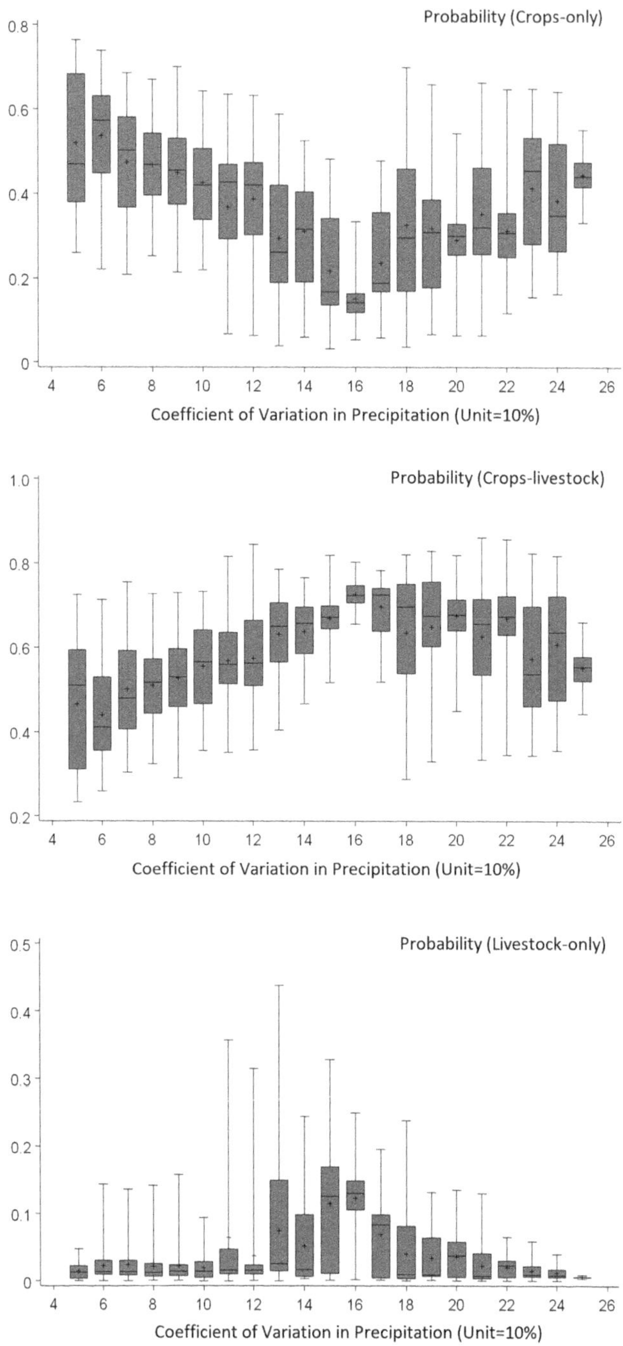

Fig. 3.4 Adoption of agricultural systems across coefficient of variation in precipitation in Sub-Saharan Africa

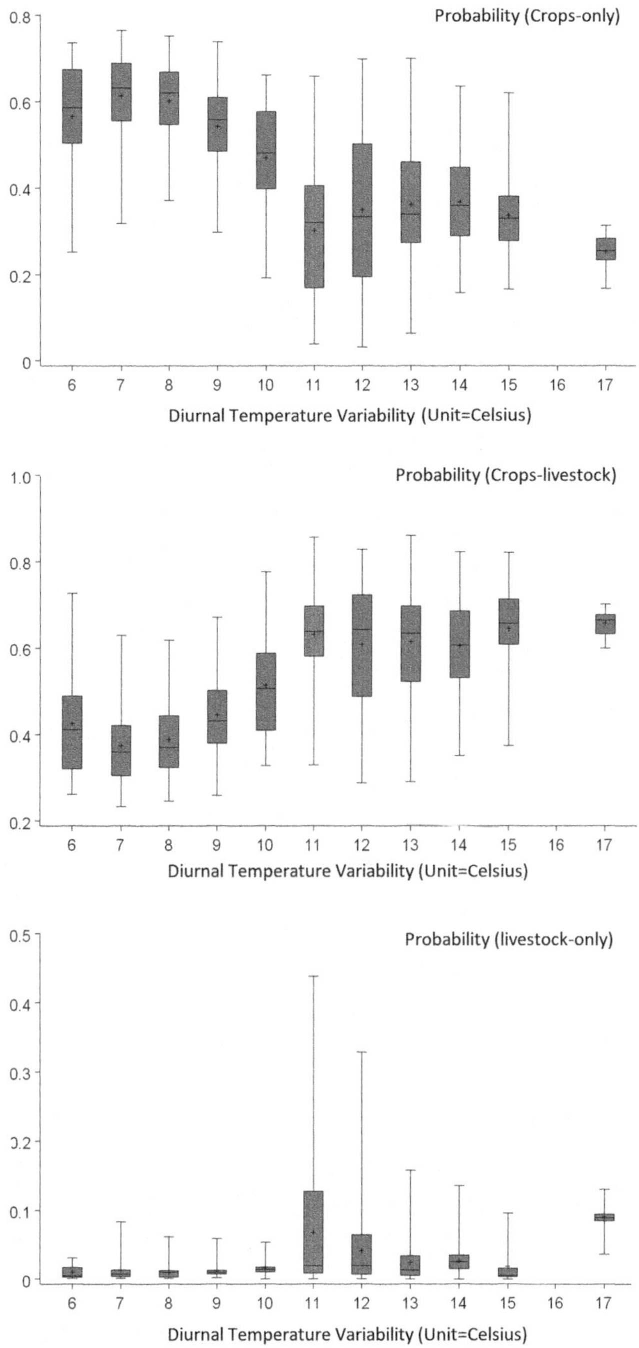

Fig. 3.5 Adoption of agricultural systems across diurnal temperature variability in Sub-Saharan Africa

explanation is that animals can endure temperature variability better than crops owing to bodily thermal regulation mechanisms or an intensive indoor management through, e.g., barns and cowsheds (Hahn 1981; Mader and Davis 2004; Hahn et al. 2009).

3.4.4 *Public Adaptation*

When it comes to an analysis of behavioral changes, two types of actions can be distinguished: private actions and public actions. A private adaptation action is one that is taken by an individual while a public adaptation action is one that is taken by a community (group) of individuals. A public action is needed when the good upon which the action is taken is a public good or the commons.

With regard to this distinction, it can be asked whether a community adaptation action is different from a private adaptation action. Further, when the good can be provided either privately or by a community, we can ask whether the community provision reveals any salient features different from the private provision of the good.

Such a privately and publicly provided good exists in the agriculture sector at the present time. Irrigation is an important input for crop farming and other agricultural activities. There are many methods for irrigating a farmland, some of which are privately supplied and others are supplied by a public agency. Private irrigation methods include, among other things, surface water irrigation, ground water irrigation, sprinkler irrigation, drip irrigation, and other precision irrigation techniques (Schoengold and Zilberman 2007).

An examination of the South American farm households' irrigation methods reveals that 65% of the farms rely on rain-fed agriculture, that is, with no irrigation. Of the 35% that irrigate, 21% relied on a public water scheme and 14% relied on private irrigations (Seo 2011a). These statistics can be compared with that of African agriculture where only 3% of the farmlands is irrigated (World Bank 2008, 2009b; Kurukulasuriya et al. 2011).

An analysis of adoption of an irrigation scheme is conducted using a mixed Logit analysis and the three irrigation options: rainfed, private, and public. Fascinatingly, stark differences are observed in the ways that the two irrigation schemes are supplied. First, while a private irrigation scheme increases as the temperature normal becomes hotter, a public irrigation scheme declines as the temperature normal becomes hotter.

Second, both a private irrigation and a public irrigation are adopted more frequently when the region receives smaller amounts of rainfall. However, the rate of adoption is drastically different between the two irrigation schemes. As the region becomes more arid, a private irrigation increases at a slow pace (in a linear fashion) but a public irrigation increases at a much faster pace (in an exponential fashion). Put differently, the drier the region becomes, the increase of adoption of a public irrigation overwhelms the increase of adoption of a private irrigation.

These results may imply that a public irrigation scheme has been historically developed as a governmental response to frequent droughts in a region which constrain crop farming, but not as a response to higher temperatures. This finding points to inefficiency in a public provision of irrigation.

The results also point to the possibility that a public irrigation scheme will be over-supplied, that is, over-supplied in highly arid zones than the economically efficient level. This could happen if the government were determined to keep farmers to maintain a certain type of agriculture even in highly arid zones in which that type of agriculture is no longer profitable or competitive in the market. Without the governmental irrigation supply, a farmer would have switched away from water-intensive crops to water-conserving crops—for example, grasslands—or non-crop agriculture such as animal husbandry. From the vantage point of the microbehavioral economics, it can be stated that a public provision of a certain type of resources can result in a distortion in resource allocations especially when the government-picked type of resources is also provided privately in the market, that is, by individual providers.

On the other hand, when there is pervasive publicness in the good that is concerned, a private provision can result in an under-provision of the public good. In the irrigation context, a dam cannot be built by an individual. Similarly, a reservoir cannot be supplied by an individual. A public sector can only provide these public goods, that is, a dam or a reservoir which can be freely accessed when needed by farmers in the surrounding areas for irrigation. Even in this context, an efficient size of a reservoir or dam is hardly an easy task to determine because the public sector in its decision cannot take into account all possible behavioral options and consequences. The dam may turn out to be too big, harming natural habitats and wildlife. The reservoir may turn out to be too large, taking away farmlands available to the locality's residents.

The concept of a public adaptation is one of the three most important economic theories that will be described in Chap. 7 of this book. It is a critical concept that may provide a justification of a governmental intervention in a public adaptation project. It may justify the Green Climate Fund's funding allocations to a range of projects internationally. A series of empirical evaluations of the GCF-funded projects on this regard is provided in Chap. 7.

3.4.5 Natural Resource Intensive Enterprises

The microbehavioral model of adaptation to global warming can be conceptualized in a broader framework than agriculture by way of natural resource intensive enterprises (Seo 2010c). Other than crop farming and livestock management, the basket of forest-related activities accounts for a large percentage of rural income and rural labor hours, for example, as much as 22% of rural income in Latin America (Vedeld et al. 2007).

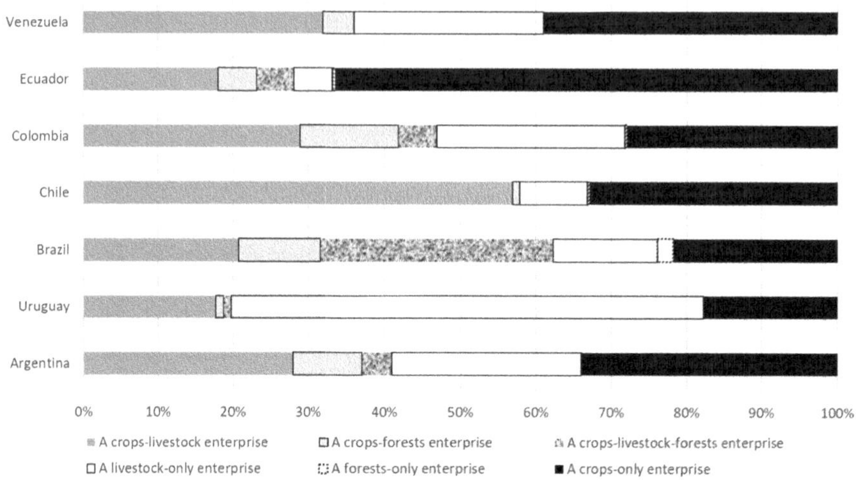

Fig. 3.6 Adoption rates of natural resource enterprises in Latin America

A natural resource sector model of the microbehavioral economics was developed with Latin American rural data (Seo 2012a). In the model, a rural natural resource manager is faced with the following set of seven alternative portfolios: a crops-only, a livestock-only, a forest-only, a crops-livestock, a crops-forests, a livestock-forests, and a crops-livestock-forests.

Adoption rates of these enterprises in each of the seven Latin American countries are summarized in Fig. 3.6. The three countries from the top are Andean countries: Ecuador, Colombia, and Venezuela. The four countries from the bottom are Southern Cone countries: Argentina, Brazil, Chile, and Uruguay.

A livestock-only enterprise is dominant in Uruguay with more than 60% of the rural enterprises adopting this type of enterprise. It is the most favored in Uruguay. This enterprise is also chosen by a large fraction of farms in Colombia, Venezuela, and Argentina, all known for the aforementioned iconic Latin American grasslands. In these countries, a crops-livestock enterprise is also adopted by a large fraction of rural farms.

A crops-only enterprise is dominant in Ecuador with more than 60% of the rural enterprises adopting this enterprise, owing to the large and sudden altitude variations across the country in the Andes Mountain range. A crops-livestock-forests enterprise is chosen most frequently by Brazilian farms, Colombian and Ecuadorian farms. A crops-forests enterprise is most favored by Colombian and Brazilian farms. A forests-only enterprise is found most often in Brazil. Note that Brazil and Colombia are the main hosts of the Amazon basin and rainforest.

How will these behavioral choices be altered in the future if the present climate system were to be changed? For future simulations, the two AOGCM scenarios are assumed, which are summarized in Table 3.4. One is the HadCM3 (Hadley Center Coupled Model) A2 scenario by the UKMO (United Kingdom Met Office) and the

Table 3.4 Climate change scenarios for the Latin American sample by mid-century

Scenarios		Temperature normals (celsius)		Precipitation normals (mm/month)	
AOGCM	Institution	Summer	Winter	Summer	Winter
HadCM3 A2	United Kingdom Meteorology Office (UKMO)	+2.77	+2.38	−12.91	+2.89
GISS ER A2	Goddard Institute for Space Studies (GISs)	+2.07	+1.97	+7.82	−0.74

Note: (1) HadCM3 means the Hadley center Coupled Model 3; (2) GISS ER means the Goddard Institute of Space Studies ModelE-R

Table 3.5 Changes in adoptions of natural resource enterprises by mid-century

	Portfolios					
	Diversified		Integrated	Specialized		
Scenarios	Crops-livestock	Crops-forests	Crops-livestock-forests	Livestock-only	Forests-only	Crops-only
Baseline probability	37.2%	6.8%	5.1%	20.4%	0.75%	29.8%
% point changes under HadCM3 UKMO A2	0.87%	−1.18%	0.10%	1.75%	−0.49%	−1.06%
% point changes under ER GISS A2	−0.23%	−0.67%	0.45%	2.00%	0.01%	−1.56%

other is the GISS (Goddard Institute of Space Studies) ModelE-R A2 scenario (Gordon et al. 2000; Schmidt et al. 2005). The A2 scenario is one family of storylines proposed by the IPCC, which foretells a high degree of warming owing to the fast rate economic growth fueled by extensive fossil fuel uses and divergent regional growth (Nakicenovic et al. 2000).

The two AOGCMs are chosen since they predict qualitatively different changes in precipitation for the region. The HadCM3 predicts a hotter future than the GISS ER: The former predicts about 2.5 °C increase in temperature while the latter predicts about 2.0 °C increase in temperature by about 2060. The precipitation predictions are more different. The HadCM3 predicts a 13 mm/month decrease in summer precipitation while the GSS ER predicts a 8 mm/month increase in summer precipitation.

Assuming the two climate models' predictions, changes in choices of the natural resource enterprises are calculated in Table 3.5 (Seo 2016b). Of the six enterprises, three are a specialized portfolio and the other three are a diversified portfolio. With the hotter and more arid HadCM3 scenario, the livestock-containing enterprises all are predicted to be chosen more frequently: The livestock-only, the crops-livestock, the crops-livestock-forest enterprise. The other three enterprises, a combination of crops and forests, are predicted to be less frequently adopted, that is, the crops-only, the crops-forests, and the forests-only enterprise.

If the GISS ER's A2 scenario were to come to pass by the middle of twenty-first century, the forests-only enterprise is not predicted to be less frequently adopted, owing to a summer precipitation increase. A similar response was also found in African agriculture (Seo 2010c, 2014a). On the other hand, the crops-livestock enterprise is expected to be less often adopted, owing again to a summer precipitation increase which negatively affects livestock management through disease outbreaks (Ford and Katondo 1977).

It is noticeable from Table 3.5 that the livestock-only enterprise and the crops-livestock-forests enterprise are predicted to be increasingly adopted under both of the climate scenarios. This may indicate that farm animals are more resilient to a hotter world than other farm assets, mainly, crops, because of internal heat regulation mechanisms as well as animals' mobility (Hahn et al. 2009).

This may also indicate that diversification and integration will work under a warming world. According to Table 3.5, an integrated rural portfolio of crops, livestock, and forests may make a natural resource enterprise more resilient against a particular natural shock such as a heat wave, a severe drought, or a heavy rainfall. Through the integration, a natural resource manager may be able to reduce the risk in the short-term and ensure the long-term profitability of the enterprise (Markowitz 1952; Udry 1995).

3.5 Microbehavioral Economics Implications for the GCF Funds Allocations

The results from the microbehavioral economic models described in the previous section offer ample insights on the central question of where the GCF funds should be directed and how much. They inform policy-makers about which agricultural system or natural resource enterprise in what region is most vulnerable to a warming climate system. The bottom line should be that the international funds should be directed to those negatively affected and not to those who will benefit from a changing climate.

A decision-maker who is given the responsibility of allocating the US$ 100 billion annually to a basket of selected projects from the large pool of proposals submitted by various policy areas in different geopolitical regions is bound to ask how much damage is expected in those enterprises that are identified to be among the most vulnerable. S/he may attempt to tie the size of the individual GCF grant to the magnitude of the potential damage expected of the potential recipients. We will have to wait until Chap. 7, however, to assess the actual decision rules determined by the GCF Board.

The microbehavioral economic model offers pivotal information in answering this important policy question. In Table 3.6, the present author summarizes the predicted changes in the long-term profits of the three agricultural systems, assuming the above-described ModelE-R's A2 scenario by the GISS.

Table 3.6 Changes in land values of natural resource enterprises owing to climate change scenarios

Scenarios	Enterprises		
	Crops-only	Crops-livestock	Livestock-only
Baseline land values ($/ha)	2418	1640	1058
GISS ER A2 scenario: absolute changes ($/ha)	−230	−57	629
GISS ER A2 scenario: % changes	−9.5%	−3.5%	59.4%

The long-term profit of the agricultural system can be measured by the present value of the stream of future rents, that is, net incomes, generated from the land, which is defined to be the land value according to David Ricardo (Ricardo 1817). It can be also thought of as the value of a capital asset which is defined, by Irving Fisher, as the present value of the stream of net incomes generated from the asset discounted by expected interest rates (Fisher 1906, 1930).

The changes in the land values of the three natural resource enterprises caused by the climatic change according to the above-referenced GISS ModelE-R A2 scenario are shown in Table 3.6 (Seo 2016a). The land value of the crops-only portfolio is predicted to fall by 230$/ha, which corresponds to 9.5% of the baseline land value, that is, the present land value.

Predicted changes in the land values of the other portfolios under the same climate scenario are quite dissimilar. The crops-livestock portfolio is predicted to lose only 57$/ha in its land value by the middle of this century, which is a 3.5% loss from the baseline.

The changes in the land value of the livestock-only portfolio are predicted to be beneficial, assuming the same climate scenario. There will be an increase in the land value by 629$/ha, which corresponds to 59.4% of the baseline land value of this portfolio. According to the microbehavioral model, the specialized livestock enterprise will benefit from the predicted climate change by the Goddard Institute.

The information presented in Table 3.6 contains one of the key pillars of the knowledge body needed for a justifiable distribution of the GCF funds allocations. More concretely, it may be argued that the GCF grants/loans should be directed to the most vulnerable sector of the economy as well as the most vulnerable enterprise of the sector. The crops-only enterprise is identified in Table 3.6 as the most vulnerable enterprise within the agricultural and natural resource sector. If the grants were to be given at all to a particular group, it should be allocated to a loser or victim group. It would not make sense for the GCF to give free money to the groups that are winners from climate change.

With the impact of climate change on each of the natural resource enterprises measured, what is the aggregate impact of climate change on the natural resource sector as a whole, that is, inclusive of all the enterprises? A prediction of the aggregate sectoral impact would require that the individual enterprises be allowed to switch from one to another enterprise as the changes in the climate system would unfold gradually in the future.

Table 3.7 Changes in the expected land value of the natural resource sector by mid-century

Scenarios	Absolute changes ($/ha)	% changes	95% confidence intervals ($/ha)
Baseline land value ($/ha)	1857		
GISS ER, A2 ($/ha)	−13	−0.7%	(−25, −1)
UKMO HadCM3, A2 ($/ha)	−62	−3.3%	(−80, −44)

The aggregate impact predicted by the microbehavioral model is presented in Table 3.7 after accounting for both changes in choices of enterprises and changes in enterprise-specific land values (Seo 2016a). If the GISS ER scenario were to unravel, the aggregate impact would be modest at 13$/ha of land. This is less than 1% of the baseline land value. If the HadCM3 scenario were to come to pass, the aggregate impact is predicted to be 62$/ha of land. This corresponds to 3.3% decrease from the present land value. All predictions are made for the middle of this century, about year 2060.

Table 3.7 provides another key element of the knowledge body needed for a proper distribution of the GCF funds. An interpretation of the result in the table is that although the crops-only portfolio is quite vulnerable to the future climate change scenario with about 10% loss of annual income by the middle of this century, the agricultural and natural resource sector as a whole would not be much vulnerable, as predicted in the table. Owing to a host of efficient adaptations, including gradual switches from one enterprise to another, the sector as a whole would suffer only a minor loss, if at all.

Another interpretation of Table 3.7 is that the aggregate impact would depend on which climate model prediction comes to pass eventually. A hotter and more arid climate scenario such as the HadCM3 would harm the natural resource sector more heavily. The results imply that the hotter the Earth's climate becomes beyond the current state, the more difficult and costly adaptation actions become.

The key takeaway message from Table 3.7 for the GCF policy-makers is that an allocation decision of the GCF funds should be made after fully accounting for adaptive behavioral changes that can be taken privately. Without considering adaptation behaviors of individuals, it can turn out to be in the worst-case scenario that the GCF fund be allocated to the entities who massively benefit from a warmer world. In another similarly perverted case, the GCF funding may encourage an enterprise to remain fixated on the identified vulnerable portfolio (Seo 2015b).

This essential message can be carried further on to an analysis with a specification of a more detailed natural resource sector. For example, a GCF grant allocation to any of the following natural resource enterprises can be thought of in a similar manner: A crops-only, a livestock-only, a forests-only, a crops-forests, a crops-livestock, a livestock-forests, a crops-livestock-forest (Seo 2012a, 2016b).

Besides the Latin American natural resource enterprises, a microbehavioral economic analysis applied to other continents can reveal similarly fundamental information which is needed for allocation decisions of the GCF policy-makers (Seo 2014a). An example is an analysis of African agriculture which showed changes in adoptions in response to climatic changes of agricultural portfolios among a crops-only, a crops-livestock, a crops-forests, and a crops-livestock-forests (Seo 2010c). A hotter and more arid climate in the future would force African farmers to switch to a crops-livestock mixed portfolio. On the other hand, a hotter and more humid climate in the future would force African farmers to adopt a crops-forests mixed portfolio.

3.6 Additional Insights of the Microbehavioral Economic Models

The description of the microbehavioral economic models of adaptation behaviors up to this point by and large rely on a certain classification of the agricultural and natural resource sector. It was referred to as different names in this chapter: a set of agricultural system, a set of natural resource enterprises, or a set of agricultural portfolios. Accordingly, a recommendation for a GCF allocation was also made in the framework of the chosen classification, e.g., at the level of an agricultural system.

However, it does not mean that a microbehavioral economic model should always be conducted within the frame of the chosen classification, say, an agricultural system. Neither does it mean that a GCF funding should be framed at the level of the agricultural system. It is possible that a series of microbehavioral models points to a certain natural resource product or system which is particularly vulnerable or resilient, as such, should be recommended for a GCF funding.

Goat ownership in India can serve as an example in this context. The Indian climate is characterized by a severe monsoon regime: An extremely large amount of rainfall during the monsoon season and an extremely scarce rainfall during the non-monsoon season. The higher the monsoon rainfall variability index which was defined in the theory section (Refer to Eq. 3.14 and Fig. 3.2), the more likely that an Indian farmer owns goats (Seo 2016c).

The Indian goat study indicates that it might be possible for climate researchers to search and find a magical product for the warming world at least on a regional basis. It doesn't have to be a livestock species. It could be a fish species, a forest product, or even a particular crop variety which turns out to be particularly resilient to changes in the climate regime (Evenson and Gollin 2003; NRC 2015).

The microbehavioral economic literature in this direction is very sparse. Climate researchers have by and large neglected this important research area where there is big potential that can be realized. A few notable examples from the scientific literature outside the microbehavioral economics would give readers a better idea. Animal scientists reported that a kangaroo can be an alternative meat food to cattle and sheep since the former is not a ruminant, so does not emit as much methane, a potent greenhouse gas (Kempton et al. 1976). An Asparagopsis, a type of seaweed, or a 3NOP (3-Nitrooxypropanol) can be used as a feed additive for ruminant animals to affect enteric fermentation of the ruminants such as cattle and sheep to reduce the amount of methane emissions (Hristov et al. 2015; CSIRO 2018).

In concluding the chapter, a few clarifications and qualifications should help readers ascertain subtler points of arguments. First, the present author would like to add that many of the arguments presented in this chapter with regard to the microbehavioral economics of global warming may not be conveyed at first to the readers as lucidly as they are written. These points will become clearer when the readers have had the opportunity to read through the rest of the scientific traditions elaborated in this book since many points in this chapter are written with reference to the alternative scientific traditions. The agro-economic modeling will be explained in Chap. 4; the statistical techniques for estimation of yield and productivity changes will be followed in Chap. 5; the ecosystem modeling will come at Chap. 6.

Another clarification is needed with regard to the rationale for supporting a particular GCF project funding. Although a microbehavioral economic research can point to the vulnerable sectors and enterprises of the economy to climate change, it does not provide directly a rationale for providing a support for a particular sector or an enterprise. This is because a vulnerable enterprise under a particular climate change scenario may be able to adapt on its own to avoid the brunt of the damage from the predicted climate change. If such is the case, there would be no need for the GCF policy-makers to allocate the GCF funding to the enterprise.

The present author will come back to this critical question in Chaps. 7 and 8 where theoretical foundations for the governmental or global-scale allocations of the global funds to selected projects are laid out. The foundational theories will include the theory of public expenditures (Samuelson 1954), of efficient resource allocations (Ricardo 1817; von Thunen 1826), of public adaptation, of crowdfunding (Mollick 2014), and the theory of bitcoin and cryptocurrency (Bohme et al. 2015).

References

Adams R, Rosenzweig C, Peart RM, Ritchie JT, McCarl BA, Glyer JD, Curry RB, Jones JW, Boote KJ, Allen LH (1990) Global climate change and US agriculture. Nature 345:219–224

Adams R, McCarl BA, Segerson K, Rosenzweig C, Bryant K, Dixon BL, Conner R, Evenson RE, Ojima D (1999) The economic effects of climate change on US agriculture. In: Mendelsohn R, Neumann J (eds) The impact of climate change on the United States economy. Cambridge University Press, Cambridge

Ainsworth EA, Long SP (2005) What have we learned from 15 years of free-air CO2 enrichment (FACE)? A meta-analytic review of the responses of photosynthesis, canopy properties and plant production to rising CO2. New Phytol 165:351–371

Amundson JL, Mader TL, Rasby RJ, Hu QS (2006) Environmental effects on pregnancy rate in beef cattle. J Anim Sci 84:3415–3420

Anselin L (1988) Spatial econometrics: methods and models. Kluwer Academic Publishers, Dordrecht

Baethgen WE (1997) Vulnerability of agricultural sector of Latin America to climate change. Clim Res 9:1–7

Basist A, Peterson N, Peterson T, Williams C (1998) Using the special sensor microwave imager to monitor land surface temperature, wetness, and snow cover. J Appl Meteorol 37:888–911

Berkson J (1944) Application of the logistic function to bio-assay. J Am Stat Assoc 39:357–365

Beron KJ, Vijberberg WPM (2004) Probit in a spatial context: a Monte Carlo approach. In: Anselin L, Florax RJGM, Rey SJ (eds) Advances in spatial econometrics: methodology, Tools and Applications. Springer, Heidelberg, pp 169–192

Biondi F, Gershunov A, Cayan DR (2001) North pacific decadal climate variability since 1661. J Clim 14:5–10

Bliss CI (1934a) The method of probits. Science 79:38–39

Bliss CI (1934b) The method of probits. Science 79:409–410

Bohme R, Christin N, Edelman B, Moore T (2015) Bitcoin: economics, technology, and governance. J Econ Perspect 29:213–238

Bourguignon F, Fournier M, Gurgand M (2004) Selection bias corrections based on the multinomial logit model: Monte-Carlo comparisons. DELTA working paper no. 20. De'partement et Laboratoire d'Economie The'orique et Applique'e (DELTA), Paris

Byerlee D, Eicher CK (1997) Africa's emerging maize revolution. Lynne Rienner Publishers Inc., Boulder

Chung CE, Ramanathan V (2006) Weakening of North Indian SST gradients and the monsoon rainfall in India and the Sahel. J Clim 19:2036–2045

Commonwealth Scientific and Industrial Research Organization (CSIRO) (2018) Future feed. CSIRO, Canberra. Accessed from https://research.csiro.au/futurefeed/

Curtis S, Adler RF, Huffman GJ, Nelkin E, Bolvin D (2001) Evolution of tropical and extratropical precipitation anomalies during the 1997 to 1999 ENSO cycle. Int J Climatol 21:961–971

Dahl GB (2002) Mobility and the returns to education: testing a Roy model with multiple markets. Econometrica 70:2367–2420

Dai A, Trenberth KE, Qian T (2004) A global dataset of palmer drought severity index for 1870–2002: relationship with soil moisture and effects of surface warming. J Hydrometeorol 5:1117–1130

Danielson JJ, Gesch DB (2011) Global multi-resolution terrain elevation data 2010 (GMTED2010). US Geo-logical Survey Open-File Report 2011–1073, 26 p

Darwin RF, Tsigas M, Lewandrowski J, Raneses A (1995) World agriculture and climate change: economic adaptations. U.S. Department of Agriculture (USDA). Economic Research Service, Washington, DC

Denman KL, Brasseur G, Chidthaisong A, Ciais P, Cox PM, Dickinson RE, Hauglustaine D, Heinze C, Holland E, Jacob D, Lohmann U, Ramachandran S, da Silva Dias PL, Wofsy SC, Zhang X (2007) Couplings between changes in the climate system and biogeochemistry. In: Solomon S et al (eds) Climate change 2007: the physical science basis. The Intergovernmental Panel on Climate Change. Cambridge University Press, Cambridge

Deschenes O, Greenstone M (2007) The economic impacts of climate change: evidence from agricultural output and random fluctuations in weather. Am Econ Rev 97:354–385

Dinar A, Hassan R, Mendelsohn R, Benhin J (2008) Climate change and agriculture in Africa: impact assessment and adaptation strategies. EarthScan, London

Driessen P, Deckers J, Nachtergaele F (2001) Lecture notes on the major soils of the world. Food and Agriculture Organization (FAO), Rome

Dubin JA, McFadden DL (1984) An econometric analysis of residential electric appliance holdings and consumption. Econometrica 52(2):345–362

Dudal R (1980) Soil-related constraints to agricultural development in the tropics. International Rice Research Institute, Los Banos

Easterling DR, Evans JL, Groisman PY, Karl TR, Kunkel KE, Ambenje P (2000) Observed variability and trends in extreme climate events: a brief review. Bull Am Meteorol Soc 81:417–425

Efron B (1981) Nonparametric estimates of standard error: the jackknife, the bootstrap and other methods. Biometrika 68:589–599

Evenson R, Gollin D (2003) Assessing the impact of the Green Revolution 1960–2000. Science 300:758–762

FAO (2005) Global agro-ecological assessment for agriculture in the twenty-first century (CD-ROM). FAO land and water digital media series. FAO, Rome

FAO (2012) FAO STAT. FAO Statistics Division, Rome. Available at http://faostat.fao.org

Fischer G, Shah M, Tubiello FN, van Velhuizen H (2005) Socio-economic and climate change impacts on agriculture: an integrated assessment, 1990–2080. Philos Trans R Soc B 360:2067–2083
Fisher I (1906) The nature of capital and income. Macmillan, New York
Fisher I (1930) The theory of interest. Macmillan, New York
Fisher FM (1966) The identification problem in econometrics. McGraw-Hill, New York
Food and Agriculture Organization (FAO) (1978) Report on agro-ecological zones (vol. 1): methodology and results for Africa. FAO, Rome
Food and Agriculture Organization (FAO) (2003) The digital soil map of the world (DSMW) CD-ROM. FAO, Rome
Ford J, Katondo KM (1977) Maps of tsetse fly (Glossina) distribution in Africa, 1973, according to subgeneric groups on a scale of 1: 5000000. Bull Anim Health Prod Afr 15:187–193
Fox NJ, Marion G, Davidson RS, White PCL, Hutchings MR (2012) Livestock helminths in a changing climate: approaches and restrictions to meaningful predictions. Animals 2:93–107
Geweke J (1989) Bayesian inference in econometric models using Monte Carlo integration. Econometrica 57:1317–1339
Gitay H, Brown S, Easterling W, Jallow B (2001) Ecosystems and their goods and services. In: McCarthy JJ et al (eds) Climate change 2001: impacts, adaptations, and vulnerabilities. The IPCC. Cambridge University Press, Cambridge
Gordon C, Cooper C, Senior CA, Banks HT, Gregory JM, Johns TC, Mitchell JFB, Wood RA (2000) The simulation of SST, sea ice extents and ocean heat transports in a version of the Hadley Centre coupled model without flux adjustments. Clim Dyn 16:147–168
Green Climate Fund (GCF) (2011) Governing instrument for the Green Climate Fund. GCF, Songdo City
Green Climate Fund (GCF) (2018a) Annex III: investment framework. GCF, Songdo City
Green Climate Fund (GCF) (2018b) Status of pledges and contributions made to the Green Climate Fund, status date: 29 January 2018. GCF, Songdo City
Green Climate Fund (GCF) (2018c) Projects + Programmes. GCF, Songdo City. Accessed from https://www.greenclimate.fund/what-we-do/projects-programmes
Hahn GL (1981) Housing and management to reduce climate impacts on livestock. J Anim Sci 52:175–186
Hahn GL (1999) Dynamic responses of cattle to thermal heat loads. J Anim Sci 77:10–20
Hahn GL, Gaughan JB, Mader TL, Eigenberg RA (2009) Chapter 5: Thermal indices and their applications for livestock environments. In: DeShazer JA (ed) Livestock energetics and thermal environmental management. American Society of Agricultural and Biological Engineers, St. Joseph
Hajivassiliou V, McFadden D (1998) The method of simulated scores for the estimation of LDV models. Econometrica 66:863–896
Hansen J, Sato M, Reudy R (2012) Perception of climate change. Proc Natl Acad Sci U S A 109: E2415–E2423
Harris I, Jones PD, Osborn TJ, Lister DH (2014) Updated high-resolution grids of monthly climatic observations – the CRU TS3.10 Dataset. Int J Climatol 34:623–642
Heckman J (1979) Sample selection bias as a specification error. Econometrica 47:153–162
Hotelling H (1931) The economics of exhaustible resources. J Polit Econ 39(2):137–175
Hristov AN, Joonpyo OH, Fabio Giallongo F, Tyler W, Frederick TW, Michael T, Harper MT, Holley L, Weeks HL, Antonio F, Branco AF, Peter J, Moate PJ, Matthew H, Deighton MH, SRO W, Kindermann M, Duval S (2015) An inhibitor persistently decreased enteric methane emission from dairy cows with no negative effect on milk production. Proc Natl Acad Sci 112 (34):10663–10668
Hulme M, Doherty RM, Ngara T, New MG, Lister D (2001) African climate change: 1900–2100. Clim Res 17:145–168
Indian Institute of Tropical Meteorology (IITM) (2012) Homogeneous Indian monthly rainfall data sets & Indian regional monthly surface air temperature data set. IITM, Pune
Intergovernmental Panel on Climate Change (IPCC) (1990) Climate change: the IPCC scientific assessment. Cambridge University Press, Cambridge

Intergovernmental Panel on Climate Change (IPCC) (2014) Climate change 2014: the physical science basis. The fifth assessment report of the IPCC. Cambridge University Press, Cambridge
Janowiak JE (1988) An investigation of interannual rainfall variability in Africa. J Clim 1:240–255
Johnson HD (1965) Response of animals to heat. Meteorol Monogr 6:109–122
Johnston J, DiNardo J (1997) Econometric methods, 4th edn. McGraw-Hill, New York
Kafle K, Winter-Nelson A, Goldsmith P (2016) Does 25 cents more per day make a difference? The impact of livestock transfer and development in rural Zambia. Food Policy 65:62–72
Kahneman D, Tversky A (1979) Prospect theory: an analysis of decision under risk. Econometrica 47:263–291
Kaiser HM, Riha SJ, Wilks DS, Rossiter DG, Sampath R (1993) A farm-level analysis of economic and agronomic impacts of gradual climate warming. Am J Agric Econ 75:387–398
Kala N (2015) Ambiguity aversion and learning in a changing world: the potential effects of climate change from Indian agriculture. Ph.D. dissertation, Yale University, New Haven
Kazianga H, Udry C (2006) Consumption smoothing? Livestock, insurance, and drought in rural Burkina Faso. J Dev Econ 79:413–446
Keane M (1994) A computationally practical simulation estimator for panel data. Econometrica 62:95–116
Kempton TJ, Murray RM, Leng RA (1976) Methane production and digestibility measurements in the grey kangaroo and sheep. Aust J Biol Sci 29:209–214
Koopmans TC (1951) Efficient allocation of resources. Econometrica 19:455–465
Koopmans TC (1965) On the concept of optimal economic growth. Academiae Scientiarum Scripta Varia 28(1):1–75
Kurukulasuriya P, Mendelsohn R, Hassan R, Benhin J, Diop M, Eid HM, Fosu KY, Gbetibouo G, Jain S, Mahamadou A, El-Marsafawy S, Ouda S, Ouedraogo M, Sène I, Maddision D, Seo SN, Dinar A (2006) Will African agriculture survive climate change? World Bank Econ Rev 20:367–388
Kurukulasuriya P, Kala N, Mendelsohn R (2011) Adaptation and climate change impacts: a structural Ricardian model of irrigation and farm income in Africa. Clim Change Econ 2:149–174
Le Treut H, Somerville R, Cubasch U, Ding Y, Mauritzen C, Mokssit A, Peterson T, Prather M (2007) Historical overview of climate change. In: Solomon S et al (eds) Climate change 2007: the physical science basis. The fourth assessment report of the Intergovernmental Panel on Climate Change. Cambridge University Press, Cambridge
Lee LF (1983) Generalized econometric models with selectivity. Econometrica 51:507–512
Mader TL, Davis MS (2004) Effect of management strategies on reducing heat stress of feedlot cattle: feed and water intake. J Anim Sci 82:3077–3087
Mader TL, Frank KL, Harrington JA, Hahn GL, Nienaber JA (2009) Potential climate change effects on warm-season livestock production in the Great Plains. Clim Chang 97:529–541
Magrin G, Garcia CG, Choque DC, Gimenez JC, Moreno AR, Nagy GJ, Nobre C, Villamizar A (2007) Latin America. In: Parry ML, Canziani OF, Palutikof JP, van der Linden PJ, Hanson CE (eds) Climate change 2007: impacts, adaptation, and vulnerability. Intergovernmental Panel on Climate Change. Cambridge University Press, Cambridge, pp 581–615
Markowitz H (1952) Portfolio selection. J Financ 7:77–91
Mata LJ, Campos M et al (2001) Latin America. In: McCarthy JJ, Canziani OF, Leary NA, Dokken DJ, White KS (eds) Climate change 2001: impacts, adaptation, and vulnerability. Intergovernmental Panel on Climate Change. Cambridge University Press, Cambridge
May W (2011) The sensitivity of the Indian summer monsoon to a global warming of 2 degrees C with respect to pre-industrial times. Clim Dyn 37:1843–1868
McFadden DL (1974) Conditional logit analysis of qualitative choice behavior. In: Zarembka P (ed) Frontiers in econometrics. Academic, New York, pp 105–142
McFadden D (1999) Sampling and selection. Lecture note. University of California, Berkeley
McFadden D, Train K (2000) Mixed MNL models for discrete response. J Appl Econ 15:447–470
Meehl GA, Hu A (2006) Megadroughts in the Indian monsoon region and southwest North America and a mechanism for associated multidecadal Pacific sea surface temperature anomalies. J Clim 19:1605–1623

Mendelsohn R, Nordhaus W, Shaw D (1994) The impact of global warming on agriculture: a Ricardian analysis. Am Econ Rev 84:753–771
Mendelsohn R, Kurukulasuriya P, Basist A, Kogan F, Williams C (2007) Measuring climate change impacts with satellite versus weather station data. Clim Chang 81:71–83
Mollick E (2014) The dynamics of crowdfunding: an exploratory study. J Bus Ventur 29:1–16
Moran PAP (1950a) Notes on continuous stochastic phenomena. Biometrika 37:17–23
Moran PAP (1950b) A test for the serial dependence of residuals. Biometrika 37:178–181
Nakicenovic N, Davidson O, Davis G, Grübler A, Kram T, La Rovere EL, Metz B, Morita T, Pepper W, Pitcher H, Sankovski A, Shukla P, Swart R, Watson R, Dadi Z (2000) Emissions scenarios. A special report of Working Group III of the Intergovernmental Panel on Climate Change. IPCC, Geneva
Nash J (1951) Non-cooperative games. Ann Math 54(2):286–295
National Aeronautics and Space Administration (NASA) (2018) Missions: Earth Observing System (EOS). NASA, Washington, DC. Accessed from https://eospso.nasa.gov/mission-category/3
National Research Council (NRC) (2015) Critical role of animal science research in food security and sustainability. National Academies Press, Washington, DC
New M, Lister D, Hulme M, Makin I (2002) A high-resolution data set of surface climate over global land areas. Clim Res 21:1–25
Nordhaus WD (1973) The allocation of energy resources. Brook Pap Econ Act 1973(3):529–576
Porter JH, Parry ML, Carter TR (1991) The potential effects of climatic change on agricultural insect pests. Agric For Meteorol 57:221–240
Rahmstorf S, Coumou D (2011) Increase of extreme events in a warming world. Proc Natl Acad Sci U S A 108:17905–17909
Ricardo D (1817) On the principles of political economy and taxation. John Murray, London
Samuelson P (1938) A note on the pure theory of consumers' behaviour. Economica 5:61–71
Samuelson P (1954) The pure theory of public expenditure. Rev Econ Stat 36:387–389
Schlenker W, Roberts M (2009) Nonlinear temperature effects indicate severe damages to crop yields under climate change. Proc Natl Acad Sci U S A 106(37):15594–15598
Schlesinger WH (1997) Biogeochemistry: an analysis of global change, 2nd edn. Academic, San Diego
Schmertmann CP (1994) Selectivity bias correction methods in polychotomous sample selection models. J Econ 60:101–132
Schmidt GA, Ruedy R, Hansen JE et al (2005) Present day atmospheric simulations using GISS ModelE: comparison to in-situ, satellite and reanalysis data. J Clim 19:153–192
Schoengold K, Zilberman D (2007) The economics of water, irrigation, and development. In: Evenson R, Pingali P (eds) Handbook of agricultural economics, vol 3. Elsevier, Amsterdam, pp 2933–2977
Seo SN (2006) Modeling farmer responses to climate change: climate change impacts and adaptations in livestock management in Africa. Ph.D. dissertation, Yale University, New Haven
Seo SN (2010a) A microeconometric analysis of adapting portfolios to climate change: adoption of agricultural systems in Latin America. Appl Econ Perspect Policy 32:489–514
Seo SN (2010b) Is an integrated farm more resilient against climate change? A micro-econometric analysis of portfolio diversification in African agriculture? Food Policy 35:32–40
Seo SN (2010c) Managing forests, livestock, and crops under global warming: a microeconometric analysis of land use changes in Africa. Aust J Agric Resour Econ 54(2):239–258
Seo SN (2011a) An analysis of public adaptation to climate change using agricultural water schemes in South America. Ecol Econ 70:825–834
Seo SN (2011b) A geographically scaled spatial analysis of adaptation to climate change by agricultural systems in Africa. J Agric Sci 149:437–449
Seo SN (2012a) Adapting natural resource enterprises under global warming in South America: a mixed logit analysis. Economia 12:111–135
Seo SN (2012b) Decision making under climate risks: an analysis of sub-Saharan farmers' adaptation behaviors. Weather Climate Soc 4:285–299
Seo SN (2013a) An essay on the impact of climate change on US agriculture: weather fluctuations, climatic shifts, and adaptation strategies. Climate Change 121:115–124

Seo SN (2013b) Refining spatial resolution and spillovers of a microeconometric model of adapting portfolios to climate change. Mitig Adapt Strateg Glob Chang 18:1019–1034
Seo SN (2014a) Evaluation of agro-ecological zone methods for the study of climate change with micro farming decisions in sub-Saharan Africa. Eur J Agron 52:157–165
Seo SN (2014b) Adapting sensibly when global warming turns the field brown or blue: a comment on the 2014 IPCC report. Econ Aff 34:399–401
Seo SN (2015a) Micro-behavioral economics of global warming: modeling adaptation strategies in agricultural and natural resource enterprises. Springer, Cham
Seo SN (2015b) Helping low-latitude, poor countries with climate change. Regul Winter 2015–2016:6–8
Seo SN (2016a) Modeling farmer adaptations to climate change in South America: a micro-behavioral economic perspective. Environ Ecol Stat 23:1–21
Seo SN (2016b) The micro-behavioral framework for estimating total damage of global warming on natural resource enterprises with full adaptations. J Agric Biol Environ Stat 21:328–347
Seo SN (2016c) Untold tales of goats in deadly Indian monsoons: adapt or rain-retreat under global warming? J Extreme Events 3:1650001. https://doi.org/10.1142/S2345737616500019
Seo SN (2016d) Microbehavioral econometric methods: theories, models, and applications for the study of environmental and natural resources. Academic Press (Elsevier), Amsterdam
Seo SN (2017) The behavioral economics of climate change: adaptation behaviors, global public goods, breakthrough technologies, and policy-making. Academic Press (Elsevier), Amsterdam
Seo SN (2018) Natural and man-made catastrophes: theories, economics, and policy designs. Wiley-Blackwell, Hoboken
Seo SN (2019) Will farmers fully adapt to monsoon climate? J Agric Sci (under revision)
Seo SN, Mendelsohn R (2008a) Measuring impacts and adaptations to climate change: a structural Ricardian model of African livestock management. Agric Econ 38:151–165
Seo SN, Mendelsohn R (2008b) A Ricardian analysis of the impact of climate change impacts on South American farms. Chilean J Agric Res 68:69–79
Seo SN, Mendelsohn R, Dinar A, Hassan R, Kurukulasuriya P (2009) A Ricardian analysis of the distribution of climate change impacts on agriculture across agro-ecological zones in Africa. Environ Resour Econ 43:313–332
Seo SN, McCarl B, Mendelsohn R (2010) From beef cattle to sheep under global warming? An analysis of adaptation by livestock species choice in South America. Ecol Econ 69:2486–2494
Shanahan TM, Overpeck JT, Anchukaitis KJ, Beck JW, Cole JE, Dettman DL, Peck JA, Scholz CA, King JW (2009) Atlantic forcing of persistent drought in West Africa. Science 324:377–380
Shaw MR, Zavaleta ES, Chiariello NR, Cleland EE, Mooney HA, Field CB (2002) Grassland responses to global environmental changes suppressed by elevated CO2. Science 298:1987–1990
Shiller RJ (2003) From efficient markets theory to behavioral finance. J Econ Perspect 17:83–104
Shiller RJ (2005) Irrational exuberance, 2nd edn. Princeton University Press, Princeton
Steiger C (2006) Modern beef production in Brazil and Argentina. Choices Mag 21:105–110
Strzepek K, McCluskey A (2006) District level hydroclimatic time series and scenario analyses to assess the impacts of climate change on regional water resources and agriculture in Africa. CEEPA discussion paper no. 13. Centre for Environmental Economics and Policy in Africa, University of Pretoria, Pretoria
Sutherst RW (1991) Pest risk analysis and the greenhouse effect. Rev Agric Entomol 79:1177–1187
Taylor KE, Stouffer RJ, Meehl GA (2012) An overview of CMIP5 and the experiment design. Bull Am Meteorol Soc 93:485–498
Thaler RH (2015) Misbehaving: the making of behavioral economics. W.W. Norton Company, New York
Train K (2003) Discrete choice methods with simulation. Cambridge University Press, Cambridge
Tubiello FN, Ewert F (2002) Simulating the effects of elevated CO2 on crops: approaches and applications for climate change. Eur J Agron 18:57–74

Tubiello FN, Amthor JA, Boote KJ, Donatelli M, Easterling W, Fischer G, Gifford RM, Howden M, Reilly J, Rosenzweig C (2007) Crop response to elevated CO2 and world food supply. Eur J Agron 26:215–223
Udry C (1995) Risk and saving in Northern Nigeria. Am Econ Rev 85:1287–1300
United Nations Framework Convention on Climate Change (UNFCCC) (1992) United nations framework convention on climate change. UNFCCC, New York
United Nations Framework Convention on Climate Change (UNFCCC) (2011) Report of the transitional committee for the design of Green Climate Fund. UNFCCC, New York
United Nations Framework Convention on Climate Change (UNFCCC) (2015) The Paris agreement. Conference of the Parties (COP) 21. UNFCCC, New York
United States Animal Health Association (USAHA) (2008) Foreign animal diseases-the gray book, 7th edn. Committee on Foreign and Emerging Diseases of the USAHA, St. Joseph
United States Department of Agriculture (USDA) (2012) Census of agriculture 2012. USDA, Washington, DC. Accessed from https://www.agcensus.usda.gov/Publications/2012/#full_report
University of Alabama in Huntsville (UAH) (2018) Global temperature record. The Earth System Science Center, UAH. https://www.nsstc.uah.edu/climate/
Vedeld P, Angelsen A, Bojø J, Sjaastad E, Kobugabe GK (2007) Forest environmental incomes and the rural poor. Forest Policy Econ 9:869–879
von Neumann J, Morgenstern O (1947) Theory of games and economic behavior, 2nd edn. Princeton University Press, Princeton
von Thünen JH (1826) Der Isolierte Staat in Beziehung auf Landwirtschaft und Nationalökonomie, Hamburg, Perthes. (English trans. by Wartenberg CM (1996) The Isolated State. Pergamon Press, Oxford)
Welch JR, Vincent JR, Auffhammer M, Moya PF, Dobermann A, Dawe D (2010) Rice yields in tropical/subtropical Asia exhibit large but opposing sensitivities to minimum and maximum temperatures. Proc Natl Acad Sci U S A 107:14562–14567
White N, Sutherst RW, Hall N, Whish-Wilson P (2003) The vulnerability of the Australian beef industry to impacts of the Cattle Tick (Boophilus microplus) under climate change. Clim Chang 61:157–190
World Bank (2008) World development report 2008: agriculture for development. World Bank, Washington, DC
World Bank (2009a) Africa Infrastructure and Country Diagnostics (AICD). World Bank, Washington, DC. Available from: http://www.infrastructureafrica.org/aicd/
World Bank (2009b) Awakening Africa's sleeping giant: prospects for commercial agriculture in the Guinea Savannah Zone and beyond. World Bank and FAO, Washington, DC
World Bank (2017) Climate change knowledge portal for development practitioners and policy makers: Thailand. World Bank, Washington, DC
World Resources Institute (WRI) (2005) World resources 2005: the wealth of the poor—managing ecosystems to fight poverty. WRI, Washington, DC
Ziska LH (2003) Evaluation of yield loss in field-grown sorghum from a C3 and C4 weed as a function of increasing atmospheric carbon dioxide. Weed Sci 51:914–918

Chapter 4
Agro-Economic Models for Measuring the Impact of Climate Change on Agriculture

Abstract This chapter explains the scientific tradition of the Agro-Economic Model (AEM) which was developed purposefully for measuring the impacts of climate change on agriculture. The theory and methodology of the initial AEM and its extensions are elaborated in detail. The extensions and applications of the initial AEM include an application to a non-US country in the Sahel, a global-scale AEM framework, and a reliance on alternative measures of productivity than crop yields. This chapter also summarizes the results from the agronomic crop simulation models and the FACE (Free Air CO2 Enhancement) experiments which form the basis for an economic analysis in the AEM modeling. The chapter discusses ad hoc adaptations that are incorporated into the AEM modeling. The AEM models reported severe impacts of climate change on agriculture and consumers, especially when adaptations are not incorporated.

Keywords Agro-Economic Model · Major grains · Crop simulation · FACE · Green Climate Fund · ad hoc adaptation

4.1 Introduction

The second scientific tradition that will be reviewed in this chapter is broadly referred to as the Agro-Economic Model (AEM), whose analytical framework has been widely applied to examine the impacts of climate change on agriculture (Adams et al. 1990). The AEM was also the first scientific framework presented to the climate change community for measuring the impacts of climate change on agriculture.

In comparison with the microbehavioral model of adaptation behaviors presented in the previous chapter whose focus is on an individual manager who makes numerous management and adaptation decisions, the AEM framework is anchored at major staple crops that are selected by the AEM modelers based on the production history of the region under examination (Seo 2015a).

S. N. Seo, *The Economics of Global Allocations of the Green Climate Fund*,
https://doi.org/10.1007/978-3-030-18274-8_4

The AEM model is called that way because it combines an agronomic crop model with an economic model (Adams et al. 1999). With the use of an agronomic crop model, AEM researchers learn the impact of carbon doubling in the atmosphere and temperature increase on the yields of major crops through field or laboratory experiments.

In the second stage of the AEM, an economic model is carefully calibrated and used to extrapolate the locally-based agronomic crop model results to the national economy under examination, e.g., the US agriculture, and to transform the yield change results to changes in the economic variables such as producer income and consumer surplus.

There are multiple variants of the AEM. To name a few, an AEM variant can be built for a global agricultural economy by linking national economies assuming numerous parameters on agricultural trade relations and future changes of them (Parry et al. 2004). For another, the AEM framework can be used to calculate other variables of interests such as the number of people in hunger or malnutrition (Rosenzweig and Parry 1994; Hertel and Rosch 2010). Further, it has also advanced to include productivity changes in farm animals that are obtained from controlled experiments, albeit in a rudimentary way (Butt et al. 2005).

For the purposes of this book, the results from the AEMs applied to a country, many countries, or a continent would provide important information which can be utilized by the present author and others to evaluate the allocations of funds from the Green Climate Fund. The results from the field/laboratory experiments on yield changes of a major crop, say, maize, would provide basic information for such evaluations. The results on producer income changes or changes in the number of people in hunger owing to a future climate regime would also offer an essential input for the evaluation of the GCF-funded projects (GCF 2011, 2018a).

4.2 Theory

An Agro-Economic Model starts from the field experiments designed to examine the effects of elevated carbon dioxide concentration in the atmosphere on selected crops. An experiment is carried out on a selected grain such as wheat, maize, soybeans, cotton, and rice which holds major significance to the national economy under examination by an agronomist or climate scientist.

The basic scientific tools that are relied upon during the crop experiment stage are the principles of experimental designs and analyses that can be traced far back to the crop experiments by Ronald A. Fisher during the early twentieth century (Fisher 1935; Kuehl 1999). In a randomized design, factors that affect the crop's growth and yields are randomized. Then, the changes in the yield of the concerned crop are measured by comparing the yield achieved at the present level of carbon dioxide with that achieved at the elevated level of carbon dioxide (Jones and Kiniry 1986; Williams et al. 1989; Tubiello and Ewert 2002).

The experiment can be carried out in a laboratory setting, but an outdoor field experiment can replicate the field conditions more accurately. In a laboratory setting, experiments are conducted using a controlled experimental chamber, or a greenhouse, or closed-up or open-top field chambers in which all non-climate factors are, including soils and winds, in principle perfectly controlled (Tubiello and Ewert 2002). The climate factors can be either carbon dioxide concentration or temperature or both.

In an open-air field experiment, after randomizing other factors of crop growth than carbon dioxide concentration, climate scientists elevate CO_2 level through the pre-installed pipes placed around the crop plant in the experimental plots. They observe and record the changes in the yield (or growth rate) of the crop. For the measurement of temperature effects, the exact same experiment may be repeated at two different temperature zones (Ainsworth and Long 2005).

The yield changes of the crop observed from the experimental plots placed in a controlled experimental setting can still diverge from the field observations of the yield changes at the actual farms of the crop (plant) that have occurred over the past several decades owing to the effects of a temperature increase and a carbon dioxide concentration elevation (Lobell and Field 2008).

The open-air field experiment is referred to as the FACE (Free-Air CO2 Enrichment) experiment, which has been conducted extensively since the first application in the early 1990s. A thorough review and meta-analysis of the FACE literature is available from multiple sources including the Ainsworth and Long (2005). A thorough review and meta-analysis for the agronomic crop simulation models can also be found from multiple sources including the Tubiello and Ewert (2002). Both literatures are also extensively showcased by the Intergovernmental Panel on Climate Change (IPCC)'s official reports (Reilly et al. 1996; Gitay et al. 2001; Easterling et al. 2007; Porter et al. 2014). The predicted changes in the yields of the crops from the two experimental methods will be explained and compared in the next sections of this chapter.

In the second stage of the AEM, the locally-based experimental outcomes of crop yield changes are summarized in a certain way for the national-level outcomes which are then inserted into a national agricultural model constructed to be representative of the agricultural economy of the country under examination (Adams et al. 1990; Reilly et al. 2003; Butt et al. 2005). This is not a simple task but rather a complex but also highly controversial procedure, which may be summarized by the following basic five points of contention.

First, the most detailed descriptions of the AEM methodology are found in the works of Adams and his coauthors who are the original developers of the methodology (Adams et al. 1999). In the seminal AEM study of the US agriculture, researchers relied on the so-called Agricultural Sector Model (ASM) in which the agro-economy of the country is divided into 63 homogeneous agricultural production regions in the 48 contiguous United States (Adams et al. 1990, 1999). To feed agronomic/experimental results into the ASM, the authors first choose 17 representative farms (sites) across the country representative of 17 major agro-climatic regions in the US. Each representative enterprise is to represent all the farms in the corresponding agro-climatic region in terms of, *inter alia*, assets, geography, soils, education, markets, which is obviously a high-level of simplification.

Second, the experimental results on the selected crops obtained from each of the 17 selected sites are then fed into each of the 17 representative enterprises to calculate the changes in crop yields for each of the 17 representative enterprises attributable to site-specific changes in the climate factors. The AEM modelers must choose the set of crops to be examined in the second-stage economic modeling, which can be different from the family of crops that agronomists or climate scientists choose to conduct experiments on, because the representative enterprise for one of the 17 U.S. regions would not include minor crops that are grown in the region on marginal lands by a small number of farmers. The calibrations of the 17 representative agricultural enterprises may not capture U.S. agriculture accurately (Kaiser et al. 1993).

Third, representative enterprise-level outcomes for the 17 enterprises are then extrapolated to the national agricultural economy, relying on the ASM's specification of the 63 homogenous agricultural production regions. In order to simulate the impact of climate change on US agriculture, researchers must account for different conditions across the 63 production regions such as land uses, water availability, and irrigation, that is, in addition to the yield effects of climate change obtained from the first-stage crop experiments/simulations. For the effects of climate change on land uses, water availability, irrigation, and other processes, the AEM needs to outsource from other scientific traditions, at the least (Seo 2010a, b, 2011, 2016a, b; Kurukulasuriya et al. 2011).

Fourth, an economic evaluation further requires assumptions on prices of crops as well as costs of inputs. From these assumptions, an AEM modeler can calculate the baseline net revenue of the US agriculture, that is, the net revenue before a climatic change. Changes in prices and costs of all agriculture-related assets and products are difficult to be projected for a century time horizon, to say the least (Evenson 2002).

Fifth, calibrations of prices and costs cannot be done without assumptions on international trade of agricultural products. The AEM should assume a certain pattern of exports and imports of the crops, livestock, and livestock products that are considered in the model. In addition, owing to complementarity and substitutability, there needs to be a certain pattern of exports and imports of the substitutes and complements (Reilly et al. 1994). In addition, changes in the existing agricultural trade regime will alter production decisions by domestic farmers.

Assuming that all the above matters are resolved satisfactorily, the impacts of climate change on US agriculture are calculated by measuring the difference between the baseline net revenue (profit) and the future net revenue (profit). Other economic indicators of the impacts of climate change are also often derived by AEM researchers with a list of assumptions, which include the total grain production, the number of people in hunger, the number of mal-nourished people, and consumer/producer surplus (Resenzweig and Parry 1994; Parry et al. 2004; Hertel and Rosch 2010).

For measuring the future agricultural profit, a future climate scenario is chosen by the AEM researchers from the range of future climate change scenarios offered by the climate models referenced by the Intergovernmental Panel on Climate Change (IPCC)'s scientific reports (IPCC 2001, 2014). The IPCC reports as well as the AGGCM climate models have relied on a family of storylines (scenarios) and make scenario-dependent predictions of the future climate system. A storyline (scenario) is a set of assumptions on social, economic, and technological changes in the future

(Nakicenovic et al. 2000), but the recent revision of the IPCC makes these storylines meaningless and instead adopts an emissions-based RCP (Representative Concentration Pathways) approach (IPCC 2014).

Assuming the demand, technologies, and a host of other aforementioned parameters, AEM modelers calculate the baseline yields, prices, consumer surplus, producer surplus, and economic welfare for the current period, that is, without climate change. These measures are recalculated for the future period, that is, with the climate changed. Researchers can either assume that the demand, technologies, and other parameters remain fixed or assume that these parameters are updated over time following a certain formula selected by the modeler.

The welfare change is calculated as follows. For each crop, the area between the baseline demand curve and the supply curve is defined as the baseline economic welfare for the crop. The future economic welfare for the crop is defined in the same way. For each crop, the impact of the assumed climate change is then calculated as the difference between the new economic welfare and the baseline economic welfare (Adams et al. 1999). The total economic impact of climate change in the agricultural sector is then calculated by summing the welfare impacts across all crops included in the model.

4.3 Empirical Results: FACE Versus Agronomic Simulations

From this point on, major findings and predictions from the tradition of the Agro-Economic Modeling will be presented, with an eye on their applicability to the assessments of the Green Climate Fund grant decisions.

The AEM studies start with the basic experimental results on yield or productivity changes that agronomists or climate scientists obtain through a controlled experiment setting. As explained previously, the entire range of experiments can be categorized into either an agronomic crop simulation model or a large-scale field experiment. The agronomic crop simulation models are preferred by agronomic modelers because the simulation models are calibrated to explicitly account for all the factors that affect crop growth and yields, that is, in addition to atmospheric CO_2 concentration (Jones and Kiniry 1986; Williams et al. 1989; Tubiello and Ewert 2002). These other factors are, *inter alia*, soils, fertilizer applications, water applications, pest controls. Refer to, for example, Zilberman (1998); Ruttan (2002) for factors of production in agriculture.

The FACE experiment, on the other hand, is preferred by climate scientists and researchers because the experiment's focus is entirely on climate factors, more specifically, carbon dioxide concentration after randomizing all other factors, and it is superior in replicating the open-air field conditions than the laboratory-based crop simulations. The FACE experiment was first conducted at the Duke Forest in the early 1990s and since then many FACE experiments have followed (Schlesinger 1997; Reich et al. 2018).

In Table 4.1, the major outcomes from the 15-year experiences of the FACE experiments on major crops are summarized (forth column), along with the recently reported long-term experimental results (sixth column) (Ainsworth and Long 2005; Reich et al. 2018). The FACE experiments' results are also compared with the agronomic crop simulation models' results juxtaposed at the fifth column of the Table (Tubiello et al. 2007).

The table summarizes the results by three plant categories: grains, non-grain crops, and non-crop plants and trees. The crops for which the FACE experiments are conducted most often are rice, wheat, cotton, sorghum, and maize. A FACE experiment on millet, another major grain especially in Africa, is rare (Seo et al. 2009). Non-grain crops are cotton and soybeans/legumes. Again, scientific experiments on many non-grain crops are rare, for example, lentil, cassava, tobacco (Seo and Mendelsohn 2008a). The FACE experiments are frequently conducted also on non-crop plants, although agronomic studies focus on crops. Three non-crop plants included in the table are C3 grasses, shrubs, and trees. Again, there may be differences among the trees in their responses to carbon dioxide or temperature increase and some trees are overlooked by scientists, for example, rubber trees in Sri Lanka (Seo et al. 2005).

In the case of agronomic crop simulation models, the indictor of the impact of climate change is exclusively the yield of the crop upon which experiments are conducted. By contrasts, multiple indictors besides the crop yield are measured in the FACE experiments: Plant height, dry matter production (DMP), leaf area index (LAI), and stem diameter (SD).

With the carbon dioxide concentration doubling from the current level at the time of the experiments, a meta-analysis of the results from the FACE experiments show that the yields of crops are enhanced with little uncertainty (Table 4.1). On average, the rice yield is predicted to increase by 10%, the wheat yield to increase by 15%, the sorghum yield to increase by 40% under no additional stress and by 5% under additional stresses.

The yield increases would occur owing to the fertilization effect of elevated atmospheric carbon dioxide which increases the rate of photosynthesis of a crop plant through which carbohydrates are produced (Denman et al. 2007).

For non-grain crops, the increase in yield is predicted to be even larger. Cotton yield increases by as much as 42%, on average of all the experimental results, under the elevated CO2 condition created by the FACE experiments. The dry matter production (DMP) of soybean/legumse is predicted to increase on average by 24%.

For non-crop plants, the dry matter production of trees is predicted to increase on average by 28% while the DMP of C3 grasses to increase on average by 10%. The leaf area index (LAI) of trees is predicted to increase on average by 21% while the LAI of C3 grasses to increase on average by 10%. For shrubs, the plant height is predicted to increase by 23% under elevated CO2 condition, averaging all available experimental results.

In the above, C3 plants are the plants that use a three-carbon photosynthetic pathway, which is contrasted to C4 plants for a four-carbon pathway. Most crops are C3 crops. Notable C4 crops are maize, millet, sugar cane, and sorghum. The theory and evidence show that C3 plants are more responsive than C4 plants to elevated CO2 concentration (Reich et al. 2018).

Table 4.1 Predictions from the FACE, FACE long-term, agronomic crop simulation models

			FACE experiments (Ainsworth and Long 2005)	Agronomic crop simulation models (Tubiello et al. 2007)	FACE long-term (20-year) experiments (Reich et al. 2018)
Classifications	Plants	Indicators	Average change	Average change	Average change
Grains	Rice	Yield (kg/plot)	Around +10%	+10%(AEZ); +17%(CERES-C3); +19% (EPIC-C3)	
	Wheat	Yield (kg/plot)	+15%∗	+11%(AEZ); +17%(CERES-C3); +19% (EPIC-C3)	
	Sorghum	Yield (kg/plot)	Around +5% +40% (under no stress)	+6%(CERES-C4); +8% (EPIC-C4)	
	Maize	Yield (kg/plot)		+4%(AEZ); +6%(CERES-C4); +8% (EPIC-C4)	
	Millet	Yield (kg/plot)			
Non-grain crops	Cotton	Yield (kg/plot)	+42%∗		
	Soybeans (legumes)	Yield (kg/plot)		+16%(AEZ)	
		Dry matter production (DMP)	+24%∗		
Non-crop plants	Trees	Dry matter production (DMP)	About +28%		
		Leaf area index (LAI)	About +21%		
	C3 grass	Dry matter production (DMP)	About +10%		
		Leaf area index (LAI)	About +10%		
		Total biomass			−2%
	C4 grass	Total biomass			+24%
	Shrubs	Plant height	About +23%		

Note: (1) ∗ denotes the 95% Confidence Level; (2) The FACE is for Free-Air CO_2 Enrichment; (3) The AEZ is for Agro-Ecological Zone FAO (2005); (4) The CERES is for Crop Environment Resource Synthesis; (5) The EPIC is for Erosion Productivity Impact Calculator

The FACE experiments reviewed by Ainsworth and Long relied on short-term experiments, that is, less than a decade. A recent FACE study suggests that a long-term experiment may yield different predictions, based on 20-year experiments of four C3 grasses and four C4 grasses on 80 plots in Minnesota, USA. The authors report that, as predicted by the theory, the total biomass of C3 plants were enhanced by CO2 elevation more than that of C4 plants during the first 12 years of the experiments. For the final 8 years, the authors find a reversal of the responses: The total biomass of C4 plants were vastly enhanced but not the total biomass of C3 plants (Reich et al. 2018).

This new study seems to explain the behavioral changes reported by the microbehavioral models in Sub-Sahara and South America where farmers are observed to choose millet, sorghum, sugarcane more frequently in a hotter climate condition (Seo and Mendelsohn 2008a; Seo et al. 2008). The long-term FACE study also signals that the beneficial impacts of climate change on crop yields may have been underestimated by short-term experiments.

The yield changes predicted from the FACE experiments are compared with the predictions from the agronomic crop simulation models in the fifth column of Table 4.1. The meta-analysis results from the three agronomic crop models are presented: The AEZ (Agro-Ecological Zone) model, the CERES (Crop Environment Resource Synthesis) model, and the EPIC (Erosion Productivity Impact Calculator) model (Tubiello et al. 2007).

Notably, the agronomic crop models, more specifically, the CERES and the EPIC, predict larger increases in crop yields than the FACE experiments (Long et al. 2006). For example, the CERES predicts a 17% increase and the EPIC predicts a 19% increase in rice yield, which can be contrasted with a 10% increase in rice yield predicted by the FACE. A similar point can be observed in the wheat yield predictions. All statistics are averages of the range of predictions used by the authors.

However, if an additional correction is forced to the above agronomic crop models to account for the Agro-Ecological Zones (AEZ) in which the simulations are set up, yield predictions from the agronomic crop models correspond well to the predictions from the FACE experiments. In the case of rice, for example, the FACE experiments and the AEZ models lead to about 10% increase in the yield.

4.4 Empirical Results: Extrapolations from the Plot Level to the National Level

For the AEM researchers to assess the national-level or global-level welfare changes, they must extrapolate the plot-level experimental results to the national level. This extrapolation requires a complicated modeling work on the characteristics of the national agricultural economy and geography. Owing to the size and heterogeneity of the territory and economy of the United States, it is not at all a handy task.

The national agricultural model developed for this purpose is often referred to as the Agricultural Sector Model (ASM) in the United States (Adams et al. 1990, 1999). For the study of Mali, a Sahelian country in Africa, the national model was called the Mali Agricultural Sector Model (MASM) (Butt et al. 2005).

As explained in the theory section, the ASM of the United States relies on the two key concepts: 17 agro-climatic regions of the country and 63 homogenous production regions. Agronomic crop experiments are conducted in each of the 17 agro-climatic regions, likely at multiple sites within each region, and averaged for each region.

Each of the 63 homogenous production regions in the lower 48 States is characterized by its own agricultural portfolio as well as its own agro-climatic characteristics. The agricultural portfolio of a production region specifies the types and quantities of the crops the region produces. The designation of a production region also defines irrigation characteristics, water uses, and land uses for the region, which calls inevitably for drastic simplifications of the field-level variations.

The calibrations of the production regions in this way make it easy, although much controversial, for the AEM researchers to calculate the changes in productions of selected crops in each of the production regions. The production changes at the production region level can then be added up to calculate the national level changes in productions of the selected crops.

In Table 4.2, national-level predicted changes in the yields of major selected crops for the U.S. agriculture are presented, which was reported by Adams, McCarl, and Mearns (2003). Five major crops selected for modeling by the authors are corn, cotton, soybeans, wheat, and sorghum. For wheat, four different varieties of wheat are examined: hard red spring wheat, hard red winter wheat, soft white wheat, and Durham wheat.

Two climate models are used for future projections: The CSIRO and the RegCM. The CSIRO is the climate prediction from the Australian Commonwealth Scientific and Industrial Research Organization (CSIRO). The RegCM is a nested model within the

Table 4.2 National percentage changes in yields of selected crops (kg/ha)

		Rainfed		Irrigated	
		CSIRO: Blanket resolution	RegCM: Finer resolution	CSIRO: Blanket resolution	RegCM; Finer resolution
Corn		0%	−6%	−16%	−15%
Cotton		25%	22%	46%	45%
Soybeans		33%	5%	42%	42%
Wheat	Hard red spring wheat	−11%	−2%	−19%	−12%
	Hard red winter wheat	−2%	4%	5%	0%
	Soft white wheat	−9%	−5%	−12%	−5%
	Durham wheat	41%	22%	23%	20%
Sorghum		−7%	−6%	0%	−1%

CSIRO model with a finer spatial resolution, abbreviated for a regional climate model (Giorgi et al. 1993a, b). The CSIRO model is almost a blanket prediction for the United States while the RegCM gives differentiated climate changes at different US regions. For both models, the future climate regime is the doubling of the atmospheric carbon dioxide concentration from the pre-industrial level.

The table reveals that the national-level predicted changes in yields under the RegCM prediction are very different from those under the CSIRO in the case of rainfed agriculture, especially for soybeans and Durham wheat. However, predicted changes in yields seem to be almost identical under the two climate models when it comes to irrigated agriculture.

Focusing on the RegCM and the rainfed agriculture, the cotton yield is predicted to increase by 22% and the soybean yield to increase by 5%. The corn yield is predicted to decrease by 6%, so is the sorghum yield. Predictions of changes in wheat yields vary across the different varieties: The hard red spring wheat yield is predicted to decrease by 2%, the hard red winter wheat yield is to increase by 4%, the soft white wheat yield is to decrease by 5%, and the Durham wheat yield is to increase by 22%.

The table indicates that the impact of climate change on irrigated lands would be quite different from the impact on rainfed agriculture. In the irrigated lands, the increase in the yields of non-grain crops is more than twice larger than the increase predicted in the rainfed lands. For example, the cotton yield is predicted to increase by 45% and the soybean yield is to increase by 42%, both in the irrigated lands. For grain crops, on the other hand, the corn yield is predicted to fall by 15% and the hard red spring wheat yield is predicted to fall by 12%, both in the irrigated lands.

4.5 Empirical Results: National Economic Welfare Changes

With Tables 4.1 and 4.2, the supply-side changes under the future climate regimes can be established. It should be noted, however, that the predictions like those in Table 4.2 do not tell us all the things that are needed for a complete analysis: Whether non-selected crops for the modeling or currently-not-grown crops would become more important in the future climate regime or whether the supply-side changes can occur additionally through international markets and technological changes.

Leaving these issues aside for the moment, the AEM modelers need to set up demand-side changes in order to make economic welfare calculations of the impacts of climate change. Further, owing to the long temporal horizon that needs to be considered, as long as a century, they need to calibrate or assume technology changes to unfold in the future.

Assuming that the demand and technology remain unchanged from the baseline year, say, the present agricultural economy, the AEM modelers can quantify the changes in agricultural prices caused by the shifts of agricultural supplies caused by the change in the climate regime. The price changes are quantified by the changes in

Table 4.3 Changes in fisher price index and quantity index

Climate models and scenarios	Fisher ideal price index	Agricultural quantity index
Baseline	1.00	1.00
% changes under GISS (Goddard Institute for Space Studies); 2∗CO2	−17%	+9%
% changes under GFDL (Geophysical Fluid Dynamics Laboratory); 2∗CO2	+34%	−20%

the agricultural price index which is encompassing prices of all agricultural products, say, crops or grains, but again selected and currently-grown crops and grains only.

In Table 4.3, predicted changes in the Fisher price index are calculated given two climate change scenarios: The GISS (Goddard Institute for Space Studies) scenario and the GFDL (Geophysical Fluid Dynamics Laboratory) scenario (Adams et al. 1990). The Fisher price index, named after Irving Fisher, is known as the Fisher ideal price index because it is defined as the geometric average of the Laspeyres price index and the Paasche price index (Fisher 1920; Eurostat 2018). In other words, the Fisher price index takes into account changes in the basket of goods and services in both the base period and the future period, unlike the other price indexes.

Under the GFDL climate change scenario, the agricultural quantity index is predicted to fall by 20%, i.e., from 1 to 0.8, owing to climate change-caused yield reductions of the modeled grains. With the constant demand for agricultural products, the agricultural price is predicted to increase by 34%, i.e., from 1 to 1.34 in the Fisher price index.

Under the GISS scenario, these predictions are reversed. The price index is predicted to fall by 13% from 1 to 0.87 owing to climate change-caused yield increases of modeled crops. The agricultural quantity index is predicted to increase by 9% from 1.00 to 1.09.

With the price and supply changes estimated as in Table 4.3, the AEM modelers calculate the change in the total economic welfare as the sum of the changes in the three components: producer surplus, consumer surplus, and foreign surplus. In a closed-economy model, there is no foreign surplus in the model.

The total economic welfare change predictions by the two AEMs are shown in Table 4.4. The first set of predictions is from the 1990-AEM version and the second set is from the 1999-AEM version by Adams and his coauthors. For the 1990-AEM version, the predictions are made assuming the 1981–1983 baseline economy. The baseline economy's consumer surplus was US$ 77 billion and the producer surplus was US$ 18 billion, summing up to the total surplus of US$ 95 billion.

According to the AEM-1990 version, the authors predicted 11.4% gain of total economic welfare under the GISS model's carbon dioxide doubling scenario. This economic gain from the assumed climate change results from the large increase in the agricultural production and the decrease in the agricultural price, as shown in Table 4.3. These changes lead to the increase of the producer surplus by 9% due to the larger production, and the increase in consumer surplus by 12% owing to the lower price.

Table 4.4 Changes in economic welfare caused by CO2 doubling

	Consumer surplus	Producer surplus	Foreign surplus	Total surplus
The 1990-AEM version⁺				
Baseline	US$ 77.3 billion	US$ 17.8 billion		US$ 95.2 billion
% changes under the GISS model	+12.0%	+8.9%		+11.4%
% changes under the GFDL model	−18.0%	+19.9%		−10.9%
The 1999-AEM version⁺⁺				
% changes under the GISS model	US$ +20.6 billion	US$ +45.4 billion	US$ +50.6 billion	US$ +116.6 billion (~ +7% of total value of agricultural production)
% changes under the GFDL model	US$ −65.7 billion	US$ +52.2 billion	US$ −3.4 billion	US$ −16.9 billion (~ −1% of total value of agricultural production)

Note: (⁺) The model assumes 1981–1983 baseline agricultural economy and assumes demand and technology remain fixed; (⁺⁺) The model assumes the baseline 2060 economy, technology, and demand

If the GFDL scenario were to come to pass, it was predicted that the total economic welfare loss would amount to 10.9% of the baseline economic surplus. This loss would occur owing to the large decrease in the agricultural production and the increase in agricultural price, as shown in Table 4.3. These changes lead to the increase in producer surplus by almost 20% owing to higher sale prices, but the decrease in consumer surplus by 18% owing again to higher purchase prices.

The initial 1990-AEM paper was further developed to address major weaknesses of the model perceived by climate researchers, both theoretical and empirical. First of all, it was obvious that the analysis does not capture US agriculture comprehensively since it focuses on selected major grains such as corn, wheat, soybeans, cotton. The 1999-version of the model extends the range of examined crops to the crop groups of cotton-sorghum, tomatoes-citrus-potatoes, and forage-livestock production (Adams et al. 1999).

However, either the agronomic crop simulation models or the FACE experiments for these additional groups of agricultural products are not established as well. This is especially the case for forage and livestock production (Mader et al. 2009; Hahn et al. 2009). In addition, the sets of agricultural products analyzed in the 1999-AEM paper, although somewhat richer than that from the 1990 AEM version, cannot be said to cover comprehensively agricultural activities in the United States.

As measured by the annual value of total agricultural production, animal husbandry accounts for almost half of the total value of production in the United States (NASS 2017). Notwithstanding, the AEM modelers have largely failed to capture the livestock economy in the model from the very beginning of the modeling tradition (Seo 2006, 2010a; Seo and Mendelsohn 2008b).

For the 1999 revision of the AEM, the authors relied on forage yield changes and cattle appetite depressing as indicators of the impact of climate change on farm animals. The forage yield changes were obtained from the EPIC crop simulation model for the Southeast US and separately from the CENTURY model for the western US (Parton et al. 1992). From the pasture yield changes, the number of acres required per head of cattle for grazing was estimated under the changed climate conditions by the authors. Direct effects of climate change on farm animals were obtained from cattle food intake changes due to appetite depressing which were estimated using the NUTBAL (Nutritional Balance) model (Stuth et al. 1999).

Despite these progresses, some of which will be further explained in the following section, many aspects of the climate change impacts on livestock remain poorly understood. The knowledge gap includes physical adaptive capacities of different animal species which may differ, heat tolerant animals, animals that are resilient to rainfall changes and shocks, effects on conception, effects of climate-related management practices, and interactions with the other agricultural sectors such as crop farming and forestry (Hahn 1981; Mader and Davis 2004; Amundson et al. 2006; Seo 2014a, 2016c; NRC 2015).

For the 1999 revision of the AEM, the economic welfare change predictions are summarized at the bottom panel of Table 4.4. Under the GFDL model's carbon doubling scenario, the total welfare change is about 1% loss of the annual value of total agricultural production in the United States. The increase of the producer surplus by US$ 52 billion can be ascribed to the price increases of crops. Under this scenario, consumers lose because of the price increases, both domestic consumers (US$ – 67 billion) and foreign consumers (US$ −3.4 billion).

Under the other scenario by the GISS, the total economic welfare change is estimated to be about 7% benefit of the annual value of agricultural production in the United States from the assumed climate change. Under this climate model scenario, both the producer surplus and the consumer surplus are increased due to climate change-caused yield increases and price decreases. The big beneficiary of such changes is foreign consumers.

4.6 Empirical Results: International Applications

The AEM framework has been extended enthusiastically by agronomists and climate scientists to an international policy context, which can be summarized as two directions. On the one direction, the AEM methodology was applied to examine the impact of climate change on agriculture of the country other than the United States, especially those countries perceived to be particularly vulnerable to future climatic changes. On the other direction, a global-scale AEM assessment was constructed by linking multiple national AEMs with critical assumptions on international trade relations and extrapolations of the national AEMs.

For the development in the first direction, an exemplary study is found in the application of the AEM to Mali, an arid zone country located in the Sahel (Butt et al. 2005). Based on the above-described U.S. Agro-Economic Model, researchers made

Table 4.5 The impacts of climate change on Mali agriculture: Mali-AEM

			HadCM 2030	CGCM 2030
Classification	Products	Indicators		
Crops	Maize	Yield (kg/ha), % change	−11.2%	−13.5%
	G-nuts	Yield (kg/ha), % change	−10.0%	−13.4%
	Cotton	Yield (kg/ha), % change	+6.2%	+3.5%
	Sorghum	Yield (kg/ha), % change	−11.5%	−17.1%
	Millet	Yield (kg/ha), % change	−6.3%	−11.5%
	Cowpea	Yield (kg/ha), % change	−8.4%	−12.2%
Forage	Forage-north	Yield (kg/ha), % change	−18%	−26%
	Forage-central	Yield (kg/ha), % change	−5%	−12%
	Forage-west	Yield (kg/ha), % change	−26%	−36%
Livestock	Cattle	Animal feed intake, % change	−12.8%	−13.3%
	Sheep	Animal feed intake, % change	−3.4%	−5.9%
	Goats	Animal feed intake, % change	−4.1%	−4.7%
	Cattle	Animal weight (kg), % change	−13.6%	−15.7%
	Sheep	Animal weight (kg), % change	0%	0%
	Goats	Animal weight (kg), % change	0%	0%

several key modifications. For example, the researchers developed the country-specific agricultural sector model referred to as the Mali Agricultural Sector Model (MASM). For another, although the authors hoped to apply the results from the agronomic crop simulations and the FACE experiments that were conducted in Mali, such results were not available. Hence, the simulation and experiment results on crops, forage, and livestock obtained from the United States were applied with minimal adjustments.

The predictions of the impact of climate change on Mali agriculture made by the Mali-AEM are summarized in Table 4.5. Included in the analysis of the Mali-AEM are six crops such as maize, ground nuts, cotton, sorghum, millet, and cowpea; forage in the three regions of the country; three farm animals such as cattle, sheep, and goats.

The projection period is 2030, so the predictions in Table 4.5 cannot be directly compared with the US results in the preceding sections whose projection period corresponds to the time when carbon dioxide concentration reaches 2∗CO2 concentration. Given the relatively near projection period, the harmful effects on crops, plants, and animals presented in the Mali-AEM are eye-catching. The maize yield (kg/ha) is predicted to fall by 11.2% under the HadCM (Hadley center Coupled Model) climate model and by 13.5% under the CGCM (Coupled General Circulation Model) model. The yields of other crops are also predicted to fall. Exception is the cotton yield which is predicted to increase by 6%.

Interestingly, the decreases in forage yields are even larger, according to the authors. The North, Central, and Western regions of the country are predicted to experience the decrease of forage yield by 26%, 12%, and 36% by the year 2030. These predictions are contradictory to the agronomic and FACE experiment results presented in in Table 4.1.

Table 4.5 summarizes the prediction results of the impact of climate change on animal performances for three livestock species: Cattle, sheep, and goats. Two performance indicators are used. The first is the changes in food intake rate en route animal appetite depression owing to increased heat stress. The second indicator is animal weight.

For cattle, it is predicted that food intake should get reduced by about 13% and cattle weight get reduced by 14–16%. For sheep, food intake is predicted to decrease by 3–6%. For goats, food intake is predicted to decrease by 4–5%. The animal weight of sheep would not be affected by the assumed climate change scenarios, so is that of goats. Again, given the relatively near future projected in the model, these predictions foretold severe damages on animals from climate change, especially on cattle.

Be reminded that all these predictions of climate change impacts were made for the future period of 2030, that is, not for the future period of 2100 or the time when the carbon dioxide concentration reaches 2∗CO2 concentration. Based on the physical impacts of climate change in Table 4.5, the authors calculated the economic welfare changes for Mali. By 2030, they predicted staggering damage to Mali agriculture. Owing to the reductions in agricultural productivities shown in Table 4.5, the authors predicted 76% increase in the agricultural price index under the HadCM scenario and 145% increase under the CGCM scenario. Owing to the price spike, Mali consumers were predicted to lose 29% in consumer surplus under the HadCM scenario and 45% under the CGCM scenario (Butt et al. 2005).

In the second direction of the AEM modeling's applications to the global-scale policy setting, researchers labored to build a global-scale agricultural impact assessment based upon the national AEMs. When first conceived, it seemed straightforward to link one national AEM to another national AEM, which is the methodological basis for the whole family of the global AEM works (Rosenzweig and Parry 1994).

However, it soon dawned on the AEM researchers that there are many and tall obstacles to overcome for a successful global assessment model. To the readers who had the patience to follow through the review of this tradition up to this point, it would not be very difficult to conjecture those obstacles. In short, the limitations of the national AEM multiplies, say, a hundred times in the global AEM assessment.

First of all, there is the question of how one national AEM should be linked to another national AEM or to multiple national AEMs. The AEM researchers relied on an agricultural trade model or another. For example, Parry and coauthors relied on the Basic Linked System (BLS) for the world food trade (Parry et al. 2004). However, the global agricultural trade models including the BLS are not built with climate factors as one of the key variables in the world trade system, therefore, it is impossible to trace climate change-induced alterations in the complex relationships embedded in the trade models (Fischer et al. 1996, 2001; Hertel and Rosch 2010).

Second, the global trade models are built with an emphasis on primary crops such as maize, rice, wheat, and cattle, which hold major international importance in the today's climate system. The trade models neglect or may omit some non-staple crops as well as some minor farm animals which nonetheless provide essential nutrients and income to many farms. Further, the global trade models overlook crop varieties and livestock species that are nationally or locally important, although not a big part in international trades (Steiger 2006; Nin et al. 2007; Seo and Mendelsohn 2008b).

Third, another big, rather obvious, challenge is that there are simply too many national AEMs to build. A global AEM analysis calls for a national AEM analysis for more than 100 countries in the world. As we saw in the above in the AEM assessment for the US agriculture, it is quite complicated to build one national AEM even for a country with a well-established research tradition and well-compiled transparent historical data. In many countries in the world, it may not even be possible to build an AEM assessment.

AEM modelers sought to resolve this issue by grouping a number of countries into one cluster and building a limited number of the cluster-level AEMs which were then linked (Rosenzweig and Parry 1994; Parry et al. 2004; Fischer et al. 2005). However, it is hard to justify such a grouping approach because each cluster would comprise many different and disparate nations. Further, it would be difficult to provide a solid rationale for a chosen criterion for such grouping out of many possible criteria.

Fourth, there are a large array of crops, farm animals, plants, trees, and forest products that are frequently managed by farmers worldwide. There are probably several 100 s–1000 s' species of them that must be accounted for in a global AEM assessment. Moreover, for each species, there are multiple varieties cultivated or raised somewhere around the world. According to the Food and Agriculture Organization (FAO), there are 50,000 edible plant species in the world, although of which 15 crop plants provide almost 90% of the world's food energy intake (FAO 1995; Ziska 2003).

The vast number of agricultural products to be included in the analysis has posed a severe constraint to the global AEM analysis because the basic unit of the AEM analysis is, by conception, an agricultural product, that is, a crop variety, or a livestock species, or a tree species. A crop that is very important in one local area in the globe may have no importance in the rest of the world, but cannot be ignored in the global AEM assessment.

The predictions from one of the global AEMs are summarized in Fig. 4.1 where three indicators are presented: global cereal production, global cereal price, and the number of people at risk of hunger (Parry et al. 2004). The predictions are made by the authors assuming the range of the scenarios from the IPCC's special report on emission scenarios (Nakicenovic et al. 2000). Two scenarios in the figure are A2a and B2b scenarios. The former is the high-end emissions scenario due to a high rate of economic growth and the latter is the low-end emissions scenario owing to a low rate of economic growth.

The predicted damage from the assumed climate change, as shown in Fig. 4.1, is striking. By 2080, the changed climate regime under A2a scenario is predicted to cause 560 million people to be at the risk of hunger. By 2050, over 200 million people would be at the risk of hunger. Even under a much milder climate change scenario B2b, the people at hunger would reach 200 million by 2080.

The stunning predictions are purely from the reduction in total cereal production caused by the climatic change. Under the A2a scenario, the total cereal yield is predicted to fall by over 400 million tons per year, which is then predicted to cause the price increase of 350% by 2080. By 2050, the total cereal production is predicted to fall by about 180 million tons per year, which is accompanied by the price increase of 100%. For reference, the global cereal production stood at about 2 billion tons in 2000 (FAO 2018).

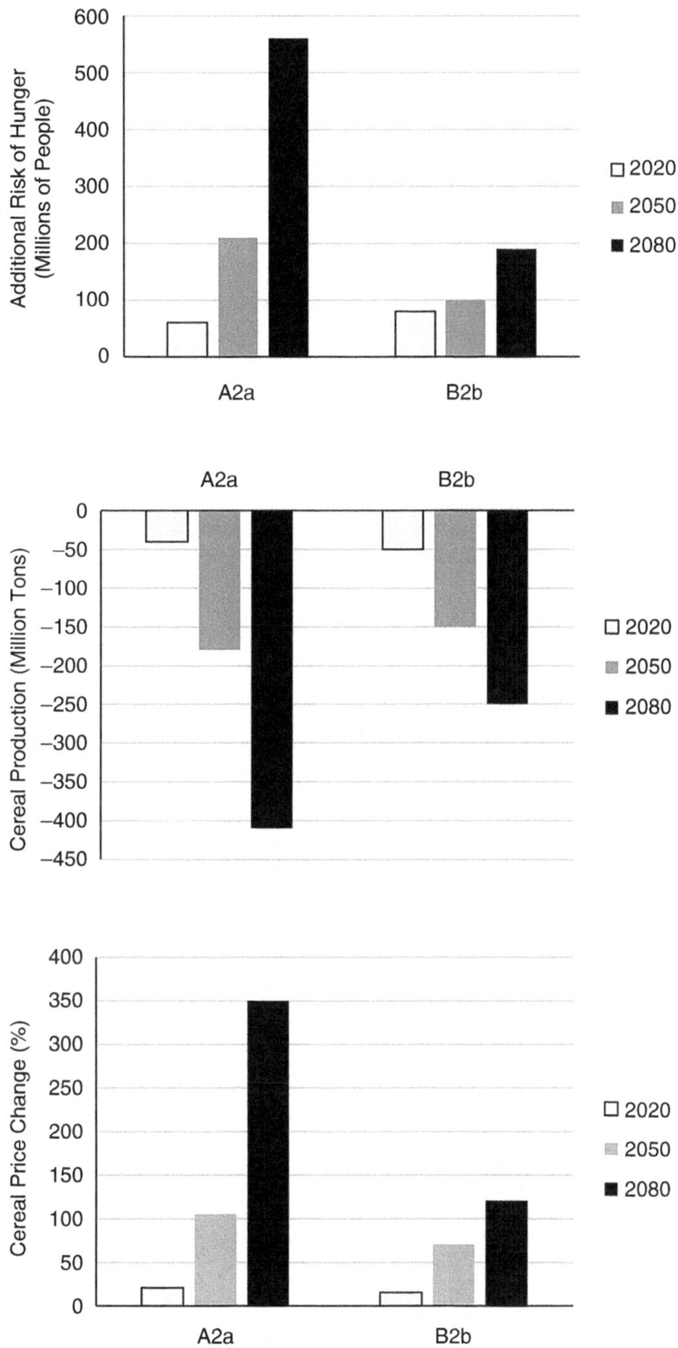

Fig. 4.1 The predictions from the global AEM: Hunger (top), Cereal production (middle), Cereal price (bottom)

For the reasons described above, AEM researchers were not able to measure in any accuracy either the global-level impact of climate change on each agricultural product included in the worldwide agricultural portfolio or the global agricultural portfolio. Faced with the daunting challenges, some researchers tried to go for an alternative measure of productivity. For example, Reilly and coauthors rely on the concept of Net Primary Product (NPP) at the vegetation level (Reilly et al. 2007). The NPP change in the cultivated land was used as an indicator of yield changes of crops, which is in no way a reliable indicator of the productivity of crops.

The NPP is the net assimilation of CO2 into plant tissues, that is, plant growth (Reilly et al. 2007). More formally, the gross primary production (GPP) is the amount of chemical energy as biomass that producers produce in a fixed period. Of the GPP, some fraction is used for respiration by the producers. The NPP is the remaining energy:

$$NPP = GPP - Respiration. \tag{4.1}$$

For another, researchers went for the Normalized Difference Vegetation Index (NDVI) as an alternative indictor of crop/plant productivity or a similar satellite-based indicator of greenness of the land under investigation (Barrett et al. 2017). Again, the NDVI does not capture differences between green grasses, green crops, green shrubs, and green trees. Further, in locations with drastic altitude variations, the NDVI can mistake altitude drops for changes in green productivities.

The NDVI is measured as follows. The Earth-orbiting satellites run by the National Aeronautics and Space Administration (NASA) or the National Oceanic and Atmospheric Administration (NOAA) have a microwave sensor with the channels that measure reflectance of visible wavelengths (VIS) light or near-infrared wavelengths (NIR) light (NASA 2015). Owing to the cell structure of leaves that perform photosynthesis, the more leaves there are, the more NIR light are reflected. The NDVI is defined as the following ratio that ranges from −1 to +1:

$$NDVI = \frac{NIR - VIS}{NRI + VIS}. \tag{4.2}$$

4.7 Ad Hoc Approach on Adaptations

The AEM modeling tradition has been the most widely distributed and reported scientific tradition, especially by the Intergovernmental Panel on Climate Change (IPCC), of all the traditions reviewed in this book with regard to the issues of climate change and agriculture and food productions. The IPCC has produced five scientific reports since its establishment in 1988, once per 5–7 years (IPCC 1990). The most dire but controversial reports by the IPCC on agriculture include the initial 1990 report and the most recent 2014 report (IPCC 2014; Seo 2014b).

Each rendition of the IPCC assessment report is composed of the three working group reports in which the second working group report contains the chapter on climate change and agriculture and food productions. The first assessment report (FAR) of the IPCC entitled "The IPCC Scientific Assessment" was published in 1990, which has turned out to lay out the framework to be followed by the subsequent IPCC reports (IPCC 1990). The most critical 'contribution' of the initial assessment, when it comes to the impact modeling of climate change, has turned out to be that the FAR introduced agriculture and food production as the most vulnerable sector under future climate changes, along with the coastal cities faced with the threats from sea level rises.

Since then, five reports were written on agriculture and food productions by various groups of authors and each report with a slightly different focus: Parry et al. (1990), Reilly et al. (1996), Gitay et al. (2001), Easterling et al. (2007) and Porter et al. (2014). The full lists of authors for these reports can be verified from the references given at the end of this chapter, from which you will find that many of them have been already referred to in this chapter as AEM authors and/or crop scientists.

The critical weaknesses of the AEM modeling as a scientific methodology for predicting what will happen in the economic value of agriculture and food productions under a new climate system have been pointed out by a host of researchers since the early 1990s. Many of their critiques were already incorporated into the review in the preceding sections, but one critique deserves special and additional attention.

The critique is that, despite the grueling as well as opaque modeling efforts, there is a fundamental drawback in the AEM modeling with regard to the possible mechanisms through which farmers' behavioral adaptations can be taken into account in the AEM (Mendelsohn et al. 1994; Seo 2014a, b). I say this a fundamental flaw because the AEM is built up from each crop. In other words, the basic unit of analysis for the AEM is a crop, not a farmer, that is, a decision-maker who makes behavioral changes.

The AEM framework makes it difficult to account for the effects of adaptation possibilities on the final estimate of the economic impact of climate change on agriculture (Mendelsohn et al. 1994). An even more fundamental issue is that the AEM modeling structure is not equipped with the scientific tools that can reveal how adaptation decisions of farmers in response to climatic changes are made and will unfold (Seo 2006, 2016a, b, d). The large array of behavioral adaptations that should be incorporated in any impact modeling of climate change on agriculture was elaborated at length in Chap. 3.

The AEM researchers have made astronomical efforts to address this well-perceived shortcoming widely shared by climate researchers (Antle et al. 2004; McCarl 2007; Challinor et al. 2007; Parry et al. 2009; Hillel and Rosenzweig 2010). The extensive efforts, some of the most pertinent of which were explained in the preceding sections, were less than satisfactory and perhaps didn't move the needle at all on this critical point. The AEM's adaptation modeling can be summed up as an ad hoc approach. In essence, the modeling committee (of one) determines which of the vast list of adaptation strategies and practices should be included in which crop, or livestock, or forest, and by what degrees (Seo 2015a).

The AEM researchers can cite the sources for their specific adaptation assumptions in the AEM modeling. But, the outsourced results from the external sources are not internally consistent with the AEM modeling processes, nor are the outsourced results meant to be consistent with each other. Further, many behavioral adaptation assumptions that cannot be left out for the AEM construction cannot be referred back to because of the lack of research on such behaviors since they are not taken by a decision-maker independently of other considerations (Seo 2014b, 2015a).

Let me elaborate on this point with an actual example. Besides the crop simulation and experimental research as shown in Table 4.1, there is another group of crop research which is directed to the quantification of the effects on crops of increased climate variabilities or extreme weather realizations (Easterling et al. 2000; Rosenzweig et al. 2001; Aggarwal and Mall 2002; Porter and Semenev 2005; Challinor et al. 2007). This line of crop research has responded to the reports of climate research that the changes in the climate system may unfold through increased variability of temperature or precipitation or realizations of extreme temperatures and precipitations (Rahmstorf and Coumou 2011; Hansen et al. 2012).

For the empirical results from this line of crop research to be validated, it would take perhaps more than a decade because field/laboratory experiments on a large number of agricultural products in different situations have to be performed and verified for a large number of possible climate risk increase scenarios. If such validations were to materialize, the AEM modeler would then proceed to incorporate the experimental results to the AEM modeling, which is another grueling process that may take another decade for the reasons amply highlighted in this chapter. However, even such an AEM framework were to be developed at last, the AEM modeler would not be able to incorporate a large set of behavioral changes made by decision-makers in response to the changes in climate variabilities or extremes or both.

By contrasts, the microbehavioral model of adaptation behaviors explained in Chap. 2 has already revealed plentiful findings in this regard (Seo 2018a). Sub-Saharan farmers were shown to adjust their farm portfolios in response to increased climate risks through increases in precipitation variabilities and increases in diurnal temperature ranges (Seo 2012). Such behavioral adaptations are found in Latin America where farmers responded to El Nino induced precipitation patterns and in South Asia where farmers adapted to the monsoon precipitation pattern and its variability (Seo 2014c, 2016c, 2018b).

4.8 Applications to the Green Climate Fund's Decisions

Having clarified the AEM machine and all the nuts and bolts of it, we can now answer the primary question of this book: To what extent, should the AEM modeling and their findings and predictions be utilized for the funding decisions by the Green Climate Fund Board (GCF 2018a, b)?

The part of the AEM framework can inform decision-makers on which crops are particularly vulnerable to an assumed climate change scenario and which regions or countries are more vulnerable than others. The information shown in Tables 4.1 and

4.5 are pertinent to the former question and the information shown in Table 4.4 are pertinent to the latter question. Nonetheless, this does not mean at all that the information in these tables can be relied upon directly.

Of the host of food security projects to which funds were allocated by the GCF, the Namibia's project entitled "Climate Resilient Agriculture in three of the Vulnerable Extreme northern crop-growing regions (CRAVE)" provides a good case study in this regard (GCF 2018c). It can be stated that the GCF decision-makers certainly recognized the three northern regions of Namibia as well as crop-growers there as the most vulnerable group when they approved a 5-year grant of US$ 10 million. The basis for such recognition by the GCF decision-makers may have come from the results and predictions made available by the AEM modeling.

However, such a grant can increase, not decrease, the economic risk of climate change on the northern crop-growing regions in Namibia. A grant would create an incentive for the crop farmers there to continue to grow the most vulnerable crops and a monetary incentive for the northern farmers not to migrate to more favorable climate regions (Seo 2015b).

This line of reasoning illustrates that the GCF funding in the form of either a grant or a loan will not be a silver bullet but can have a boomerang effect on the recipient without a highly sophisticated analysis and empirical back-ups (Seo 2017a, b). This book will come back to this central issue in Chap. 7 where each of the 93+ GCF-funded projects is evaluated.

References

Adams R, Rosenzweig C, Peart RM, Ritchie JT, McCarl BA, Glyer JD, Curry RB, Jones JW, Boote KJ, Allen LH (1990) Global climate change and US agriculture. Nature 345:219–224

Adams R, McCarl BA, Segerson K, Rosenzweig C, Bryant K, Dixon BL, Conner R, Evenson RE, Ojima D (1999) The economic effects of climate change on US agriculture. In: Mendelsohn R, Neumann J (eds) The impact of climate change on the United States economy. Cambridge University Press, Cambridge

Adams RM, McCarl BA, Mearns LO (2003) The effects of spatial scale of climate scenarios on economic assessments: an example from U.S. agriculture. Clim Chang 60:131–148

Aggarwal PK, Mall PK (2002) Climate change and rice yields in diverse agro-environments of India. II. Effect of uncertainties in scenarios and crop models on impact assessment. Clim Chang 52:331–343

Ainsworth EA, Long SP (2005) What have we learned from 15 years of free-air CO2 enrichment (FACE)? A meta-analytic review of the responses of photosynthesis, canopy properties and plant production to rising CO2. New Phytol 165:351–371

Amundson JL, Mader TL, Rasby RJ, Hu QS (2006) Environmental effects on pregnancy rate in beef cattle. J Anim Sci 84:3415–3420

Antle JM, Capalbo SM, Elliott ET, Paustian KH (2004) Adaptation, spatial heterogeneity, and the vulnerability of agricultural systems to climate change and CO2 fertilization: an integrated assessment approach. Clim Chang 64:289–315

Barrett C, Mude A, Jensen N (2017) Assessing the value of index insurance for herders: comparing NDVI-based insurance products in Ethiopia and Kenya. Feed the future: the U.S. Government's global hunger & food security initiative, U.S. government

Butt TA, McCarl BA, Angerer J, Dyke PT, Stuth JW (2005) The economic and food security implications of climate change in Mali. Clim Chang 68:355–378

Challinor A, Wheeler T, Craufurd P, Ferro C, Stephenson D (2007) Adaptation of crops to climate change through genotypic responses to mean and extreme temperatures. Agric Ecosyst Environ 119:190–204

Denman KL, Brasseur G, Chidthaisong A, Ciais P, Cox PM, Dickinson RE, Hauglustaine D, Heinze C, Holland E, Jacob D, Lohmann U, Ramachandran S, da Silva Dias PL, Wofsy SC, Zhang X (2007) Couplings between changes in the climate system and biogeochemistry. In: Solomon S et al (eds) Climate change 2007: the physical science basis. Cambridge University Press, Cambridge

Easterling DR, Evans JL, Groisman PY, Karl TR, Kunkel KE, Ambenje P (2000) Observed variability and trends in extreme climate events: a brief review. Bull Am Meteorol Soc 81:417–425

Easterling WE, Aggarwal PK, Batima P, Brander KM, Erda L, Howden SM, Kirilenko A, Morton J, Soussana J-F, Schmidhuber J, Tubiello FN (2007) In: Parry ML, Canziani OF, Palutikof JP, van der Linden PJ, Hanson CE (eds) Food, fibre and forest products. Climate change 2007: impacts, adaptation and vulnerability. Contribution of working group II to the fourth assessment report of the intergovernmental panel on climate change. Cambridge University Press, Cambridge

Eurostat (2018). Statistics explained: fisher price index. Accessed from http://ec.europa.eu/eurostat/statistics-explained/index.php/Glossary:Fisher_price_index

Evenson R (2002) Technology and prices in agriculture. In report of the consultation on agricultural commodity price problems. Food and Agriculture Organization (FAO), Rome

FAO (2005) Global agro-ecological assessment for agriculture in the twenty-first century (CD-ROM), FAO land and water digital media series. FAO, Rome

FAO (2018) FAO STAT. FAO Statistics Division, Rome. Available at http://faostat.fao.org

Fischer G, Frohberg K, Parry ML, Rosenzweig C (1996) Impacts of potential climate change on global and regional food production and vulnerability. In: Downing TE (ed) NATO ASI series, climate change and world food security, vol 137. Springer, Berlin

Fischer G, Shah M, van Velthuizen H, Nachtergaele FO (2001) Global agro-ecological assessment for agriculture in the 21st century. IIASA research report 02–02. International Institute for Applied Systems Analysis, Laxenburg, p 119

Fischer G, Shah M, Tubiello FN, van Velhuizen H (2005) Socio-economic and climate change impacts on agriculture: an integrated assessment, 1990–2080. Philos Trans R Soc B 360:2067–2083

Fisher I (1920) The purchasing power of money: its determination and relation to credit, interest and crises. The MacMillan Company, New York

Fisher RA (1935) Design of experiments. Oliver and Boyd, Edinburgh

Food and Agriculture Organization (FAO) (1995) Dimensions of need: an atlas of food and agriculture. FAO, Rome

Giorgi F, Marinucci MR, Bates GT (1993a) Development of a second generation regional climate model (RegCM): boundary layer and radiative transfer processes. Mon Weather Rev 121:2794–2813

Giorgi F, Marinucci MR, De Canio G, Bates GT (1993b) Development of a second generation regional climate model (RegCM): convective processes and assimilation of lateral boundary conditions. Mon Weather Rev 121:2814–2832

Gitay H, Brwon S, Easterling W, Jallow B (2001) Ecosystems and their goods and services. In: McCarthy et al (eds) Climate change 2001: impacts, adaptations, and vulnerabilities. Cambridge University Press, Cambridge, pp 237–342

Green Climate Fund (GCF) (2011) Governing instrument for the green climate fund. GCF, Songdo City

Green Climate Fund (GCF) (2018a) Annex III: investment framework. GCF, Songdo City

Green Climate Fund (GCF) (2018b) Annex IX: results management framework. GCF, Songdo City

Green Climate Fund (GCF) (2018c) Projects + Programmes. GCF, Songdo City. Accessed from https://www.greenclimate.fund/what-we-do/projects-programmes

Hahn GL (1981) Housing and management to reduce climate impacts on livestock. J Anim Sci 52:175–186

Hahn GL, Gaughan JB, Mader TL, Eigenberg RA (2009) Chapter 5 thermal indices and their applications for livestock environments. In: DeShazer JA (ed) Livestock energetics and thermal environmental management. ASABE. Copyright 2009 American Society of Agricultural and Biological Engineers. ASABE #801M0309. ISBN 1-892769-74-3, St. Joseph, pp 113–130

Hansen J, Sato M, Reudy R (2012) Perception of climate change. Proc Natl Sci Acad USA 109: E2415–E2423. https://doi.org/10.1073/pnas.1205276109

Hertel TW, Rosch SD (2010) Climate change, agriculture, and poverty. Appl Econ Perspect Policy 32:355–385

Hillel D, Rosenzweig C (eds) (2010) Handbook of climate change and agroecosystems: impacts, adaptation, and mitigation. Imperial College Press, London

Intergovernmental Panel on Climate Change (IPCC) (1990) Climate change: the IPCC scientific assessment. Cambridge University Press, Cambridge

Intergovernmental Panel on Climate Change (IPCC) (2001) Climate change 2001: the physical science basis. The fifth assessment report of the IPCC. Cambridge University Press, Cambridge

Intergovernmental Panel on Climate Change (IPCC) (2014) Climate change 2014: the physical science basis. The fifth assessment report of the IPCC. Cambridge University Press, Cambridge

Jones CA, Kiniry JR (1986) CERES-maize: a stimulation model of maize growth and development. Texas A&M University Press, College Station

Kaiser HM, Riha SJ, Wilks DS, Rossiter DG, Sampath R (1993) A farm-level analysis of economic and agronomic impacts of gradual climate warming. Am J Agric Econ 75:387–398

Kuehl RO (1999) Design of experiments: statistical principles of research design and analysis, 2nd edn. Duxbury Press, Pacific Grove

Kurukulasuriya P, Kala N, Mendelsohn R (2011) Adaptation and climate change impacts: a structural Ricardian model of irrigation and farm income in Africa. Clim Chang Econ 2:149–174

Lobell D, Field CB (2008) Estimation of the carbon dioxide (CO2) fertilization effect using growth rate anomalies of CO2 and crop yields since 1961. Glob Chang Biol 14:39–45

Long SP, Ainsworth EA, Leakey ADB, Nösberger J, Ort DR (2006) Food for thought: lower than-expected crop yield stimulation with rising CO2 concentrations. Science 312:1918–1921

Mader TL, Davis MS (2004) Effect of management strategies on reducing heat stress of feedlot cattle: feed and water intake. J Anim Sci 82:3077–3087

Mader TL, Frank KL, Harrington JA, Hahn GL, Nienaber JA (2009) Potential climate change effects on warm-season livestock production in the Great Plains. Clim Chang 97:529–541

McCarl BA (2007) Adaptation options for agriculture, forestry and fisheries. A report to the UNFCCC secretariat financial and technical support division. Texas A&M University, College Station

Mendelsohn R, Nordhaus W, Shaw D (1994) The impact of global warming on agriculture: a Ricardian analysis. Am Econ Rev 84:753–771

Nakicenovic N, Davidson O, Davis G, Grübler A, Kram T, La Rovere EL, Metz B, Morita T, Pepper W, Pitcher H, Sankovski A, Shukla P, Swart R, Watson R, Dadi Z (2000) Emissions scenarios. A special report of working group III of the intergovernmental panel on climate change. IPCC, Geneva

National Aeronautics and Space Administration (NASA) (2015) Taking a global perspective on Earth's climate. Available at http://climate.nasa.gov/nasa_role/

National Agricultural Statistics Service (NASS) (2017) Census of agriculture 2017. NASS USDA, Washington, DC. Accessed from https://www.nass.usda.gov/AgCensus/

National Research Council (NRC) (2015) Critical role of animal science research in food security and sustainability. National Academies Press, Washington, DC

Nin A, Ehui S, Benin S (2007) Livestock productivity in developing countries: an assessment. In: Evenson R, Pingali P (eds) Handbook of agricultural economics, vol. 3. North Holland, Oxford, pp 2467–2532

Parry ML, Medzhulin GV, Sinha S, Duinker PN, Morison JIL, Porter JH, Reilly J, Wright LJ (1990) Agriculture and forestry. In impacts assessment of climate change. Cambridge University Press, Cambridge

Parry ML, Rosenzweig CP, Iglesias A, Livermore M, Fischer G (2004) Effects of climate change on global food production under SRES emissions and socioeconomic scenarios. Glob Environ Chang 14:53–67

Parry M, Arnell N, Berry P, Dodman D, Fankhauser S, Hope C, Kovats S, Nicolls R, Sattherwaite D, Tiffen R, Wheeler T (2009) Assessing the costs of adaptation to climate change: a critique of the UNFCCC estimates. The International Institute for Environment and Development and The Grantham Institute for Climate Change, London

Parton WJ, McKeown B, Kirchner V, Ojima D (1992) CENTURY users manual. Natural Resource Ecology Laboratory, Colorado State University, Fort Collins

Porter JR, Semenev M (2005) Crop responses to climatic variation. Philos Trans R Soc B 360:2021–2035

Porter JR, Xie L, Challinor AJ, Cochrane K, Howden SM, Iqbal MM, Lobell DB, Travasso MI (2014) Food security and food production systems. In: Field CB et al (eds) Climate change 2014: impacts, adaptation, and vulnerability. Cambridge University Press, Cambridge/New York

Rahmstorf S, Coumou D (2011) Increase of extreme events in a warming world. Proc Natl Acad Sci U S A 108:17905–17909

Reich PB, Hobbie SE, Lee TD, Pastore MA (2018) Unexpected reversal of C3 versus C4 grass response to elevated CO2 during a 20-year field experiment. Science 360:317–320

Reilly J, Hohmann N, Kane S (1994) Climate change and agricultural trade. Glob Environ Chang 4:24–36

Reilly J, Baethgen W, Chege F, Van de Geijn S, Enda L, Iglesias A, Kenny G, Patterson D, Rogasik J, Rotter R, Rosenzweig C, Sombroek W, Westbrook J (1996) Agriculture in a changing climate: impacts and adaptations. In: Watson R, Zinyowera M, Moss R, Dokken D (eds) Climate change 1995: impacts, adaptations, and mitigation of climate change. Intergovernmental Panel on Climate Change (IPCC). Cambridge University Press, Cambridge

Reilly J, Tubiello T, McCarl B, Abler D, Darwin R, Fuglie K, Hollinger S, Izaurralde C, Jagtap S, Jones J, Mearns L, Ojima D, Paul E, Paustian K, Riha S, Rosenberg N, Rosenzweig C (2003) U.S. agriculture and climate change: new results. Clim Chang 59:43–69

Reilly J, Paltsev S, Felzer B, Wang X, Kicklighter D, Melillo J, Prinn R, Sarofim M, Sokolov A, Wang C (2007) Global economic effects of changes in crops, pasture, and forests due to changing climate, carbon dioxide, and ozone. Energy Policy 35:5370–5383

Rosenzweig C, Parry M (1994) Potential impact of climate change on world food supply. Nature 367:133–138

Rosenzweig C, Iglesias A, Yang XB, Epstein PR, Chivian E (2001) Climate change and extreme weather events: implications for food production, plant diseases, and pests. Glob Chang Hum Health 2(2):90–104

Ruttan VW (2002) Productivity growth in world agriculture: sources and constraints. J Econ Perspect 16:161–184

Schlesinger WH (1997) Biogeochemistry: an analysis of global change, 2nd edn. Academic, San Diego

Seo SN (2006) Modeling farmer responses to climate change: climate change impacts and adaptations in livestock management in Africa. Ph.D. dissertation, Yale University, New Haven

Seo SN (2010a) A microeconometric analysis of adapting portfolios to climate change: adoption of agricultural systems in Latin America. Appl Econ Perspect Policy 32:489–514

Seo SN (2010b) Managing forests, livestock, and crops under global warming: a micro-econometric analysis of land use changes in Africa. Aust J Agric Resour Econ 54(2):239–258

Seo SN (2011) An analysis of public adaptation to climate change using agricultural water schemes in South America. Ecol Econ 70:825–834

Seo SN (2012) Decision making under climate risks: an analysis of sub-Saharan farmers' adaptation behaviors. Weather Clim Soc 4:285–299

Seo SN (2014a) Evaluation of agro-ecological zone methods for the study of climate change with micro farming decisions in sub-Saharan Africa. Eur J Agron 52:157–165

Seo SN (2014b) Adapting sensibly when global warming turns the field brown or blue: a comment on the 2014 IPCC report. Econ Aff 34:399–401

Seo SN (2014c) Coupling climate risks, eco-systems, and anthropogenic decisions using south American and sub-Saharan farming activities. Meteorol Appl 21:848–858
Seo SN (2015a) Micro-behavioral economics of global warming: modeling adaptation strategies in agricultural and natural resource enterprises. Springer, Cham
Seo SN (2015b) Helping low-latitude, poor countries with climate change. Regulation 38:6–8
Seo SN (2016a) Modeling farmer adaptations to climate change in South America: a micro-behavioral economic perspective. Environ Ecol Stat 23:1–21
Seo SN (2016b) The micro-behavioral framework for estimating total damage of global warming on natural resource enterprises with full adaptations. J Agric Biol Environ Stat 21:328–347
Seo SN (2016c) Untold tales of goats in deadly Indian monsoons: adapt or rain-retreat under global warming? J Extreme Events 3(1):1650001. https://doi.org/10.1142/S2345737616500019
Seo SN (2016d) Microbehavioral econometric methods: theories, models, and applications for the study of environmental and natural resources. Academic Press, Amsterdam
Seo SN (2017a) Beyond the Paris agreement: climate change policy negotiations and future directions. Reg Sci Policy Prac 9:121–140
Seo SN (2017b) The behavioral economics of climate change: adaptation behaviors, global public goods, breakthrough technologies, and policy-making. Academic Press, Amsterdam
Seo SN (2018a) Natural and man-made catastrophes: theories, economics, and policy designs. Wiley-Blackwell, Hobokken
Seo SN (2018b) Will farmers fully adapt to monsoon climate? J Agric Sci (under revision)
Seo SN, Mendelsohn R (2008a) An analysis of crop choice: adapting to climate change in south American farms. Ecol Econ 67:109–116
Seo SN, Mendelsohn R (2008b) Measuring impacts and adaptations to climate change: a structural Ricardian model of African livestock management. Agric Econ 38:151–165
Seo SN, Mendelsohn R, Munasinghe M (2005) Climate change and agriculture in Sri Lanka: a Ricardian valuation. Environ Dev Econ 10:581–596
Seo SN, Mendelsohn R, Dinar A, Hassan R, Kurukulasuriya P (2008) Differential adaptation strategies of African cropland across agro-ecological zones, World Bank Policy Research Working Paper 4601. World Bank, Washington, DC
Seo SN, Mendelsohn R, Dinar A, Hassan R, Kurukulasuriya P (2009) A Ricardian analysis of the distribution of climate change impacts on agriculture across agro-ecological zones in Africa. Environ Resour Econ 43:313–332
Steiger C (2006) Modern beef production in Brazil and Argentina. Choices Mag 21:105–110
Stuth JW, Freer M, Dove H, Lyons RK (1999) Nutritional management for free-ranging livestock. In: Jung H (ed) Nutrition of herbivores. American Society of Animal Science, Savoy
Tubiello FN, Ewert F (2002) Simulating the effects of elevated CO2 on crops: approaches and applications for climate change. Eur J Agron 18:57–74
Tubiello FN, Amthor JA, Boote KJ, Donatelli M, Easterling W, Fischer G, Gifford RM, Howden M, Reilly J, Rosenzweig C (2007) Crop response to elevated CO2 and world food supply. Eur J Agron 26:215–223
Williams JR, Jones CA, Kiniry JR, Spaniel DA (1989) The EPIC crop growth model. Trans Am Soc Agric Eng 32:497–511
Zilberman D (1998) Agricultural and environmental policies: economics of production, technology, risk, agriculture and the environment. The State University of New York, Oswego
Ziska LH (2003) Evaluation of yield loss in field-grown sorghum from a C3 and C4 weed as a function of increasing atmospheric carbon dioxide. Weed Sci 51:914–918

Chapter 5
Statistical Methods for Estimating Yield Changes Attributable to Climate Change

Abstract This chapter provides a review of the literature of statistical methods for estimating yield/productivity changes of major crops, plants, and other resources that are attributable to climate change. A host of statistical yield estimation techniques were developed as an effort to describe yield changes of staple crops of the world such as corn, wheat, rice, soybeans, and cotton. The statistical techniques were originated because climate researchers wanted to make use of the readily available national yield data such as the US agricultural census. In contrast to the field/laboratory experiments reviewed in the previous chapter, applications of the statistical techniques have resulted in the predictions of extremely severe losses of yields of grains caused by future climate changes. This chapter explains underlying factors that may explain the discrepancies between the predictions from this tradition and those from other traditions. The empirical results from the yield statistical studies were interpreted in the context of providing a rationale for the GCF's funding allocations into vulnerable major crops.

Keywords Statistical methods · Crop yields · Staple crops · Crop growth · Field experiment · GCF

5.1 Introduction

The third scientific tradition in the literature of climate change that will be introduced in this chapter is the tradition of statistical methods whose primary goal is to estimate the changes of yields of the world's staple crops attributable to climatic shifts. From the estimated statistical relationships, statisticians make predictions of future yields of the crops assuming a climate scenario or another.

The motivation for the emergence of the statistical tradition was clear: There was the inevitable need for explaining the impacts of climate change on major crops of significance in the world to the general public and policy-makers. If you had wanted to tell the impacts of climate change on agriculture, the most straightforward way should

S. N. Seo, *The Economics of Global Allocations of the Green Climate Fund*,
https://doi.org/10.1007/978-3-030-18274-8_5

have been to pick the most salient crop in the region that people are concerned about and tell how much yield loss of the concerned crop the region is going to suffer from future climatic changes.

Major staple crops are what everyone in the world cognizes easily and consumes almost every day. Out of 50,000 edible plants globally, nearly two-thirds of world food energy intake is provided by three staple crops: rice, maize, wheat (FAO 1995). Other major staple crops are millet and sorghum. Other staple crops that are particularly salient in Sub-Saharan Africa include tubers and root crops such as potatoes, cassava, yams, and taro (Seo et al. 2008). Another class of staple crops particularly prominent in South Asia are pulses and beans (Kala 2015). There are cash crops that are important for farm income such as cotton in the US, coffee and tobacco in Latin America, cocoa and gum in Sub-Saharan Africa, and tea and rubber in South Asia (Adams et al. 2003; Seo et al. 2005; Seo and Mendelsohn 2008a, b).

The motivation of the tradition of statistical methods is analogous to that of the crop simulation models and the FACE experiments explained in the previous chapter. That is, both traditions are motivated by the concern on potentially deleterious impacts of climatic shifts on the most dominant food crops. A differential factor that has motivated the statistical tradition over the crop simulation and experimental tradition is that the crop simulation models and the FACE experiments are expensive to build and take much time and human resources to complete (Reich et al. 2018). By contrast, yield data of major staple crops are readily available as part of the national statistics on agriculture in all developed countries. In the United States, the National Agricultural Statistics Service (NASS) maintains the yield data, including the data from the U.S. census of agriculture taken once every 5 years (NASS 2018).

The yield data can also be accessed from the international agricultural organizations, e.g., the FAOSTAT platform provided by the Food and Agriculture Organization (FAO) of the United Nations (FAO 2017). The data accessed at the international organizations are processed ones from the data submitted by national agricultural organizations of the United Nations' member countries.

In addition, for most developed countries, there is no cost charged for obtaining the crop yield and other agricultural data. In China, an access to the national agricultural data is limited to certain government-approved groups, that is, communist party-approved groups. In India, the data can be accessed by researchers but only after a payment for the data (Open Government Data Platform of India 2018). The statistical tradition of climate change and agriculture has arisen in order to take advantage of these readily available, consistently measured historical data.

The development of this tradition during the past two decades, however, proved that the readily available public data do not guarantee any success in climate change and agriculture research, but instead lead to other data needs and theoretical challenges that turned out to be too tall to overcome.

This chapter will provide a review of the modeling development of this research tradition and highlight major empirical outcomes and methodological limitations. The present author then discusses how the future yield predictions from this tradition can be applied to the funding allocation decisions of the Green Climate Fund (GCF 2018c).

5.2 Statistical Methods

A national agricultural economy is made up of a large portfolio of grains, non-grain crops, and other non-crop products. Of the full range of agricultural portfolios, a handful of crops account for a dominant fraction of the national agricultural income and production. Globally, just 15 major crop plants account for about 90% of the world food energy intake (FAO 1995). In the United States, the most important revenue-generating crops are wheat, maize, soybeans, and cotton.

The grain and non-grain crops are commonly categorized into multiple sub-groups. For example, they can be classified into grains (corn, rice, wheat, millet, sorghum, barley, etc.), vegetables (lettuce, broccoli, cabbage, avocado, onions, garlic, etc.), root and tuber crops (potatoes, yam, taro, carrot, radish, cassava, etc.), cash crops (cotton, coffee, tea, tobacco, canola, sugar cane, etc.), and tree crops (citrus, tangerines, apples, coconuts, dates, persimmon, Brazil nuts, etc.) (Seo 2016c).

A national agricultural organization, for example, the National Agricultural Statistics Service of the United States Department of Agriculture (USDA), is given the responsibility to maintain precise national data on major crops, especially, of national importance (NASS 2018). The designated organization maintains annual production data of these crops, which are of high quality in many developed countries.

The national data are maintained, in nearly all cases, at the level of political units (Nordhaus 2006). In the United States, it means the 2nd administrative unit, that is, a county. There are about 3000 counties in the United States (Mendelsohn et al. 1994). In many other countries, the 2nd administrative unit is a district while the 1st administrative unit is a province (Seo 2013a). Often in developing countries, national agricultural data are available only at the provincial level, that is, the first administrative division (Auffhammer et al. 2006).

From this point on, I will elaborate a couple of statistical models that were most widely referred to of all the empirical studies that belong to this tradition and that captured the spirit of this econometric tradition best. The first can be called the yield growth function approach and the second the growing-season degree days approach (Deschenes and Greenstone 2007; Schlenker and Roberts 2009).

Let's start with the US crop yield growth study (Schlenker and Roberts 2009). Of the country's crop portfolio, the authors picked three crops: corn, soybeans, and cotton. They 'intentionally' picked corn for the grain family, soybeans for the bean family, and cotton for cash crops. For each of the three crops, the authors estimate the yield of the crop as a function of climate variables and other control variables.

Let's pick corn (maize) to explain their empirical model. Let y_{it} be the yield of corn for county i and year t, expressed in tons per hectare of land. Let h be heat (temperature) during the growing season of corn. Let z_{it} be other explanatory variables than the heat, for example, precipitation. Let φ_i be the fixed effect of the county, that is, the effect of time-invariant factors that belong to the county, e.g., soils and geography.

With this set-up, the authors estimated the following yield function:

$$y_{it} = \int_{\underline{h}}^{\bar{h}} g(h)\tau_{it}(h)dh + z_{it}\beta + \varphi_i + \varepsilon_{it}, \tag{5.1}$$

where the last term on the right-hand side is the error term which is assumed to be spatially correlated and is estimated by the routine proposed by Conley (1999).

The only novel thing in Eq. 5.1, at the time of publication, was the first term on the right-hand side, hence needs to be explained here. The $\tau_{it}(h)$ is the time distribution of heat for the county and the year during the growing season of corn. More concretely, it is a probability density function whose range is from $_{h\_}$.

The $g(h)$ is a yield growth function for corn, that is, the growth of corn yield at a given heat level, h. The growth function is applied to all counties and years, so there are no subscripts. By integrating the growth function over the range of heat using the probability density function, researchers can calculate the corn yield that depends on heat.

The authors approximate the integral in Eq. 5.1, that is, the first term on the right-hand side, numerically by using 1 degree (°C) temperature interval and the cumulative distribution function Φ_{ih}:

$$\sum_{h=-5}^{49} g(h + 0.5)(\Phi_{ih}(h + 1) - \Phi_{ih}(h)) \tag{5.2}$$

For the estimation of the yield growth function, $g(h)$ in the above equation, the authors tried three different statistical specifications: (1) a step function with a different growth rate in each 3 ° C temperature interval; (2) an eighth order Chebychev polynomial; and (3) a piecewise linear function.

There are several conspicuous drawbacks in the yield growth function model in Eqs. 5.1 and 5.2, which were pointed out sporadically by various researchers. First, the heat (temperature) distribution for region i would be different from one year to another. The focus on a single year cannot establish climate normals. Further, weather variability, also called climate variability normal, is not accounted for by the above model. This is one form of the critique about the inability to differentiate the impacts of climate changes on agriculture from those of weather variations (Seo 2013b).

Second, the yield growth rate function is independently estimated in any of the three non-parametric methods mentioned above. More clearly, the estimation of the yield growth function does not control, that is, is assumed to be independent of other variables such as precipitation, diurnal temperature changes, rainfall variability, and monsoon variability which have all been quantified to be important factors for crop management and growth (Seo 2010b, 2012b, 2014b, 2016d). This will lead to biased estimates of the yield growth parameters.

Third, the estimation of the yield growth function does not control non-climate factors such as soils, geography, education, experience, extension services, transportation, among many other important socio-economic variables that the literature of climate change and agriculture has identified (Seo 2014a, 2016c). Be remined that these non-climate factors are carefully simulated in the crop simulation models that were explained extensively in the previous chapter (Jones and Kiniry 1986; Tubiello and Ewert 2002).

Another way to state this problem is that the yield growth function should be different from one county to another, from which only one yield growth function is estimated by the authors. More accurately, it would vary from one plot to another. Notwithstanding, the estimation of the yield growth function is performed assuming independences of the above-mentioned non-climate as well as socio-economic factors from the heat.

The readers will have another opportunity to evaluate the yield growth model approach when the empirical results are presented shortly in the next sections of this chapter. For now, we will move on to the second modeling approach, which is similar to the first modelling approach in many aspects. In fact, the second modeling method preceded in time to the first modeling approach.

The second modeling work is similarly developed to predict the impact of climate change on US agriculture using the county level yield and profit data (Deschenes and Greenstone 2007, 2012). To be fair to the authors, their focus was on the agricultural profit model while the crop yield model was included as an additional analysis that supports their arguments. The crop model is written:

$$y_{ct} = \sum_i \alpha_i f_i(W_{ict}) + H_{ct}\beta + \delta_c + \delta_t + \varepsilon_{ct}. \tag{5.3}$$

The authors used the yield data of major crops obtained from the USDA agricultural survey in 1987, 1992, 1997, and 2002. They also constructed farm profits using farm revenues and production expenditures at the level of US county. There were 2342 counties and a total of 9368 observations included in their regression analysis.

The model in Eq. 5.3 is similar to the model in Eq. 5.2 in many aspects: H_{ct} denotes non-weather explanatory variables at a specific county (c) and a specific year (t), δ_c county fixed effects, δ_t year fixed effects, ε the error term.

What differentiates the two models is the first term on the right-hand side. The W denotes a vector of weather variables, including precipitation and temperature, and the subscript i denotes a specific weather variable. The weather variables are not directly used in the equation for estimation, but as a function of the weather variables. More concretely, the authors construct the growing-season degree days from the daily average temperature data. The growing season is defined as the period during which the average daily temperature is between 46.4 °F (8 °C) and 89.6 °F (32 °C). The growing-season (GS) degree days, DDs, is defined as follows:

$$DDs = \sum_{j \in GS} \left(T_j - 46.4^{\circ} F\right). \tag{5.4}$$

Once the climate-yield relationship in Eq. 5.1 or Eq. 5.3 is estimated, the impact of climate change is calculated as the difference between the future yield given the future climate and the baseline (present) yield given the baseline (present) climate. There are additional challenges involved in this procedure.

For the second methodology (Eq. 5.3), the change in the growing season degree days should be estimated from a selected climate change scenario from the family of scenarios and climate models proposed by climate scientists (Taylor et al. 2012; IPCC 2014a). Since climate scientists can only make predictions of annual (or monthly) temperature changes for the future time period, e.g., year 2100, yield statisticians should determine how to use the annual climate information to estimate the change in growing season degree days which requires daily temperature data for year 2100. The choice of one method for this estimation over other methods alters both the growing season and the growing season degree days.

For the first methodology in Eq. 5.1, the task of predicting future yields is even harder. The researchers should assume a particular heat distribution in the future based on the climate change predictions expressed as the changes in annual (or monthly) mean temperatures. To be more specific, a temperature change by 4 °C at the end of twenty-first century which is one of the climate change predictions made by climate scientists does not tell you how the annual heat distribution will be altered. That is, $\tau_{it}(h)$ in Eq. 5.1 at the end of twenty-first century.

5.3 Agricultural Census

Statisticians who applied a sophisticated statistical technique or another to estimate yield changes of important crops induced by climatic shifts were in favor of not wasting time in obtaining and constructing the base data that must be plugged into statistical models. They relied on readily-available, public agricultural data that have been historically compiled by a national agency.

The most prominent data for this research tradition was the agricultural census by the United States Department of Agriculture (USDA). By the USDA, the census of agriculture has been conducted every 5 years. The initial year of the census goes back to 1840 (NASS 2018). The researchers used the USDA censuses conducted in 'recent' years to construct the panel data of crop yield, for example, in 1982, 1987, 1992, 1997, 2002, 2007, and 2012 (Deschenes and Greenstone 2007; Schlenker and Roberts 2009).

According to the National Agricultural Statistics Service (NASS), the census of agriculture aims for a complete count of U.S. farms and ranches and the people who operate them. For the 2017 census, the NASS mailed questionnaires to farmers and ranchers on December 2017 and completed forms were due by February 5, 2018. The census questionnaire asks farmers and ranchers about land use and ownership, operator characteristics, production practices, income, and expenditures (NASS 2018). All responses are counted. Both rural and urban farms are counted. Even the small plots of land growing some vegetables or raising some animals are counted in the census.

Notwithstanding the widespread belief, there are many reasons to conclude that the census of agriculture may not precisely count the US agricultural sector. First, some farms are not on the census mail list (CML), which tend to be small farms in acreage, production, and sales. Second, the responses by the farms and ranches in the CML that raised and sold less than $1000 in the census year were excluded from counting in the census of agriculture. Third, the census response rate of those in the CML is not 100%: The response rate for 2007 census was 85.2%, that for the 2002 census 88%, and that for the 1997 census 86.2% (US Census Bureau 2012; USDA 2012). Fourth, there can be inaccuracies in the editing and imputing process of the completed forms by the NASS staff. For example, when there is a missing unreported item in the submitted questionnaire, the NASS staff imputes the value from a recent NASS survey.

An additional discrepancy can occur due to the way that a farm's outputs are assigned to a county. Completed forms are processed by the Census Bureau's National Processing Center and edited further by the NASS staff when necessary. The data are reported at the US county level using the 'principal county' method. That is, a farm that operates in multiple counties is classified into the principal county.

For the statistical yield studies, yield data for corn, soybeans, and cotton are obtained from the USDA's National Agricultural Statistics Service. The yield information is given and used in the model at the US county level. The period of yield data availability covers many decades, e.g., from 1950 to 2015 (Schlenker and Roberts 2009). The yield of a crop is defined as the total county-level production divided by the acres harvested.

Yield researchers made another adjustment by limiting their analysis to the counties located east of 100° meridian (excluding Florida) for the study of corn and soybeans. They argued that the grain productions in the Western states are different because of the irrigated production systems which are subsidized heavily by the governments (Schenkler et al. 2005). For the analysis of cotton yields, all counties in the U.S. were included.

5.4 Empirical Results: Studies in U.S. Agriculture

The yield models by statisticians attracted much attention from both academic researchers and policy circles on climate change. It might even be correct to say that the yield studies have been the most influential climate research, of the four traditions explained in this book, on global warming activists and policy negotiators at the United Nations. The outsized interest and influence of this tradition is due to the severe and dreadful impacts of climate change on crop yields forecast by the tradition of the statistical yield studies (IPCC 2014b; Seo 2014c).

In this section, the present author will explain major results and predictions selected from the range of statistical yield studies. There are a large number of crop yield studies in the climate and agriculture literature, but we do not need to explain all of them to understand the major empirical findings and conclusions from this tradition since the multitude of statistical yield models can be interpreted as a variant of the two yield models elaborated in the previous section.

Let me begin by the crop yield studies for U.S. agriculture reported by Deschenes and Greenstone (2007, 2012) and Schlenker and Roberts (2009). From the yield growth function approach in Eqs. 5.1 and 5.2, the yield change predictions for the three crops shown in Table 5.1 are made based on two climate change scenarios at the end of the twenty-first century. The three crops are corn, soybeans, and cotton. Two climate scenarios are from the Hadley climate center. The Hadley B1 is a mild climate change scenario attributable to a low rate of economic growth globally and the Hadley A1F1 is a severe climate change scenario attributable to a high rate of economic growth globally relying on intensive use of fossil fuels (Nakicenovic et al. 2000).

The yield loss is expected to be very large in all crops and largest for corn. If climate change were to be realized as predicted by the Hadley A1F1 scenario, the loss of corn yield is predicted to be about 82% loss from the present level by the end of the twenty-first century. Even with the mild Hadley B1 scenario, the loss of corn yield is predicted to be about 43%.

The authors also predict that the yield of soybeans would also be severely impacted, so is that of cotton. With the Hadley A1F1 scenario, the yield of soybeans is predicted to fall by about 72%. With the Hadley B1 scenario, the expected loss of soybean yield is about 35%. For cotton, the yield is predicted to decline by 72% under the Hadley A1F1 scenario and by 38% by the Hadley B1 scenario.

These predictions were in stark contrasts to the earlier reports that soybeans and cotton are more resilient to higher temperature and that these non-grain crops would be less vulnerable to the changes in carbon dioxide concentration and global warming (McKinsey and Evenson 1998; Ainsworth and Long 2005). Readers can refer back to Table 5.1 in Chap. 4 for the predictions of yield changes of the grains and the non-grain crops.

You can ask whether the predictions in Table 5.1 would hold under a different statistical yield modeling such as the alternative yield model expressed in Eqs. 5.3 and 5.4, that is, the growing-season degree days approach. The answer to this question depended on to whom you posed this question. In Table 5.2 below, yield change predictions from the two independent research groups are put together, which diverge quite dramatically.

Table 5.1 Impacts of climate change on yields of major crops in the U.S.: a yield growth function model

Climate models and scenarios (for 2070–2099)	Corn	Soybeans	Cotton
US agriculture			
Hadley III Model B1	~ −43%	~ −35%	~ −38%
Hadley III Model A1F1	~ −82%	~ −72%	~ −72%

Note: The table is based on Schlenker and Roberts (2009)

Table 5.2 Statistical yield models: impacts of climate change on yields of major crops in the U.S.

Yield models and climate scenarios (for 2070–2099)	Agricultural profit (US$)	Corn (production, yield)	Soybeans (production, yield)
[1] DG 2007 Model			
Baseline	US$ 32 B	8.67 BB	2.38 BB
Hadley II	US$ +1.29 B	+0.01BB	+0.02 BB
[2] DG 2012 Model			
Hadley II	US$ -1.7 B		
Hadley II with distributed lag	US$ -3.7 B		
CCSM III A2 with distributed lag	US$ -8.6 B		
[3] FHRS 2012 Model			
Hadley III B2	−55.99%	−42.01%	−51.59%

Note: (1) The DG is for Deschenes and Greenstone (2007, 2012). (2) The FHRS is for Fischer et al. (2012). (3) *B* Billion, *BB* Billion Bushels

Developing the growing-season degree days model in Eqs. 5.3 and 5.4, Deschenes and Greenstone made initial predictions, which is denoted as DG 2007 in the table. The authors predicted that agricultural profit would increase owing to climate change, although slightly, under the Hadley II model climate scenario. More specifically, the authors predicted that annual farm profit would increase by US$ 1.3 billion per year from the baseline level of US$ 32 billion per annum.

The increase in agricultural profit would occur because of the increases in crop yields owing to climatic changes. The authors predicted that the corn production would increase slightly (by 0.01 billion bushels) and the soybean production would also increase slightly (by 0.02 billion bushels), as shown in Table 5.2. These predictions were made assuming that the Hadley II scenario would unfold by the end of the twenty-first century.

The predictions by the DG 2007 yield model are starkly different from the predictions in Table 5.1. The other group of researchers challenged this forecast. Fisher and others, relying on the same data and methodology as the DG 2007 model, applied the Hadley III model B2 scenario and their forecasts are shown at the bottom of the table, which is designated as the FHRS 2012 model in Table 5.2 (Fisher et al. 2012). The FHRS 2012 model altered the DG 2007 model only slightly by introducing county fixed effects and different soil definitions.

The FHRS (for Fisher, Hanemann, Roberts, Schlenker) reanalysis predicted that the reduction of agricultural profit would be as large as 56% given the above-mentioned climate scenario. The loss of corn yield is predicted to be 42% and the loss of soybean yield to be 52% from the baseline, all by the 2070–2099 time period.

The two results seem at first two different farmland animals, like a dog and a hawk, that have nothing in common. The results are starkly contrasting. In response to the critique, Deschenes and Greenstone corrected several data irregularities pointed out by the latter group of researchers and recalculated their impacts (Deschenes and Greenstone 2012). The results of the DG 2012 model are also shown at the middle panel of the table.

The recalculated impacts are different, but not drastically different from their initial impact calculations, that is, the DG 2007 model. The impact on agricultural profit turns from slightly positive (+1.29 billion US$/year, +4%) to slightly negative (−1.7 billion US$/year, −5.3%), assuming the same climate change scenario, that is, the Hadley II model. Considering that these predictions are for the end of the twenty-first century and there are standard errors for the predictions, the initial predictions and the corrected predictions may not be dramatically different.

The authors provided two alternative results adding two new yield model specifications. In the first specification, the authors ran a distributed lag model in which this year's weather is assumed to affect agricultural profits in the two succeeding years in addition to the current year's profit. They argue that the counting of the lagged effects increases the damage to −3.7 billion US$ per year (−11.6%). This is again with the assumption of the Hadley II scenario.

In the second specification, they further add the assumption of more severe climate change scenario, i.e., the CCSM (Community Climate System Model) A2 scenario. The counting of both the lagged effects and the severer climate change scenario results in the damage estimate of −8.6 billion US$ per year (−26.8%), which is dramatically different from the initial prediction of +4% in the DG 2007 model.

It should be noted that the high-end climate change predictions from the host of climate models relied upon by the IPCC are known to forecast a very high degree of global warming far beyond the range of climate change predictions suggested by the IPCC assessment reports, that is, a warming in the range of 1.2–4.5 °C by the end of this century (IPCC 2014a).

Considering the wide range of climate change forecasts by the century's end, the impact estimate of −26.8% by the DG 2012 model cannot be said to be forecasting a very high damage on agriculture owing to the future global warming. To clarify, if the authors were to adopt the above-mentioned range of climate change forecasts used by the IPCC, the same statistical yield model with a distributed lag is likely to predict a much smaller damage, perhaps less than half of the 26.8% loss of profit, even with the high-end climate prediction of the IPCC range.

A more serious critique of the DG 2012 distributed lag model is that the distributed lag approach does not fundamentally answer the critique that neither the DG 2007 model nor the FHRS 2012 model does distinguish the difference between the weather impacts and the climate change impacts on agriculture (Seo 2013b). Put differently, the impact of a climate regime in a farming region would certainly have more than two lags.

5.5 Applications: International Contexts

The statistical models introduced in Eqs. 5.1 through 5.4 and their variants were applied to study agriculture in non-US countries as well as global agriculture. These studies relied on the national agricultural data of concern, the continental-scale data, or the global-scale data constructed by combining the national-level data

(Rosenzweig et al. 2001; McCarl et al. 2008; Schlenker and Lobell 2010; Lobell et al. 2011). The geographical unit of analysis is again a level in the national administrative divisions chosen by a modeler, for example, a district.

Constructing a global scale data, Lobell and coauthors calculated the impacts of the observed climate changes during the past decades on the yields of selected crops (Lobell et al. 2011). That is, they examined the effects of past climate changes on yields of crops, not the effects of future climate changes.

As commented elsewhere by the present author, what the modeling of these authors arrives at is similar to the growing-season degree days model explained in Eqs. 5.3 and 5.4 with time fixed effects (Seo 2013b). What the model reveals is the effects of yearly weather variations from the long-term climate normals on yearly yield variations from the long-term average yields. More concretely, what is estimated is the impacts of yearly temperature variations from the temperature normal on yearly corn yield variations from the average yield of corn in the region of study. For this reason, the host of similar statistical models can be called a statistical yield-weather model.

In an international context, a crop of utmost importance is often different from the dominant crops of the United States upon which the statistical yield models explained in the preceding section were developed. The most notable example is rice in South, Southeast, and East Asia. A group of researchers applied the statistical yield models to examine the impacts of the already-occurred climate changes on rice yield in South and Southeast Asia (Welch et al. 2010).

As shown in Table 5.3, the authors chose six countries: China, India, Indonesia, the Philippines, Thailand, Vietnam. The authors then chose seven farm sites located in the six countries where they obtained farm-level yield data of rice. The choices of farm locations as well as selected countries were constrained by the data availability from the International Rice Research Institute (IRRI) located in the Philippines. Therefore, the sample may not be representative, nor random.

They measured the impacts of observed climate changes on the stages of rice growth, in a similar spirit to that of the growth rate function approach in Eqs. 5.1 and 5.2. But, instead of estimating the yield growth rate function, they relied on the classification of four stages of rice growth. For example, rice growth can be divided into four phases: germination phase, vegetative phase, reproductive phase, and ripening phase. These phases are similar to, but not the same as the commonly-used four seasons: spring, summer, fall, winter (Mendelsohn et al. 1994).

For climate variables, the authors suggested to use minimum temperature, maximum temperature, and solar radiation which is altered by, among other things, the extent of cloud cover. The impacts are calculated by two seasons: High-yield season and low-yield season. There are multiple growing seasons in many areas in the selected regions owing to subtropical and tropical climate regimes.

The authors argued that the observed climate changes reduced the rice yield in Vietnam by 22.8%, in the Philippines by 19.5%, in Thailand by 11.7% (Welch et al. 2010). The large decreases in rice yield by already-occurred climate changes are almost entirely by the increase in minimum temperature. That is, a night-time minimum temperature increase owing to the trend of global warming has resulted in significant losses of rice yields.

Table 5.3 Statistical yield models: rice yields in South & Southeast Asia

	High-yield season				Low-yield season			
Country (site)	Yield change due to an increase in MINT (%)	Yield change due to an increase in MAXT (%)	Yield change due to an change in SR (%)	**Net impact on rice yield (%)**	Yield change due to an increase in MINT (%)	Yield change due to an increase in MAXT (%)	Yield change due to an change in SR (%)	**Net impact on rice yield (%)**
China (Jinhua)	0	0	0	**0**	−9.8	+14.0	0	**+4.1**
India (Aduthurai)	−6.3	+12.6	−4.1	**+2.2**	−7.2	+10.7	0	**+3.5**
Indonesia (Sukamandi)	−12	0	−0.3	**−12.3**	−7.1	0	0	**−7.1**
Philippines (Maligaya)	−21.8	0	+2.3	**−19.5**	−12.5	0	0	**−12.5**
Thailand (Suphan Buri)	−11.7	0	0	**−11.7**	0	+10.9	+6.8	**+17.7**
Vietnam (Hanoi)	−16.6	0	−6.2	**−22.8**	0	0	+11.8	**+11.8**
Vietnam (Omon)	+9.0	0	+10.7	**+19.8**	+16.9	+13.7	0	**+30.6**

Note: (1) MINT is for Minimum Temperature; (2) MAXT is for Maximum Temperature; (3) SR is for Solar Radiation; (4) The table is constructed from the data provided by Welch et al. (2010)

On the other hand, the authors argued that the increase in maximum temperature had led to the increase in rice yield. A warmer day-time maximum temperature has been beneficial. The changes in solar radiation were positive in some locations while negative in other locations, which resulted in gains or losses in rice yields.

There are a couple of salient features shown in Table 5.3. First, large negative effects of observed climatic changes can be seen in the Philippines and Indonesia but there are countries that have benefited greatly from climatic changes, for example, Vietnam (Omon), Thailand, India, and China. Second, the high-yield season effects and the low-yield season effects of observed climatic changes are uneven and even opposites in some countries such as Thailand, Vietnam (Hanoi).

5.6 Interpretations

There are not a few interpretational challenges with regard to the predictions made by the yield statisticians elaborated in this chapter. It has been noticed and pointed out by many that these predictions are starkly different from the predictions made by other scientific traditions. Let's first compare the yield predictions by the yield statisticians with those made by the experimental scientists such as the FACE and crop simulations.

In Fig. 5.1, the present author puts together two predictions. The FACE experimental results are for cotton, soybeans, and trees. The predictions are made for future carbon dioxide concentration doubling, i.e., $2*CO2$ concentration at the pre-industrial times. For soybeans and trees, the productivity indicator is dry matter production (DMP) (Ainsworth and Long 2005). The FACE results are a meta-analysis of a large number of FACE experiments conducted at some point over a 15-year time span.

The yield studies are presented for maize, soybeans, and cotton. The predictions are made for the future climate scenario, Hadley A1F1, a severe climate change scenario by the end of the twenty-first century (Schlenker and Roberts 2009). The productivity indicator is crop yield, that is, kg produced per hectare of cropland.

For cotton, the two predictions are completely opposite: About 70% loss of cotton yield by the yield statisticians and about 40% gain of cotton yield by the experimental scientists. For other crops and plants, opposite predictions are clearly striking.

An interpretation of this odd discrepancy is that heat effects may be different from carbon dioxide effects. However, the FACE experiments control the heat effects carefully and the experiments conducted across the range of climate regimes broadly support the FACE figures in Fig. 5.1. Another interpretation may be that the FACE experiments may be universal while the statistical yield studies are applicable to only U.S. agriculture.

Another commentary on the statistical yield studies has been that the yield effects of climate change should be quite different from the effects of climate change on agricultural profit—or revenue or income—as a whole. As shown in Table 5.2, the

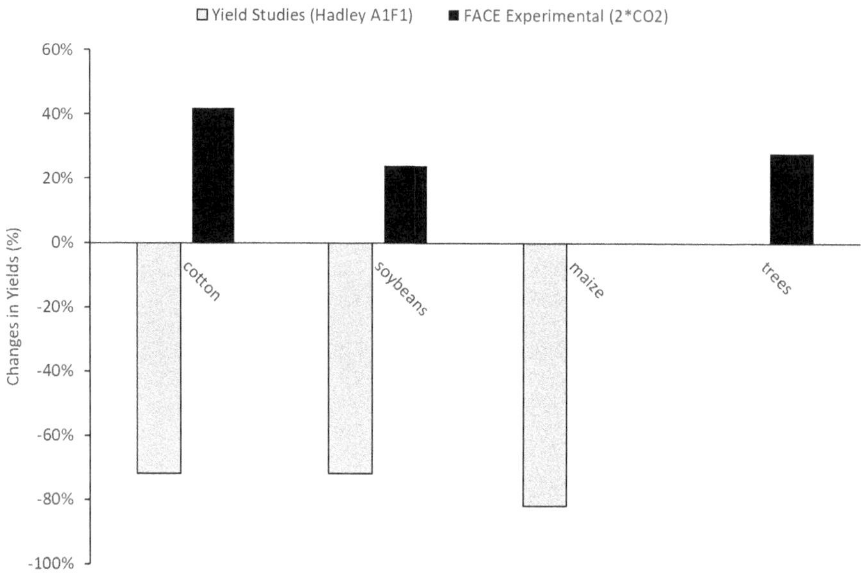

Fig. 5.1 Yield statisticians versus experimentalists

DG statistical models forecast that the effects of climate change on agricultural profit are negligible if the degree of climate change were not to be extreme.

This has been pointed out by economists from the very early period of climate change science literature, many aspects of which were extensively covered throughout Chap. 3 of this book. To mention a few studies here, Mendelsohn and coauthors predicted that the effects of climate change on US agriculture would be modest owing to input substitutions and changes in practices that take place as global temperature rises (Mendelsohn et al. 1994).

The microbehavioral studies that explicitly quantify changes in farm behaviors in response to climate changes do find that farmers' adaptation strategies and behaviors are indeed vast and effective as responses to changes in climate conditions. A microbehavioral economic model that incorporates the full array of adaptation behaviors will lead to a modest impact of climate change on agriculture (Seo 2010b, 2016a, c).

Another interpretation of the dire predictions by the statistical yield studies is that climate is perhaps quite different from weather. More pertinently, weather effects on crop growth are quite different from climate effects (Seo 2013b). Weather is defined as a short-term, daily in most cases, atmospheric phenomenon while climate is defined as a long-term weather, e.g., a 30-year average of weather events (Le Treut et al. 2007; IPCC 2014a).

One way to examine the effects of climate changes on the crop yields is to construct a normal yield, that is, a 30-year average yield of a crop. We can compare

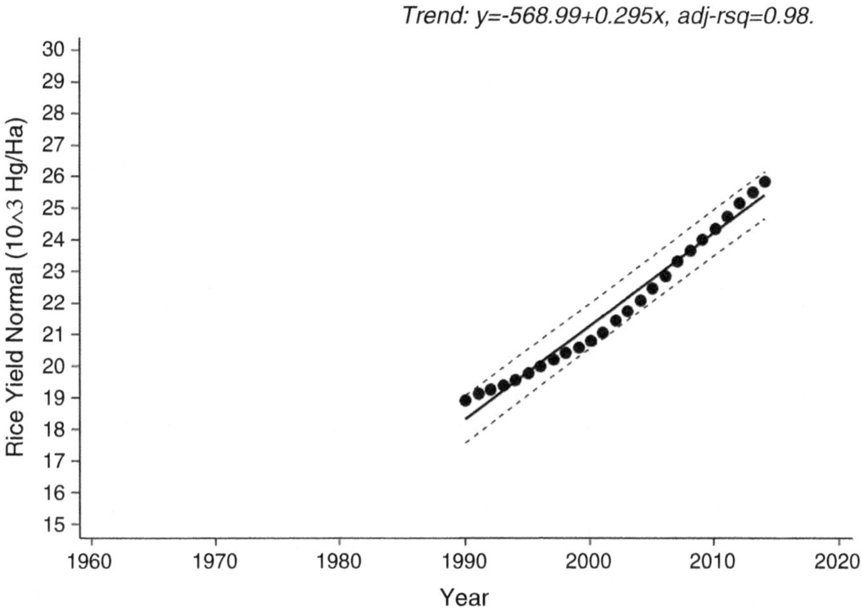

Fig. 5.2 Changes in rice yield normal in Thailand from 1960 to 2015. (Note: 10 Hg = 1 kg)

and quantify the changes in the normal crop yield against the changes in climate normals such as a 30-year average temperature or a 30-year average precipitation.

In Fig. 5.2 below, the present author constructed the time series of the rice yield normal using the Thailand agricultural data (Seo 2019). Rice is the most important agricultural product in Thailand which is the world's leading rice exporter (World Bank 2009; Welch et al. 2010). The yield data in the figure cover the period of 1960–2015. The agriculture data are from the FAOSTAT from the United Nations' organization in Rome which are compiled from the official national data submitted by the Thailand government (FAO 2017).

In a major departure from other statistical yield studies (Deschenes and Greenstone 2007; Schlenker and Roberts 2009; Lobell et al. 2011; Kala 2015), in order to separate the impact of changes in climate normals from the impact of changes in yearly weather, annual rice yields (y_t) (in hg per ha of cropland) reported from 1960 to 2015 are averaged for each 30-year period to construct rice yield normal (y_t^O):

$$y_t^O = \frac{1}{30} \sum\nolimits_{k=t-30}^{t} y_k. \tag{5.5}$$

Using the rice yield data from the 1960s, the increase in rice yield normal since 1990 in Thailand is shown in Fig. 5.2, which shows a nearly linear increase from 18,000 hg/ha to 26,000 hg/ha. For your reference, 1 hectogram (hg) is 1/10th of 1 kilogram (kg). As the overlaid trend line in the figure shows, the increase in rice yield is remarkable. The Thailand's remarkable experience was also highlighted by development economists in the World Bank report (World Bank 2009). Note that the rice yield normal (per ha) shown in Fig. 5.2 is substantially lower than the highest rice yield in major rice producing countries such as California, Japan, and China because the rice yield normal drawn in the figure is the average of 30 years as well as the average of all the regions in Thailand, that is, not the rice yield in the year 2019 (World Bank 2008).

The trend line estimated and shown in the figure tells that the normal yield has increased almost 30 kg/ha per year during this time period and 800 kg/ha in total during the period. How much of this increase can be attributable to the global warming trend? During this time period, the trend of global warming has been reported extensively by international organizations such as the Intergovernmental Panel on Climate Change (IPCC 2014a). For our continued analysis of the Thailand yield data, let's take the climate data made available to the public by the World Bank Climate Change Knowledge Portal from which we can obtain monthly temperature and monthly precipitation since the year 1900 for Thailand (World Bank 2017). The World Bank data are in turn provided by the Climate Research Unit at the University of East Anglia (Harris et al. 2014).

The World Bank climate data reveals that, for Thailand, the degree of global warming during the time span was a 0.3 °C increase in the temperature normal, which is much smaller than the global average change reported by the IPCC but closer to the low-latitude average temperature change reported by the IPCC (IPCC 2014a).

Be aware that, with only the information in Fig. 5.2, it is not possible for us to conclude that the trend of global warming was hugely beneficial for rice production in Thailand. From the trend, other causes of yield growth should be subtracted before such a conclusion is drawn, the most prominent of which could be technological changes and innovative agricultural practices developed by Thai people (Ruttan 2002; Evenson and Gollin 2003). Thais also credit the late King Bhumibol Adulayej's agricultural policy for the country's agricultural success (Seo 2019). For a full statistical analysis of the data, refer to Seo (2019).

Notwithstanding, the figure does inform us that the effects of the global warming that has occurred during the past half-century have not been fearfully ruinous. On the contrary, Thai agriculture has overcome the challenges posed by shifts in the climate system to improve the country's rice yield remarkably through a basketful of strategies.

5.7 Implications for GCF Funding Decisions

In what ways do the statistical yield studies assist the Green Climate Fund's funding decisions as well as the design of a funding rule (GCF 2011, 2018a, b)? More specifically, can the statistical yield studies be the basis of or criteria of allocations of the GCF fund to the growers of the most vulnerable crops identified by these studies?

For example, should the GCF Board support financially the maize farmers in the United States or Sub-Saharan Africa or rice farmers in India, Bangladesh, or Thailand because of potential threats of climate change, based on the statistical yield studies described in the preceding sections?

The answer may be no owing to the interpretational issues described in the previous section. The statistical yield studies have not done a good job of distinguishing climate effects from carbon dioxide effects on crop yields. Neither have they done a good job of isolating climate effects from weather effects on crop yields. Given this, it would seem unreasonable to allocate the international climate-change-purpose funds to a particular crop in a particular country, say, maize in the United States, based solely on the statistical yield studies.

An alternative perspective in answering the question is that a farmer in developing countries located in low-latitude regions tends to practice an integrated farming in which her/his portfolio of agricultural products includes a mixed bag of grains, non-grain crops, farm animals, trees, and forest products (Seo 2012a, 2014a, 2016b). What this means is that a farmer may benefit from a particular scenario of climate change even if one asset of her/his portfolio, for example, rice, were to lose big under the scenario. In such a case, a financial grant to the farmer by the GCF would be very difficult to be justified.

From this perspective, the most serious drawback of the statistical yield models as a policy relevant research endeavor is the absence of relationship in the models with the livestock sector as well as the omission of the livestock effects caused by climatic changes. The physical effects of climate change on farm animals are vastly different in their physical mechanisms and adaptations from those of grains and crops (Mader and Davis 2004; Hahn et al. 2009). The microbehavioral economics literature has emphatically shown the inter-relationships between crop management and animal husbandry as the core economic strategy of low-latitude farmers against climate constraints (Seo 2010b, 2012b, 2016a, d).

A more favorable commentary on the statistical yield studies would be that what their predictions inform the GCF decision-makers is that a specialized agriculture on a certain crop is more likely to be more severely hit from climate change (Seo 2010a, b, 2012b). A farm with a highly diversified portfolio, on the other hand, is likely to cope far better than a specialized farm with a single grain cultivation.

Should the GCF grants then be directed to a single grain farm in vulnerable developing countries? Such an allocation rule may run into the behavioral incentive problems well-studied in the economic literature, for example, adverse selection and moral hazards (Laffont and Martimort 2002). The international climate fund's grant to a single crop farm will make it harder for the grant recipient to make a transition from the beneficiary crop to a different crop or a mixed farming system. The GCF funding in this case will encourage a mal-adaptation of recipient farmers (Seo 2015).

This concludes the chapter on the statistical yield models. Along with the other modeling traditions elucidated in this book, the expositions of this chapter will serve as one of the empirical bases for the evaluations of individual GCF-funded projects conducted in Chap. 7 (GCF 2018c).

References

Adams RM, McCarl BA, Mearns LO (2003) The effects of spatial scale of climate scenarios on economic assessments: an example from U.S. agriculture. Clim Chang 60:131–148

Ainsworth EA, Long SP (2005) What have we learned from 15 years of free-air CO2 enrichment (FACE)? A meta-analytic review of the responses of photosynthesis, canopy properties and plant production to rising CO2. New Phytol 165:351–371

Auffhammer M, Ramaathan V, Vincent JR (2006) Integrated model shows that atmospheric brown clouds and greenhouse gases have reduced rice harvests in India. Proc Natl Acad Sci U S A 103:19668–19672

Conley TG (1999) GMM estimation with cross sectional dependence. J Econ 92:1–45

Deschenes O, Greenstone M (2007) The economic impacts of climate change: evidence from agricultural output and random fluctuations in weather. Am Econ Rev 97:354–385

Deschenes O, Greenstone M (2012) The economic impacts of climate change: evidence from agricultural output and random fluctuations in weather: reply. Am Econ Rev 102:3761–3773

Evenson R, Gollin D (2003) Assessing the impact of the Green Revolution 1960–2000. Science 300:758–762

FAO (2017) FAO STAT. FAO Statistics Division, Rome. Available at http://faostat.fao.org

Fisher AC, Hanemann WM, Roberts MJ, Schlenker W (2012) The economic impacts of climate change: evidence from agricultural output and random fluctuations in weather: comment. Am Econ Rev 102:3749–3760

Food and Agriculture Organization (FAO) (1995) Dimensions of need: an atlas of food and agriculture. FAO, Rome

Green Climate Fund (GCF) (2011) Governing instrument for the green climate fund. GCF, Songdo City

Green Climate Fund (GCF) (2018a) Annex III: investment framework. GCF, Songdo City

Green Climate Fund (GCF) (2018b) Annex IX: results management framework. GCF, Songdo City

Green Climate Fund (GCF) (2018c) Projects + programmes. GCF, Songdo City. Accessed from https://www.greenclimate.fund/what-we-do/projects-programmes

Hahn GL, Gaughan JB, Mader TL, Eigenberg RA (2009) Chapter 5: Thermal indices and their applications for livestock environments. In: DeShazer JA (ed) Livestock energetics and thermal environmental management. ASABE, St. Joseph, pp 113–130. Copyright 2009 American Society of Agricultural and Biological Engineers. ASABE #801M0309. ISBN 1-892769-74-3

Harris I, Jones PD, Osborn TJ, Lister DH (2014) Updated high-resolution grids of monthly climatic observations – the CRU TS3.10 Dataset. Int J Climatol 34:623–642

Intergovernmental Panel on Climate Change (IPCC) (2014a) Climate Change 2014: the physical science basis. The fifth assessment report of the IPCC. Cambridge University Press, Cambridge

Intergovernmental Panel on Climate Change (IPCC) (2014b) Climate change 2014: impacts, adaptation, and vulnerability. The fifth assessment report of the IPCC. Cambridge University Press, Cambridge

Jones CA, Kiniry JR (1986) CERES-Maize: a stimulation model of maize growth and development. Texas A&M University Press, College Station

Kala N (2015) Ambiguity aversion and learning in a changing world: the potential effects of climate change from Indian agriculture. Ph.D. dissertation, Yale University, New Haven

Laffont J, Martimort D (2002) The theory of incentives: the principal-agent model. Princeton University Press, Princeton

Le Treut H, Somerville R, Cubasch U, Ding Y, Mauritzen C, Mokssit A, Peterson T, Prather M (2007) Historical overview of climate change. In: Solomon S et al (eds) Climate change 2007: the physical science basis. The fourth assessment report of the IPCC. Cambridge University Press, Cambridge

Lobell D, Schlenker W, Costa-Roberts J (2011) Climate trends and global crop production since 1980. Science 333:616–620

Mader TL, Davis MS (2004) Effect of management strategies on reducing heat stress of feedlot cattle: feed and water intake. J Anim Sci 82:3077–3087
McCarl BA, Villavicencio X, Wu X (2008) Climate change and future analysis: is stationarity dying? Am J Agric Econ 90:1241–1247
McKinsey JW Jr, Evenson RE (1998) Technology-climate interactions: was the Green Revolution in India climate friendly? In: Dinar A, Mendelsohn R, Evenson R, Parikh J, Sanghi A, Kumar K, McKinsey J, Lonergan S (eds) Measuring the impact of climate change on Indian agriculture, World Bank Technical Paper no. 402. World Bank, Washington, DC
Mendelsohn R, Nordhaus W, Shaw D (1994) The impact of global warming on agriculture: a Ricardian analysis. Am Econ Rev 84:753–771
Nakicenovic N, Davidson O, Davis G, Grübler A, Kram T, La Rovere EL, Metz B, Morita T, Pepper W, Pitcher H, Sankovski A, Shukla P, Swart R, Watson R, Dadi Z (2000) Emissions scenarios. a special report of working group III of the intergovernmental panel on climate change. IPCC, Geneva
National Agricultural Statistical Service (NASS) (2018) Census of agriculture. US Department of Agriculture, Washington, DC. http://www.agcensus.usda.gov/index.php
Nordhaus W (2006) Geography and macroeconomics: new data and new findings. Proc Natl Acad Sci U S A 103(10):3510–3517
Open Government Data Platform of India (2018) Available at https://data.gov.in/sector/agriculture
Reich PB, Hobbie SE, Lee TD, Pastore MA (2018) Unexpected reversal of C3 versus C4 grass response to elevated CO2 during a 20-year field experiment. Science 360:317–320
Rosenzweig C, Iglesias A, Yang XB, Epstein PR, Chivian E (2001) Climate change and extreme weather events: implications for food production, plant diseases, and pests. Glob Change Hum Health 2(2):90–104
Ruttan VW (2002) Productivity growth in world agriculture: sources and constraints. J Econ Perspect 16:161–184
Schenkler W, Hanemann M, Fisher A (2005) Will US agriculture really benefit from global warming? Accounting for irrigation in the hedonic approach. Am Econ Rev 95:395–406
Schlenker W, Lobell D (2010) Robust negative impacts of climate change on African agriculture. Environ Res Lett 5:1–8
Schlenker W, Roberts M (2009) Nonlinear temperature effects indicate severe damages to crop yields under climate change. Proc Natl Acad Sci U S A 106(37):15594–15598
Seo SN (2010a) Is an integrated farm more resilient against climate change? A micro-econometric analysis of portfolio diversification in African agriculture. Food Policy 35:32–40
Seo SN (2010b) A microeconometric analysis of adapting portfolios to climate change: adoption of agricultural systems in Latin America. Appl Econ Perspect Policy 32:489–514
Seo SN (2012a) Adapting natural resource enterprises under global warming in South America: a mixed logit analysis. Economia 12:111–135
Seo SN (2012b) Decision making under climate risks: an analysis of Sub-Saharan farmers' adaptation behaviors. Weather Clim Soc 4:285–299
Seo SN (2013a) Refining spatial resolution and spillovers of a microeconometric model of adapting portfolios to climate change. Mitig Adapt Strateg Glob Chang 18:1019–1034
Seo SN (2013b) An essay on the impact of climate change on US agriculture: weather fluctuations, climatic shifts, and adaptation strategies. Climate Change 121:115–124
Seo SN (2014a) Evaluation of agro-ecological zone methods for the study of climate change with micro farming decisions in Sub-Saharan Africa. Eur J Agron 52:157–165
Seo SN (2014b) Coupling climate risks, eco-systems, and anthropogenic decisions using South American and sub-Saharan farming activities. Meteorol Appl 21:848–858
Seo SN (2014c) Adapting sensibly when global warming turns the field brown or blue: a comment on the 2014 IPCC Report. Econ Aff 34:399–401
Seo SN (2015) Helping low-latitude, poor countries with climate change. Regulation 38(4):6–8. Winter 2015–2016

Seo SN (2016a) Modeling farmer adaptations to climate change in South America: a micro-behavioral economic perspective. Environ Ecol Stat 23:1–21

Seo SN (2016b) The micro-behavioral framework for estimating total damage of global warming on natural resource enterprises with full adaptations. J Agric Biol Environ Stat 21:328–347

Seo SN (2016c) Microbehavioral econometric methods: theories, models, and applications for the study of environmental and natural resources. Academic, Amsterdam

Seo SN (2016d) Untold tales of goats in deadly Indian monsoons: adapt or rain-retreat under global warming? J Extreme Events 03:1650001. https://doi.org/10.1142/S2345737616500019

Seo SN (2019) Will farmers fully adapt to monsoonal climate changes? An analysis of rice in Thailand. J Agric Sci. (under revision)

Seo SN, Mendelsohn R (2008a) An analysis of crop choice: adapting to climate change in South American farms. Ecol Econ 67:109–116

Seo SN, Mendelsohn R (2008b) Measuring impacts and adaptations to climate change: a structural Ricardian model of African livestock management. Agric Econ 38:151–165

Seo SN, Mendelsohn R, Munasinghe M (2005) Climate change and agriculture in Sri Lanka: a Ricardian valuation. Environ Dev Econ 10:581–596

Seo SN, Mendelsohn R, Dinar A, Hassan R, Kurukulasuriya P (2008) Differential adaptation strategies of African cropland across agro-ecological zones, World Bank Policy Research Working Paper 4601. World Bank, Washington, DC

Taylor KE, Stouffer RJ, Meehl GA (2012) An overview of CMIP5 and the experiment design. Bull Am Meteorol Soc 93:485–498

Tubiello FN, Ewert F (2002) Simulating the effects of elevated CO2 on crops: approaches and applications for climate change. Eur J Agron 18:57–74

United States Census Bureau (2012) Statistical abstract of the United States: 2012. U.S. Census Bureau, Washington, DC. Accessed from https://www.census.gov/library/publications/2011/compendia/statab/131ed.html

United States Department of Agriculture (USDA) (2012) Census of agriculture 2012. USDA, Washington, DC. Accessed at https://www.agcensus.usda.gov/Publications/2012/#full_report

Welch JR, Vincent JR, Auffhammer M, Moya PF, Dobermann A, Dawe D (2010) Rice yields in tropical/subtropical Asia exhibit large but opposing sensitivities to minimum and maximum temperatures. Proc Natl Acad Sci U S A 107:14562–14567

World Bank (2008) World development report 2008: agriculture for development. World Bank, Washington, DC

World Bank (2009) Awakening Africa's sleeping giant: prospects for commercial agriculture in the Guinea Savannah zone and beyond. World Bank and FAO, Washington, DC

World Bank (2017) Climate change knowledge portal for development practitioners and policy makers: Thailand. World Bank, Washington DC

Chapter 6
Ecosystem-Based Economic Modelers of Climate Change

Abstract This chapter provides a review and discussions of the models of ecosystem changes induced by climate change as well as the ecosystem-based economic impact models. First, the global land cover database for climate studies constructed by the National Aeronautics and Space Administration (NASA) team and other satellite-based ecosystem data are explained. The land cover data are illustrated by the South American land cover map. Second, the Agro-Ecological Zone (AEZ) classification for suitability of crop agriculture developed by the international agricultural organizations is explained. The distributions of the AEZs across the African continent are illustrated. This chapter then critically reviews the ecosystem-based economic impact studies that rely on the AEZ methodology. The fundamental drawbacks of the AEZ-based economic impact methodology are highlighted through the methodology's failure of identification of agricultural activities in each AEZ as well as the failure to capture behavioral responses such as selection, diversification, and integration.

Keywords Climate change · Ecosystem · Global land cover · Agro-Ecological Zones · Economic impacts

6.1 Introduction

The fourth scientific modeling tradition that this book utilizes is the group of studies whose focus is on ecosystem or ecological changes induced by climatic shifts. Of the large array of researches in this area for the past three decades (Gitay et al. 2001; Fischlin et al. 2007; Denman et al. 2007), this chapter will highlight the ecosystem studies whose components include an economic valuation or even whose ultimate goal is an economic (monetary) valuation of climate change-induced ecosystem changes.

Even before climate change came to materialize as an international policy issue by the establishment of the Intergovernmental Panel on Climate Change (IPCC) (WMO 1985; IPCC 1990), the concept of ecological zones and ecosystems had long

S. N. Seo, *The Economics of Global Allocations of the Green Climate Fund*,
https://doi.org/10.1007/978-3-030-18274-8_6

been relied upon by agricultural and ecological experts as a tool for identifying the lands suitable or unsuitable for agricultural production activities (FAO 1978; Dudal 1980). For example, a desert ecosystem is not a suitable place for cultivating staple grains such as corn, wheat, and rice while a moist savannah zone is more suitable for these grains.

With the advance of climate change science, this concept has become one of the important climate change indictors in the literature of climate change. An increase or decrease of Sahara Desert in Africa or Gobi Desert in Central Asia observed over a sustained period of time would indicate that the regional climate, if not the global climate, is changing (Thomas and Nigam 2018; NASA 2018).

On the other hand, the ecosystems are an integral part of the carbon cycle that determines how much carbon dioxide is released into the atmosphere or is stored in the different ecosystems. An increase in the forest ecosystem or the grassland ecosystem on the surface of Earth leads to an increase in the stored carbon dioxide in the Earth's ecosystems (Schlesinger 1991; Houghton 2008; Crowther et al. 2015). In addition to the biogeochemical processes of carbon cycle, there is the zoogeochemical process that plays an important role in the carbon cycle (Schmitz et al. 2018). Animals accumulate carbon in their body, so the predator-prey relationships in the ecosystems affects the carbon accumulation in the earth ecosystems.

Putting together the multi-faceted aspects of ecological changes, researchers have attempted to devise a systematic tool or another to evaluate the ecosystem impacts of climate change, which can be referred to as the ecosystem models of climate change impacts. The ecosystem impact models are particularly pertinent to the regions and industries dominantly based on agricultural and natural resource sectors such as Sub-Saharan Africa, Latin America, and South Asia (World Bank 2008).

This chapter will review the literature of ecosystem classifications whose main drivers are climate factors. The focus of this chapter will be laid on the efforts to classify lands and resources for the suitability for various agricultural production activities. In this vein, an influential methodology among the agricultural and development experts is the Agro-Ecological Zone (AEZ) methodology (FAO 1978; IITA 2000; FAO/IIASA 2012). Developed and advanced by international agricultural research organizations, the AEZ methodology offers a transparent relationship between climate changes and agro-ecological systems.

The AEZ methodology has been extended by climate researchers beyond the study of ecosystem changes under climate change (Seo et al. 2009). For example, it was applied to the analysis of yield changes of major crops such as maize, rice, wheat, and cotton caused by climatic shifts (Fischer et al. 2005; Tubiello et al. 2007). This book explained in depth the two scientific traditions that are primarily concerned on this topic of yield changes: the crop simulation models and the FACE experiments in Chap. 4, the statistical yield studies in Chap. 5. An AEZ impact analysis couples the AEZ classification of the concerned region with the results from the crop simulation models/FACE experiments in order to provide improved predictions of yield changes.

What critical information does the AEZ methodology or other ecosystem-based impact models present regarding the money allocation decisions of the Green

Climate Fund (GCF) (GCF 2011, 2018a)? Should the GCF Board allocate the funds to the region where the region's ecosystem is predicted to turn from a productive system to a hostile or deserted system? Among other important questions raised and answered in this chapter, these are some of the key questions you will encounter in this chapter.

To give a brief preview, a cursory examination of the GCF-funded programs, 93 of them as of the end of 2018, will reveal that the concepts of ecological zones or ecosystems are pervasive in the project applications as well as the project funding documentations, especially on the policy results areas of food and water security (GCF 2018c). Take for example the following two projects in Sub-Sahara: The project entitled "Strengthening climate resilience of agricultural livelihoods in Agro-Ecological Regions I and II in Zambia" approved for Zambia and the project entitled "Improving rangeland and ecosystem management practices of smallholder farmers under conditions of climate change in Sesfontein, Fransfontein, and Warmquelle areas of the Republic of Namibia" approved for Namibia.

6.2 Global Land Cover Database for Climate Studies

Presently, many Earth-orbiting satellites are operated by the National Aeronautic and Space Administration (NASA) and other space agencies for the purpose of studying the Earth system. The NASA Earth Observing System (EOS) program was developed in the early 1990s to monitor and understand key components of the climate system and their interactions through long-term observations by polar-orbiting satellites. Of the focus areas of the EOS program are land surfaces and ecosystem processes (NASA 2018).

According to the NASA, there have been 26 satellite missions in the EOS program as summarized in Table 6.1. Many missions are completed, for example, ACRIMSAT, ICESat, ATLAS. Current missions include Aqua, Aura, and SAGE-III. Of the current missions, Landsat 7, Landsat 8, Landsat 9, and Terra satellites are the missions that are closely related with Earth's ecosystems.

An early study of the Earth's ecosystems by the researchers at the NASA classified them by the major land cover adopting a fixed spatial resolution (Matthews 1983). The author titled it as "New High-resolution Data Bases for Climate Studies" emphasizing climate change aspects of the land cover data set. By laborious examinations of national atlases and satellite imageries from the LandSat, Matthews classified the Earth surface into 32 land covers. The list of land covers is shown in Table 6.2.

The author took a grid-cell approach. That is, a major land cover is determined for each grid cell with the size of 1-degree latitude and 1-degree longitude. It is notable that her classification was not done on political units such as counties or districts, but on a geographical unit (Nordhaus 2006). A 1-degree grid cell is the cell whose size is 10 km-long and 10 km-wide.

Table 6.1 NASA missions on Earth Observing System (EOS)

Mission	Mission category	Status
Active Cavity Radiometer Irradiance Monitor Satellite (ACRIMSAT)	EOS	Completed
Advanced Earth Observing Satellite (ADEOS)	EOS, A-Train	Completed
Aqua	EOS	Current, Extended
Atmospheric Laboratory of Applications and Science (ATLAS-1)	EOS	Completed
Aura	EOS	Current, Extended
Challenging Mini-satellite Payload (CHAMP)	EOS	Completed
Glory	EOS	Launch Failure
Ice, Cloud, and land Elevation Satellite (ICESat)	EOS	Completed
Jason-1	EOS	Completed
Landsat 7	EOS	Current
Landsat 8	EOS	Current
Landsat 9	EOS	Future, Implementation
Ocean Surface Topography Mission/Jason-2 (OSTM/Jason-2)	EOS	Current, Extended
Quick Scatterometer (QuikSCAT)	EOS	Current
Radar Satellite (RADARSAT)	EOS	Completed
Sea-viewing Wide Field-of-view Sensor (SeaWiFS)	EOS	Completed
SeaWinds on the Advanced Earth Observing Satellite II (ADEOS II) (SeaWinds)	EOS	Completed
Shuttle Radar Topography Mission (SRTM)	EOS	Completed
Solar Radiation and Climate Experiment (SORCE)	EOS	Current, Extended
Spaceborne Imaging Radar-C (SIR-C)	EOS	Completed
Stratospheric Aerosol and Gas Experiment III on ISS (SAGE III-ISS)	EOS, Earth Systematic	Current
Stratospheric Aerosol and Gas Experiment-III Meteor-3 M (SAGE-III)	EOS	Completed
Terra	EOS	Current, Extended
TOPEX/Poseidon (TOPEX)	EOS	Completed
Total Ozone Mapping Spectrometer-Earth Probe (TOMS-EP)	EOS	Completed
Upper Atmosphere Research Satellite (UARS)	EOS	Completed

For determining a major land cover, Matthews relied on two source data: vegetations and land uses. For the latter, the author relied on the composite of agricultural land use data obtained from national sources across the globe. In Table 6.2, 'cultivation' is a variety of agricultural lands which is estimated to be about 17.56 million km^2 globally.

Table 6.2 Global land covers with their sizes

Formation class	Global land covers	Area (10^6 km^2)
Forest	tropical evergreen rainforest	12.29
	trop/subtropical evergreen seasonal broad-leaved forest	3.29
	subtropical evergreen rainforest	0.19
	temperate/subpolar evergreen rainforest	0.39
	temperate evergreen seasonal broadleaved forest, summer rain	0.81
	evergreen broadleaved sclerophyllous forest, winter rain	0.47
	tropical/subtropical evergreen needle-leaved forest	0.49
	temperate/subpolar evergreen needle-leaved forest	9.29
	tropical/subtropical drought-deciduous forest	2.88
	cold-deciduous forest, with evergreens	5.18
	cold-deciduous forest, without evergreens	3.99
	xeromorphic forest/woodland	2.68
Woodland	evergreen broadleaved sclerophyllous woodland	1.71
	evergreen needleleaved woodland	2.51
	tropical/subtropical drought-deciduous woodland	3.70
	cold-deciduous woodland	2.50
Shrubland	evergreen broadleaved shrubland/thick, evergreen dwarf-shrubland	1.30
	evergreen needleleaved or microphyllous shrubland/thicket	0.67
	drought-deciduous shrubland/thicket	0.83
	cold-deciduous subalpine/subpolar shrubland/dwarf shrub	0.46
	xeromorphic shrubland/dwarf shrubland	8.86
Grassland	tall/medium/short grassland, 10–40% woody cover	6.46
	tall/medium/short grassland, < 10% woody cover	3.66
	tall/medium/short grassland, shrub cover	9.34
	tall grassland, no woody cover	0.81
	medium grassland, no woody cover	0.78
	meadow, short grassland, no woody cover	6.10
Desert	forb formations	0.28
	desert	12.57
Ice	arctic/alpine tundra, mossy bog	7.34
	ice	2.44
Cultivation	cultivation	17.56

For the former source data, that is, global vegetations, the author relied on national atlases and satellite imageries referred to as the '70 published vegetation sources' by the author. The vegetation classification was based on the United Nations Educational, Scientific, and Cultural Organization (UNESCO) classification system which was developed for a global vegetation mapping and inventory (UNESCO 1973). Synthesizing several classification theories, the UNESCO classification is based on the lifeform, density, and seasonality of the underlying vegetation, with supplementary terms on altitude, climate, and vegetation architecture (Matthews 1983).

Let's take, for an illustration of the UNESCO system, the formation class of forest vegetation. The forest class is divided into three formation subclasses: evergreen, deciduous, xeromorphic. The formation subclass of evergreen is divided further into multiple formation groups such as tropical ombrophilous (meaning 'capable of withstanding or thriving in the presence of much rain'), tropical/subtropical seasonal, winter rain sclerophyllous (meaning 'thick leaved') (UNESCO 1973).

The full list of global land covers is shown in Table 6.2. Broadly, there are different formations: Forest, woodland, shrubland, grassland, deserts, ice, and cultivation. For illustration, let's take a closer look at the grassland land cover. There are six types of grasslands, three of which are with some cover and the other three of which are without any cover. For the latter, i.e., the grasslands without woody cover, there are tall grasslands, medium grasslands, and short grasslands. For the former, there are grasslands with 10–40% woody cover, the grasslands with <10% woody cover, and the grasslands with shrub cover. Summing up the sizes of the various formations of the grasslands, the total size is about 27 million km^2 globally, which is far larger than the size of cultivated lands that is 17 million km^2.

There are 12 different types of forests, which in total occupies 42 million km^2 globally. The forest vegetations are further divided, most notably, by climate zones: tropical, subtropical, temperate, subpolar, cold-deciduous, xeromorphic forests. A xerophyte—a plant having the ability to store water in leaves or stems— is a species of plants that has adapted to survive in an environment with little water such as deserts and arid zones.

The Latin American part of the above global land cover data set maintained by the NASA Goddard Institute of Space Studies (GISS) is drawn in Fig. 6.1 (Seo 2012b). The most conspicuous of the Latin American ecosystems is tropical rainforests which are marked by a blue rectangle whose size is 1-degree latitude by1-degree longitude grid cell. Tropical rainforests are expansive across northern Brazil and the Andean countries located in and around the Amazon Basin.

In the high lands of Colombia and Venezuela where the Llanos grasslands are situated, the dominant ecosystem is the grasslands with <10% woody cover. Another prominent ecosystem in the Andean high lands is tropical/subtropical drought deciduous forests.

Eastern parts of Brazil are dominated by grasslands with smaller (<10%) or larger (>10% and < 40%) woody covers as well as xeromorphic forests. Western and southern parts of Brazil, which includes the Cerrado, are covered by various types of forests such as xeromorphic forests, tropical/subtropical drought deciduous forests, seasonal forests, and evergreen forests.

In south of Brazil, Uruguay is dominantly grasslands with shrub cover. The northern part of Argentina is dominated by tall grasslands. The large areas of Argentina are arid, dominated by drought deciduous shrublands. Even further south, there exist meadows and deserts in the southern parts of Argentina.

Along the western coasts of South America, southern parts of Peru are dominantly meadow with no woody cover. Northern parts of Chile, including the Atacama Desert, are meadows with no woody cover and xeromorphic shrublands. The middle parts of Chile are xeromorphic forests and subtropical and temperate

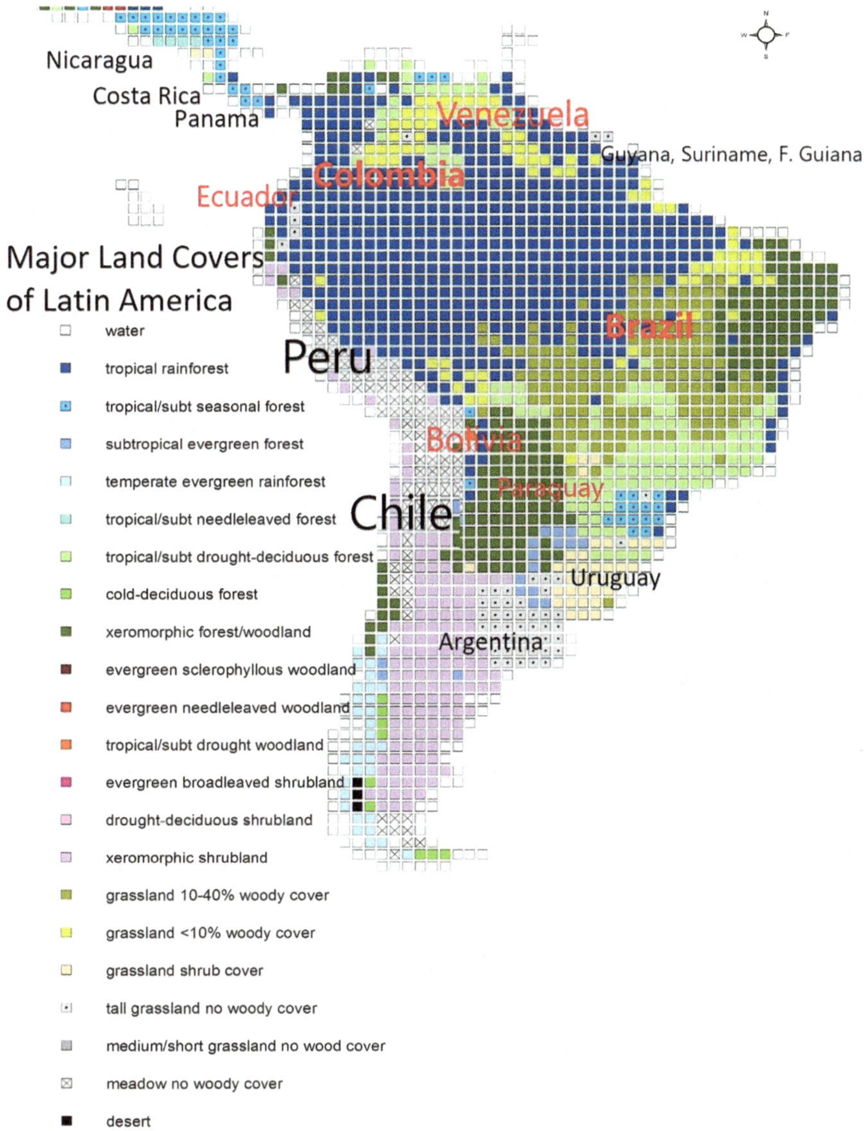

Fig. 6.1 Major land covers in Latin America

evergreen rainforests. Southern parts of Chile are temperate evergreen rainforests and cold deciduous forests.

Major land covers in Paraguay and Bolivia, two landlocked countries located in central highlands, are xeromorphic forests, tropical/subtropical seasonal forests, subtropical evergreen forests, and evergreen needle-leaved woodlands.

The Andes Mountains provide the highland mountain ecosystems that are distinct from the lowlands' ecosystems. One of the unique highland mountain ecosystems is the glaciers such as the Zongo glacier in Bolivia and the Antisana glacier in Ecuador which received much attention from climate change scientists (Rabatel et al. 2013).

The existing ecosystems in Latin America are conditioned on, among other factors, the existing climate system. Therefore, if the current climate system were to be altered owing to anthropogenic greenhouse gas emissions, the present ecosystems as shown in Fig. 6.1 would also be altered (Gitay et al. 2001; Denman et al. 2007). Possible changes in presently dominant ecosystems caused by climatic shifts are of a major concern to the world community because many anthropogenic activities, including agriculture, are conditioned on prevailing ecosystems.

This is the point where the ecosystem modeling has become prominent in the climate change literature. To be more specific, the ecosystem modeling not only tells you future changes in the Earth's ecosystems induced by climatic shifts but also gives you clues on possible changes in anthropogenic, that is, economic activities. From this point on, the present author will highlight some of the ecosystems-economy relationship using the Latin American data illustrated in Fig. 6.1 and come back later in this chapter to give a more precise analysis of the relationship.

To reveal the relationship between ecosystems and anthropogenic activities, let's first take a look at the agricultural and natural resource enterprises in South America that are presently performed in the continent. At present, about a third of the total land area in South America is utilized for agricultural lands (WRI 2005). Many different crops and their varieties are grown across South America, of which major cereals planted are wheat, maize, rice; major oil seeds planted are soybeans, peanuts, sunflowers; major vegetables are potatoes and cassavas; specialty crops include cotton, tobacco, and coffee (Mata and Campos 2001).

As displayed in Fig. 6.1, the South American continent is occupied by vast areas of different types of grassland ecosystems. Consequently, the pasturelands used for the livestock sector and farm animals dominate the total size of croplands. In Argentina, the size of pasturelands is eight times larger than that of croplands while it is four times larger in Brazil (Baethgen 1997). The dominance of the grassland ecosystems means that the weight of livestock management is far heavier in South American agriculture than elsewhere in the world (Seo 2015b).

Several production statistics are helpful for clarifying this point. A specialized livestock system accounts for about 20% of all farming households in the continent, which is huge considering only 5% of the specialized livestock system adopted in Sub-Saharan Africa (Seo 2011). Argentina and Brazil are one of the world's largest, if not the largest, beef cattle exporters and consumers of beef per head annually (Steiger 2006). The livestock species raised most often by farmers are beef cattle, dairy cattle, chickens, pigs, goats, and sheep (Seo et al. 2010).

Other than the grassland ecosystems, forest ecosystems are dominant in vast areas of South America, especially near the Equator. As can be verified in Fig. 6.1, the Amazon rainforest covers 7.5 million km^2 of land area and is the world's largest pluvial forest (Mata and Campos 2001). According to one survey supported by the World Bank, the income earned from forests and forest products may account for more than 20% of the rural income in South and Central America (Vedeld et al. 2007). South American farmers manage tree plantations for the sale of timbers, but the value of non-timber forest products is also very large (Peters et al. 1989). Most common trees in South American farms include palm, cashew, cacao, mango, pineapple, kola, eucalyptus, and Brazil nut, to name but a few (Seo 2012a).

How are agricultural and natural resource enterprises distributed across the spectrum of ecosystems in South America? In other words, given the existing, that is, present ecosystems, how are agricultural and natural resource enterprises adopted and managed in the continent? To answer these questions, we need first to define the enterprises.

Through a continental-level farm household survey conducted with the support from the World Bank, researchers asked each rural farm about the portfolio of agricultural and natural resource products that it manages (Seo and Mendelsohn 2008a). Based on the answers given, a farm that owns only crops is classified as a crops-only enterprise. A farm that owns only farm animals is classified as a livestock-only enterprise. A farm that owns trees only is classified as a forests-only enterprise. These are specialized enterprises.

Diversified enterprises are defined accordingly. A crops-livestock enterprise is one that owns both crops and farm animals. A crops-forests enterprise is one that owns both crops and trees. A livestock-forests enterprise is one that owns both farm animals and forests. A crops-livestock-forests enterprise is one that owns crops, farm animals, and forests.

This classification of agricultural and natural resource enterprises was already introduced in Chap. 3 when the microbehavioral economic model was reviewed. To summarize, the following seven agricultural and natural resource enterprises, which are defined to be mutually exclusive and exhaustive, encompass the anthropogenic/economic natural resource activities in South America:

$$\begin{aligned}
&\text{Crops} - \text{only}(\text{C});\\
&\text{Livestock} - \text{only}(\text{L});\\
&\text{Forests} - \text{only}(\text{F});\\
&\text{Crops} - \text{Livestock}(\text{C} + \text{L});\\
&\text{Crops} - \text{Forests}(\text{C} + \text{F});\\
&\text{Livestock} - \text{Forests}(\text{L} + \text{F});\\
&\text{Crops} - \text{Livestock} - \text{Forests}(\text{C} + \text{L} + \text{F}).
\end{aligned} \tag{6.1}$$

Before we formally associate the family of natural resource enterprises in Eq. 6.1 with the family of land ecosystems in Fig. 6.1 statistically, we can begin by a simpler descriptive question: Which natural resource enterprise is most favored by farm households in each of the ecosystems? Choices by South American farms of natural resource enterprises are summarized in Table 6.3 (Seo 2012b). The table shows how many farms (in percentages) located in a specific land cover (ecosystem) have chosen each of the enterprises.

There are multiple interesting findings captured in the table. First, dominance of a livestock-only enterprise in some ecosystems is noteworthy. This enterprise was

Table 6.3 Adoption percentages (%) of natural resource enterprises by land cover

	Land cover	Enterprises Crops-only	Livestock-only	Forests-only	C + L	C + F	C + L + F
Grass	tall/medium/short grassland, < 10% woody cover	21.7	41.3	4.3	8.7	19.6	4.3
	tall/medium/short grassland, shrub cover	23.8	49.0	0.7	15.4	7.0	4.2
	tall/medium/short grassland, 10–40% woody cover	17.5	21.9	0.6	23.1	7.5	19.4
	tall grassland, no woody cover	56.7	12.2	0.0	24.0	5.5	1.6
	meadow, short grass-land, no woody cover	35.0	13.8	0.0	42.5	5.0	3.8
Shrub	xeromorphic shrub-land/dwarf shrubland	36.6	30.5	0.8	19.8	10.7	1.5
Forests	trop/subtropical ever-green seasonal broad-leaved forest	3.8	0.0	11.5	23.1	3.8	57.7
	cold-deciduous forest, with evergreens	23.1	0.0	0.0	0.0	69.2	7.7
	subtropical evergreen rainforest	48.8	19.5	0.0	19.5	9.8	2.4
	temperate/subpolar evergreen rainforest	3.1	9.4	0.0	87.5	0.0	0.0
	tropical evergreen rainforest	37.2	15.5	0.4	25.9	8.8	12.2
	tropical/subtropical drought-deciduous forest	41.0	12.1	0.4	35.1	3.8	7.5
	xeromorphic forest/ woodland	30.6	19.0	0.5	35.1	4.2	10.6
Water	Water	16.1	17.8	3.4	24.6	20.3	17.8

Note: C + L = Crops + Livestock; C + F=Crops + Forests; C + L + F=Crops + Livestock + Forests

adopted by 49% of the farms in the tall/medium/short grasslands with shrub cover, a dominant land cover around Uruguay. Only 24% of the farms in this land cover chose a crops-only enterprise. A large portion of the Pampas, the iconic South American lowland grasslands that expand 750,000 km^2, is covered by this land ecosystem.

Similarly, in the tall/medium/short grasslands with sparse (<10%) woody cover, a livestock-only enterprise is favored by 41% of the households. This ecosystem is dominant in the highland grasslands in the Andes, that is, the Llanos grasslands that expand over Colombia and Venezuela. Where there is a large glade in the forested regions, Latin American farmers prefer to manage livestock and do it successfully.

Although overlooked so often by agricultural and climate researchers, this is one of the most noteworthy findings on the relationship between natural resource industries and ecological systems in the context of climate change. What it reveals is that agriculture can still be profitable in the areas where croplands no longer exist and/or where there are large open areas or glades not completely covered by forests, which is indicated by the high preference of the specialized livestock management in these ecosystems (Seo 2016a, d).

For the crops-only enterprise, tall grasslands with no woody cover, subtropical evergreen rainforests, tropical/subtropical drought-deciduous forests are favored ecological zones. The crops-only system is favored by 57%, 49%, 41% respectively of the rural households in each of these ecosystems.

The crops-livestock enterprise is adopted most often in temperate/subpolar evergreen rainforests (87.5%), meadow/short grasslands with no woody cover (42.5%), and xeromorphic forests/woodland. In the xeromorphic forests/woodland, a crops-livestock enterprise is preferred to a crops-only enterprise by 35–31% while a livestock-only is chosen by 19% of farms in this ecological zone.

A crops-forests enterprise is adopted most often by farmers in major forest zones: the tropical rainforests (8.8%), the cold-deciduous evergreen forests (69.2%), and the coastal/lake/river ecosystems (23%). A forests-only enterprise is most frequent, in the number of households that adopted this system, near the water body such as oceans, rivers, and lakes.

A crops-livestock-forests enterprise is an integrated system in that it is encompassing of crop farming, livestock management, and forestry. It is most favored by farmers located in the substantially forested zones such as tropical/subtropical evergreen seasonal broad-leaved forests (57.8%), tall/medium/short grasslands with dense (10–40%) woody cover (19.4%), and tropical rainforests (12.2%).

The description thus far of the Latin American ecosystem and economic data demonstrates the fruitfulness of the integration of the ecosystem modeling with behavioral decisions of farmers as an essential part of the literature of climate science. Ecosystem modelers have focused on the physical aspects such as a classification of ecosystems, their distributions, and climate-induced changes (Matthews 1983; Gitay et al. 2001; Fischlin et al. 2007). Economic modelers have focused on the monetary aspects such as changes in economic behaviors that are conditioned on climate factors (Seo 2010a, 2016a, b). The integration of the two scientific traditions illustrated in this section lucidly shows that they are integral components of the climate change and economy interactive loop which are often overlooked by both scientific models and economic models of climate change (Nordhaus 2008).

A simplistic climate and economy loop would look like Fig. 6.2. The literature of climate change has concentrated on the three components marked by three empty circles: climate change, ecosystem changes, carbon cycle. These are essential components of the biogeochemistry of climate change (Schlesinger 1991; Denman et al. 2007). The figure shows that there is a neglected, or hidden, component that connects between ecosystem changes and carbon cycle, marked by a shaded circle:

Fig. 6.2 A simplistic climate-economy loop

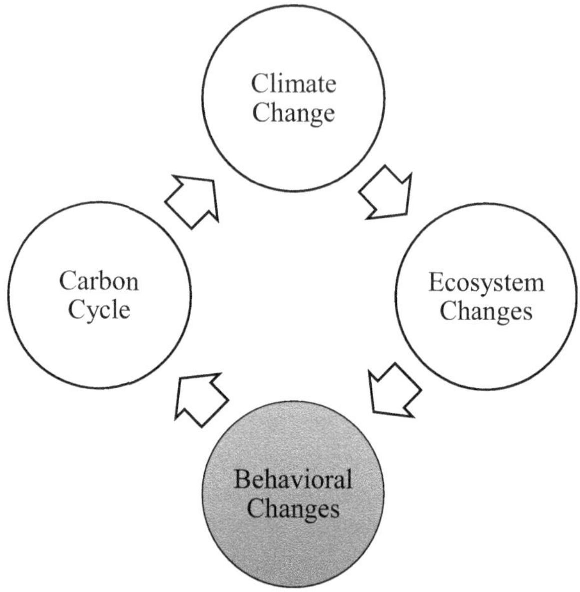

behavioral changes. The arrows indicate the directions of the cause and effect relationships among the four components.

Undoubtedly, the interactive relationships among the components are far more complex than those depicted in Fig. 6.2. To mention some of the complexities, the behavioral changes are also caused directly by climate change. For example, an increase in summer temperature in a region may force its residents to relocate to other regions. The behavioral changes are also caused directly by carbon cycle changes: For example, a carbon dioxide concentration doubling in the atmosphere can spawn a carbon-capture-and-sale industry (Seo 2017).

6.3 Agro-Ecological Zones

Another ecosystem modeling has been directed exclusively to crop agriculture with the development of the concept of the Agro-Ecological Zones (AEZ). Developed during the late 1970s by researchers from international agricultural organizations before any international climate policy meeting (WMO 1985; IPCC 1990), the AEZ methodology has been a primary tool of agricultural researchers for analyzing the land's suitability for crop farming determined by climate and soils of the land (Dudal 1980; FAO 1978).

The classification of the AEZs is made based on the concept of the Length of Growing Period (LGP) for crops (FAO 1978; Dudal 1980). The LGP is defined as the length of the period during the year when climate and soil conditions are 'conducive' to crop growth. The LGP is a similar concept to the growing season heating degree days used by the econometric models of yields surveyed in depth in the previous Chapter, although not precisely the same (Schlenker et al. 2006; Deschenes and Greenstone 2007).

Formally, the LGP is defined as the number of days in a year in which daily mean temperature is above 5 °C and soil moisture conditions are 'considered adequate.' Under rain-fed conditions, the beginning of the growing period is linked to the start of the rainy season. For crops, soil moisture that is 0.4–0.5 times the level of reference evapotranspiration is considered sufficient to meet water requirements of dryland crops, depending on the soil types and structures of the croplands considered. In turn, the soil moisture storage capacity of a soil type depends on the soil's physical and chemical characteristics, but above all on effective soil depth or volume (FAO 2005; FAO/IIASA 2012).

The AEZ classification based on the concept of the LGP was proposed originally by Dudal who was working at the Food and Agriculture Organization at the time (FAO 1978; Dudal 1980) and the basic conceptual framework remained unchanged although several refinements have been made over time (IITA 2000; FAO/IIASA 2012).

According to Dudal, the five AEZs can be classified by the LGP of the land area of concern, as illustrated in Fig. 6.3 (Dudal 1980). The desert AEZ is defined as the area whose LGP is smaller than 30 days; The arid AEZ is defined to be the area whose LGP is between 30 days and 90 days; The semi-arid AEZ is defined to be the area whose LGP is between 90 days and 180 days; The sub-humid AEZ is defined to be the area whose LGP is between 180 days and 270 days; The humid AZE is defined to be the area whose LGP is greater than 270 days.

From the FAO/IIASA data tables (2012), the spatial distribution of each of the LGP 30-day bands for the continent of Africa is drawn in Fig. 6.4. The figure also

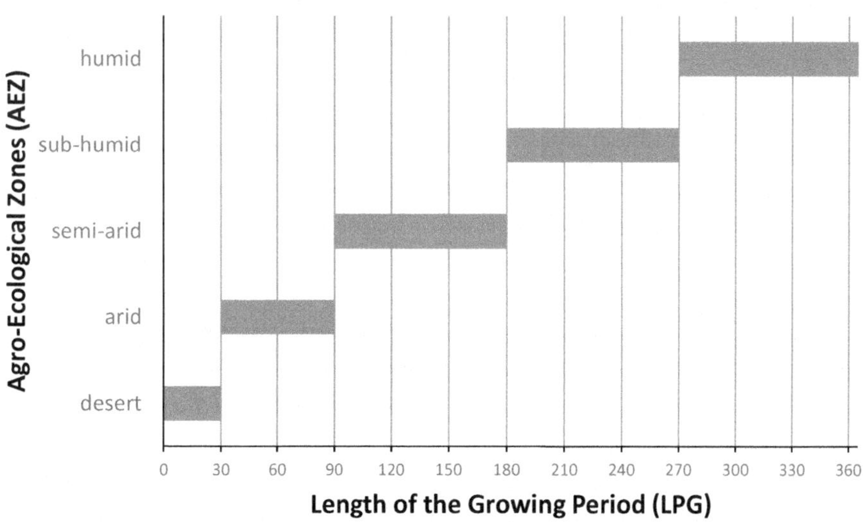

Fig. 6.3 Classification of agro-ecological zones

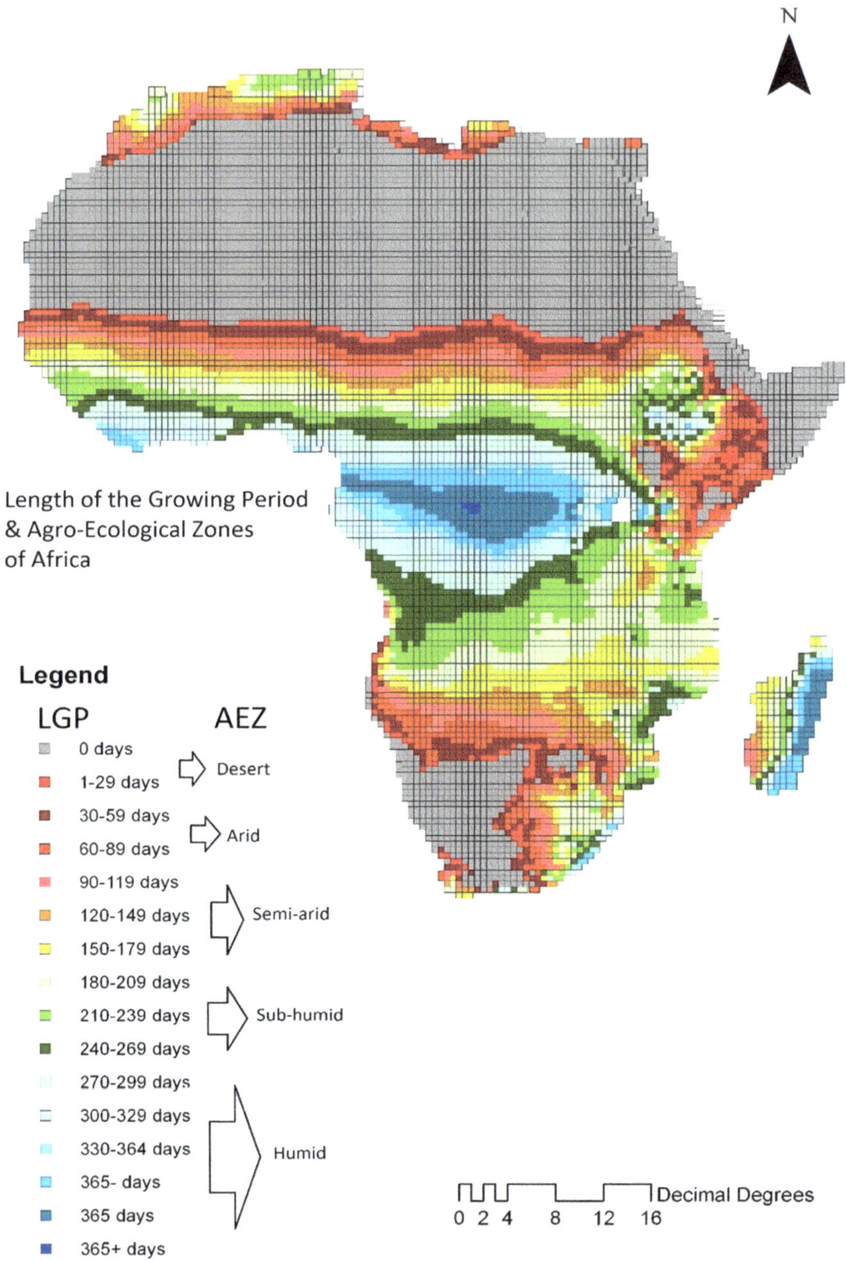

Fig. 6.4 The AEZ classification of Africa

indicates the corresponding AEZ classification of each LGP 30-day band. The spatial unit of the classification of the LGPs—and the AEZs—drawn in the figure is a grid-cell with a spatial resolution of 1/2-degree latitude by 1/2-degree longitude.

Each LGP band is 30-days long. In the figure, the first two bands of gray (a band of LGP = 0) and red (a band of LGP < 30) color are a desert AEZ. The following two "reddish" color bands are an arid AEZ. The next three "brownish and yellowish" color bands are a semi-arid AEZ. The next three "greenish" color bands are a sub-humid AEZ. The rest bands of "blueish" color are a humid AEZ.

The desert AEZ is located in the Sahara Desert in the north, the Kalahari Desert and the Namib Desert in the south of the continent, and the Horn of Africa in the east. The arid AEZ is located in the land areas that are adjacent to the major deserts, including the Sahel and the eastern highlands. The humid AEZ is located in the areas around the Equator in Gabon, Republic of Congo, Democratic Republic of Congo, Uganda, and Lake Victoria. The sub-humid AEZ is located in the areas that abut the humid AEZ. The sub-humid AEZ is also found in the eastern coastlines of the continent in Mozambique and Tanzania. The semi-arid AEZ is located in the land areas in the Sahel between the arid AEZ and the sub-humid AEZ. The semi-arid AEZ is also found in the land areas in the south around Zambia and Zimbabwe.

Taking the agro-ecological zone of the land as given, an African manager will choose an agricultural and natural resource enterprise. Let's ask the same question posed in the previous section, but this time with a different ecosystem definition, that is, the AEZ classification: Does an individual manager prefer one enterprise over another because of the agro-ecological condition of her farm?

To answer this question, we refer to the World Bank household survey conducted in 11 African countries as part of the Bank's climate change research project in Africa (Kurukulasuriya et al. 2006; Seo et al. 2009). Of the participant countries in the survey, East African countries were Ethiopia and Kenya; West African countries were Senegal, Niger, Ghana, and Burkina Faso; Central African countries were Cameroon; South African countries were South Africa, Zambia, and Zimbabwe; North African countries were Egypt.

From the household data of the 9 Sub-Saharan countries, the present author examined how many farms (in percentage) in each of the AEZ chose each of the three agricultural systems: A crops-only system, a crops-livestock system, a livestock-only system (Seo 2014a). An analysis with a more refined definition of agricultural systems, including forest-based systems, shows additional insights (Seo 2010b).

The results, shown in Table 6.4, demonstrate some usefulness of the AEZ classification based on the concept of the LGP for understanding the range of agricultural activities performed in the continent. More specifically, there is some indication that the longer the length of the growing period, the higher the adoption rate of a crops-only system. In the humid AEZ, 53% of the farms chose the crops-only system. On the other hand, 36% of the farms in sub-humid zones and 27% of the farms in semi-arid zones chose the crops-only system.

On the contrary, arid zones have a higher adoption rate of the crops-only system than the semi-arid zones. In the arid AEZ, 43% of the farms chose the crops-only system. The adoption rate of the crops-only system again declines in the desert AEZ to 33%. A higher adoption rate in the arid AEZ can signal that there is a limitation in the usefulness of the AEZ classification or that a cooler and arid climate zone is more suitable to certain crops than a hotter and semi-arid climate zone. Note in Fig. 6.4

Table 6.4 Adoption behaviors *visa vis* the AEZs: crops-only and farm animals

Adoption behaviors		AEZ				
		Desert	Arid	Semi-arid	Sub-humid	Humid
Crops						
A crops-only system	Adoption rate (%)	33.5%	42.9%	27.2%	36.1%	52.9%
Livestock						
Beef cattle	Average number owned per farm (N)	16.3	0.3	10.9	9.1	1.0
Dairy cattle	Average number owned per farm (N)	10.0	7.1	3.3	2.7	0.8
Goats	Average number owned per farm (N)	59.3	26.1	7.6	4.8	3.7
Sheep	Average number owned per farm (N)	449.3	39.1	11.3	11.8	3.3
Chickens	Average number owned per farm (N)	4.0	11.6	21.9	37.9	149.5

that the arid climate zone is farther away from the Equator than the semi-arid climate zone in the Sahel, therefore cooler.

What of the other two agricultural systems? Is the LGP/AEZ classification a good indicator of the livestock-only or the crops-livestock system of agriculture? An examination of adoption rates of these agricultural systems tells us that the AEZ/LGP classification is a poor predictor of non-cropping systems of agriculture, which is not surprising considering the original intention of the creators. As you can verify in Table 6.4, livestock-based systems are adopted more frequently in the zones where crop farming falls out of favor. The more unfavorable the agro-climate condition becomes for crops, the more enthusiastic the mixing (or switching) strategy of a crop farmer becomes.

This can be verified with the distributions of individual species of farm animals owned by sub-Saharan farmers across the AEZ zones summarized in Table 6.4 (Seo 2014a). For sheep, a rural household in the humid AEZ owns, on average, 3 sheep. On the other hand, a rural household in the arid AEZ owns, on average, 39 sheep. In the desert AEZ, a rural household owns as many as 450 sheep.

For goats, a rural household owns, on average, 3.7 goats in the humid AEZ and 4.8 goats in the sub-humid AEZ. By contrasts, the number of goats raised is much larger in arid and desert agro-ecological zones: 7.6 goats in the semi-arid AEZ, 26.1 goats in the arid AEZ, and 59 goats in the desert AEZ.

For the ownership of either beef cattle or dairy cattle, similar distributions can be found across the AEZs. But, for each of the cattle species, the differences in the numbers of each livestock species owned across the AEZs are not as dramatic as those for sheep or goats.

The distribution of the number of chickens owned per farm across the AEZs is much different from those of the above-mentioned four livestock species. It is notable that the wetter the AEZ or the longer the LGP, the larger the number of chickens owned per farm. A farmer in the desert AEZ owns, on average, 4 chickens per farm, 11.6 chickens in the arid AEZ farm, 21.9 chickens in the semi-arid AEZ farm, 37.9 chickens in the sub-humid AEZ farm, and 149.5 chickens in the humid AEZ farm.

6.4 The AEZ Impact Analysis

The discussions up to this point in this chapter showcased the usability of the two ecosystem modeling frameworks for applications in climate change analyses but also the limitations in interpreting and applying the ecosystem models to agricultural and economic questions. Because the AEZ concept has become influential in the climate change impact literature, as can be verified by the GCF project documentations (GCF 2018c), we are compelled to ask the following pivotal policy question: To what extent can climate researchers utilize the AEZ/LGP classifications for measuring the impacts of climate change on agriculture or natural resources?

The AEZ impact methodology was developed by a group of researchers, founded on the concept of the AEZ and the AEZ classifications described thus far, as an attempt to settle the contentious debate on the impact of climate change on agriculture (Fischer et al. 2005; Tubiello et al. 2007; Hillel and Rosenzweig 2010). The methodology has been cited frequently in the literature and influential in policy circles including the IPCC reports on food and agriculture (Gitay et al. 2001; Easterling et al. 2007). Notwithstanding the popularity, a critical analysis of the soundness of the AEZ impact methodology has been rarely conducted. Exceptions are the present author's two publications on the issue (Seo 2014a, 2015a).

Let me start with the description of this methodology as set up by the Fischer and his coauthors' work, the essential features of which are embedded in all similar works (Fischer et al. 2005). In the first stage, an AEZ impact modeler determines the impacts of climate change on crop growth determinants of yields, for example, length of the growing period (LGP), at the resolution of 5 arc-minute grid cell with the AEZ (Dudal 1980). A five arc-minute is approximately 1/12th of a one degree.

Since the definition of the LGP is founded on, among other things, climate and soil characteristics of the land to be defined, the first stage calls for predictions of changes in these variables in a future climate regime. The key climate variables in the first stage that determine the changes in the AEZ/LGP are daily mean temperature, soil moisture, and evapotranspiration (Dudal 1980).

In the second stage, changes in the crop production potential of a concerned crop under a future climate regime are calculated at the grid cell level, based on the information obtained in the first stage (Tubiello et al. 2007). For each AEZ, the change in the crop production potential is calculated. Notably, the primary indicator for the AEZ impact modeling is the crop production 'potential', that is, not the actual production. This calculation is based on the relationship between the crop yield and the length of growing period. A future climate change leads to a change in the length of growing period, which then leads to a change in the crop yield. This relationship is established for each AEZ.

The first and second stages are repeated for each crop considered by the modeler. In practice, the first stage and the second stage are completed through a single modeling framework such as a crop sector decision model such as the DASSAT (Decision Support System for Agrotechnology Transfer) (Tsuji et al. 1994; Rosenzweig et al. 1995) or the EPIC (Erosion Productivity Impact Calculator) crop growth model (Williams et al. 1984).

In the third stage, the yield change prediction of a crop obtained in stage 2 from the sample data under a selected climate scenario is extrapolated to the global level or the continental level using the AEZ classification of the geographical areas outside the sample. The extrapolation is again repeated for each crop considered in the modeling.

In the fourth stage, production changes obtained in stage 3 at the regional level are then fed into a global economic model called the BLS (Basic Linked System) to assess how different climate (and socio-economic) scenarios affect policy variables such as productions, demands, prices, mal-nutrition, poverty, and hunger (Rosenzweig and Parry 1994). In this stage, the modeler should assume changes in trade patterns among the nations induced by climate change, which again depend on national policy changes.

Notice that the AEZ impact methodology is not drastically different from the methodology of the Agro-Economic Models (AEM) explained exclusively in Chap. 4 of this book. First, the Basic Linked System is a less sophisticated model than the Agricultural Sector Model (ASM) of the AEM which was explained at length in Chap. 4 (Fischer et al. 1988). This is because the BLS must build a globally linked agricultural and economy model, many of which features are poorly understood by agricultural modelers.

Second, the AEMs are more forthcoming about how impact modelers acquire changes in crop yields. That is, yield predictions are based on, *in-situ* or in an experimental setting, crop simulation models and field experiments. For the AEZ impact models, crop yield changes can be obtained by other methods than the field/laboratory experiments, for example, through the statistical yield prediction models introduced in Chap. 5. Notice that the statistical yield models also rely on a similar concept such as growing season heat accumulation or growing-season degree days (Deschenes and Greenstone 2007; Schlenker and Roberts 2009). Also, the AEZ modelers can rely on some of the crop sector decision models to obtain predictions of yield changes, e.g., the DASSAT (Tsuji et al. 1994; Rosenzweig et al. 1995) or the EPIC (Williams et al. 1984).

From another perspective, the crop sector decision models can be understood as a more sophisticated crop growth simulation model under changing climate conditions. Other than the climate factors which are the primary focus of the field/laboratory-based simulation models, many decision problems should be integrated into these crop sector decision models such as, *inter alia*, land use decisions, water use decisions, system choices, crop choices, input decisions, farming practices, conversions of CO2 concentration changes to changes in temperature and precipitation. To say the least, however, these decisions cannot be obtained from these crop sector decision models internally. In turn, viewed from this vantage point, the crop sector decision models are similar in both spirit and contents to the Agro-Economic Models (AEMs) of Chap. 4 (Adams et al. 1999).

Third, the key difference between the AEZ impact model and the AEM impact model lies in the way that the AEZ impact modelers extrapolate the within-the-sample predictions to outside-the-sample. In the former, the extrapolation technique is stated explicitly, that is, an extrapolation is made from one grid cell to another grid cell if the two cells have the same AEZ. For example, the predictions obtained in the U.S. farmlands can be extrapolated to the African farmlands as long as the matching cells share the same AEZ classification. Such an extrapolation technique, however, is prone to many prediction errors because farmers' behaviors in one grid-cell can be significantly different from those in another grid-cell even though the two grid-cells share the same AEZ classification (Seo 2014a).

Applying the AEZ impact methodology, the researchers predicted changes in important food policy variables by the year 2080 under a set of climate change scenarios. To highlight some of those predictions (Fischer et al. 2005):

1. In southern Africa, an additional 11% of the total land area to be at risk of being severely constrained for crop agriculture;
2. wheat production in South Asia will fall by 20–75% and wheat production in South America will fall by 12–27%;
3. the number of people at risk of hunger is projected to increase by about 30–-175 million people from the baseline with no global warming, depending upon climate change scenarios.

These predictions can be compared with the predictions from the global AEM modeling by Parry and his coauthors which were explained in Chap. 4 (Parry et al. 2004). The AEM model predicted similarly very severe impacts from climate change by the end of the twenty-first century on crop yields. Specifically, Parry et al. (2004) predicted 250–400 million tons of reduction in the global crop yield depending upon climate change scenarios. This is commensurate with the above AEZ prediction of 20–75% loss of wheat production in South Asia.

Overall, however, the AEZ impact model predicts a far smaller loss, although still very severe, from climate change on agricultural productivities than the AEM impact model. The difference can be seen clearly in the prediction of the additional number of people at risk of hunger due to climatic changes. As summarized in Fig. 6.5, the AEM impact model predicts about 560 million additional people at risk of hunger by the end of this century assuming the high-end climate change scenario (A2a). By comparison, the AEZ impact model predicts about 175 million additional people at risk of hunger under the high-end scenario (A1F1). With the low-end climate change scenario (B2), the AEM model predicts about 190 million additional people at risk of hunger while the AEZ model predicts about 30 million people additionally at risk of hunger.

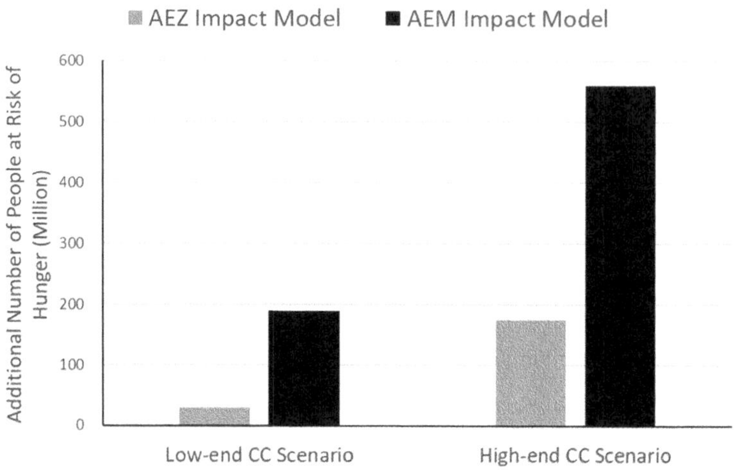

Note: 1) The AEZ impact model results are from Fischer et al. (2005); 2) The AEM impact model results are from Parry et al. (2005); 3) CC stands for Climate Change.

Fig. 6.5 Predictions from the AEZ impact model vs the AEM impact model for end of twenty-first Century. Note: (1) The AEZ impact model results are from Fischer et al. (2005); (2) The AEM impact model results are from Parry et al. (2004); (3) CC stands for climate change

As mentioned previously, the AEZ impact methodology is not fundamentally different from the AEM methodology, which is especially true when these two methodologies are applied to model the impact on a national agriculture. As such, the AEZ impact analysis suffers from the same problems that plague the AEM analysis, which were detailed in Chap. 4 of this book.

The uniqueness of the AEZ impact methodology lies in its reliance on the AEZ concept and classifications. Fundamentally, it relies on an AEZ classification for extrapolation of the within-the-sample yield prediction results to the outside-the-sample regions. Further, it relies on the LGP concept for establishing the relationships between the crops' yields and climatic shifts.

In the latter case, the LGP concept plays a substitute role for the growing season heating degree days or the cumulative heat distribution that are favored by statistical yield modelers (Deschenes and Greenstone 2007; Schlenker and Roberts 2009). The empirical problems that arise from the application of these concepts to a climate change impact analysis were clarified in Chap. 5 of this book and the same problems are unavoidable in the estimation of the LGP-based climate-yield relationships.

With regard to the former point, for the AEZ impact methodology to be scientifically meaningful, it should first be proven that the AEZ classification has the capacity to identify agricultural activities and productions that are performed in each of the AEZs (Seo 2014a). In other words, if the AEZ designation is the same for two geographical areas (grid cells), agricultural activities and productions in the two grid cells should be proven to be the same. Further, it should be proven that if the present climate system were to be shifted in the future, the two grid-cell areas with the same AEZ designation should undergo the same consequences.

As elaborated in the preceding two sections of this chapter, there are major identification failures with the AEZ/LGP concepts applied to a climate change impact model (Seo 2014a). The AEZ/LGP concepts were conceived and developed for the specifications of crop agriculture. Consequently, the AEZ/LGP classification fails to identify livestock systems from crop systems. Neither does it identify non-grain crop agriculture such as agroforestry. The consequence is that the AEZ impact methodology will get the baseline farm portfolios for each AEZ wrong which vary across the AEZs. Further, it will get the future farm portfolios under a changed climate regime wrong.

Putting it another way, the two geographical areas (grid cells) with the same AEZ designation will have the baskets of farm portfolios that are different. In this sense, the AEZ designation does not identify the basket of farm portfolios of an AEZ. The same can be said with regard to the levels of agricultural productions of an AEZ.

On this point, an even subtler analysis can be made. The effects of farmers' behavioral responses to a climatic shift cannot be accounted for by the AEZ impact models. One of such behavioral responses, which are many, is diversification. As a concrete example, let's consider a portfolio diversification by a Sub-Saharan farmer that was reported from the afore-referenced African household data (Seo et al. 2009). It was shown that the distribution of net revenue of a diversified crop portfolio is quite different from that of a specialized crop portfolio across climate zones or the AEZs.

Let's consider the net revenue from the crops-only system and that from the crops-livestock diversified system. In the humid AEZ, the net revenues earned from the two crop systems are different. In the specialized crop system, a rural household earns on average 718$ per hectare of cropland. On the other hand, in the diversified crops-livestock system, a rural household earns on average 605$ from crops per hectare of cropland, excluding the livestock profit.

In the arid AEZ, however, the positions of the two agricultural systems are even reversed. That is, the diversified crops-livestock farm earns more profit from crops than the specialized crop farm in the arid AEZ. The mixed crops-livestock farm earns on average 223$/ha of cropland from growing crops while the specialized crop farm earns 158$/ha of cropland from growing crops. This means that, in the arid AEZ where the climate regime is less favorable for crop growth, farmers have adapted to the hostile climatic condition through a diversified portfolio of crops and livestock, which has led to an increased profit even from only crops.

What these farm profit data tell us is that the effects of diversification on farms will be substantial on those who would face climatic shifts in the coming decades. It should not be missed that the behavioral strategies of farmers and natural resource managers in Sub-Sahara or elsewhere are not limited to diversification. Another prominent behavioral strategy is selection. A farmer can switch from the currently favored crop, e.g., wheat, to a heat tolerant crop such as millet or sorghum (Seo and Mendelsohn 2008c; Seo et al. 2008). Similarly, she can switch from beef cattle to a more heat tolerant animal such as sheep or goats (Seo and Mendelsohn 2008b; Seo 2016c). Still another salient adaptation strategy is integration. An integrated farm makes it resilient to changes in climatic conditions by integrating all components of agriculture such as grains, crops, farm animals, forests, and forest products (Seo 2016a, 2016d).

The AEZ impact methodology is deficient in modeling and taking into account behavioral strategies by farmers to changes in climatic conditions (Seo 2012c, 2014b). This deficiency is shared by both the AEZ impact methodology and the AEM methodology discussed in Chap. 4. Fundamentally, both methodologies are built up from major crops of interest while the microbehavioral impact models are built at the farm level where behavioral decisions are actually made.

6.5 GCF Grant Allocations into Vulnerable Ecosystems?

Having reviewed the two ecosystem modeling traditions and one ecosystem impact model, can we recommend that the Board of the GCF should allocate the funds to a particular ecosystem in a particular country because of its potentially high vulnerability? Can this approach be a decision rule for the GCF grant allocations (GCF 2011, 2018a)?

In fact, an allocation of the GCF grants on the basis of a local ecosystem's vulnerability is quite common in the approved GCF projects, which will be formally analyzed in the next chapter (GCF 2018c). The present author can handily come up with such allocation cases. A grant awarded to a West African country in the Sahel is quite common in the full list of GCF-funded projects on the grounds of the potential vulnerability of the desert and arid ecosystems thereof. For another case, a grant is given from the GCF funds to local communities in the Andes mountain ecosystems in Latin America. For the third, there are multiple GCF grants allocated to regional communities in the grassland ecosystems of Mongolia and Central Asia.

Another point of interest is that the GCF grants/loans that are awarded on the grounds of potentially vulnerable ecosystems are not handed out on the condition that the funds shall be utilized to protect the particular ecosystem. Rather, the GCF grants are given for a particular agricultural product or enterprise managed by local communities that operate in the vulnerable ecosystems: For example, maize farmers in the Sahel, cattle ranchers in Mongolia, or forest managers in the Amazon.

The risk of distorting efficient behaviors of farms by providing a GCF grant to a particular product or enterprise based on the ecosystem impact modeling can be pointed out in the same way as has been analyzed in the preceding chapters. As before, it is easy to imagine that the GCF grant to a particularly vulnerable product/enterprise yields undesirable consequences of giving an incentive to the recipients to perversely stick to the vulnerable product/enterprise, that is, giving a disincentive to the recipients to adapt efficiently to a climatic shift by switching to a more resilient enterprise (Seo 2014c, 2015c). Readers may refer back to the examples described in Chaps. 4 and 5 regarding this point.

Be also reminded that the incentive problem analyzed here can be put into the context of the principal-agent problem prominent in the broader economics literature (Laffont and Martimort 2002). The principal is the GCF policy-makers and the agent is the grant recipients. Is it possible to design a contract that outwits behavioral distortions (Hart and Holmstrom 1987)?

Proponents of the above-described decision rule may suggest an alternative rule for GCF funding decisions in order to steer clear of the incentive distortion problem: Let's fund the entire range of farm enterprises in the concerned local ecosystem and then give the farmers thereof the liberty of choosing whatever farm enterprise they find the best under the current and changed climate regimes.

A brief examination will reveal that this alternative funding model is even more problematic. Even in a local ecosystem identified to be highly vulnerable by ecosystem impact modelers, there will be certainly losers as well as winners from a climatic shift (Seo 2014a, 2016d). Under such circumstances, it would not be justifiable to hand out the same amount of money to all winners and losers alike or a lump-sum of cash transferred to a community representative. Even more seriously, the blanket grant to the specific ecosystem will force the residents in the vulnerable ecosystem not to move out of the vulnerable zones.

Proponents may offer still another amendment to correct the incentive distortion problem of the funding allocation rules: Let's make a cash transfer to a particularly vulnerable enterprise located in a particularly vulnerable ecosystem, but only on the condition that the grant money should be spent in a certain way that should be specified in any GCF funding agreement.

Even with this amended funding rule, challenges are many and very high even if we would accept the assumption that the GCF Board can correctly pick the future choice of enterprise to grant recipients. To give an example, the local ecosystem to which a GCF funding is awarded will also undergo many changes as the present climate system is altered. The enterprise designated by the GCF grant agreement that should be adopted by the grant recipients could turn out in the future to be a hard-hit enterprise by a future change in the climate system that is unpredictable today or too abrupt when it occurs in the future (NRC 2013; Titley et al. 2016). The rule is too rigid to be prepared for unpredictable climatic shifts.

Further, the rigidity of the rule means that it would be difficult for the GCF policy board to monitor how the grant is actually used by the funding recipients. They may report that the money was used for the purposes specified in the funding agreement, although they have spent the grant money for other non-specified purposes (Groves 1973). It would be impossible for the funding agency to monitor each and every expense of the recipients.

More fundamentally, there is a question of to what policy outcome and the level of performance the GCF board should award the grant and how much (GCF 2018a, b). Should it be the health of the particular ecosystem? Or should it be the income of the residents in the ecosystem? Or should it be the maintenance of yields of staple crops within the ecosystem? A choice of one of these policy outcomes by the GCF policy board will profoundly affect the behaviors of the grant recipients as well as the design of a contract between them (Hart and Holmstrom 1987).

The descriptions thus far in this section point to the challenges as well as drawbacks of the family of ecosystem-based GCF funding decision rules, should the rules actually be adopted by the GCF decision board. In the next chapter, the present author will come back to the entire list of GCF funding programs and projects that received either a grant or loan and provide an economic evaluation of each project (GCF 2018c). The present author will have an opportunity there to single out the ecosystem-based GCF funding projects and provide a more rigorous analysis of them.

References

Adams R, McCarl BA, Segerson K, Rosenzweig C, Bryant K, Dixon BL, Conner R, Evenson RE, Ojima D (1999) The economic effects of climate change on US agriculture. In: Mendelsohn R, Neumann J (eds) The impact of climate change on the United States economy. Cambridge University Press, Cambridge

Baethgen WE (1997) Vulnerability of agricultural sector of Latin America to climate change. Clim Res 9:1–7

Crowther TW, Glick HB, Covey KR et al (2015) Mapping tree density at a global scale. Nature 525:201–205

Denman KL, Brasseur G, Chidthaisong A, Ciais P, Cox PM, Dickinson RE, Hauglustaine D, Heinze C, Holland E, Jacob D, Lohmann U, Ramachandran S, da Silva Dias PL, Wofsy SC, Zhang X (2007) Couplings between changes in the climate system and biogeochemistry. In: Solomon S et al (eds) Climate change 2007: The physical science basis. Cambridge University Press, Cambridge

Deschenes O, Greenstone M (2007) The economic impacts of climate change: Evidence from agricultural output and random fluctuations in weather. Am Econ Rev 97:354–385

Dudal R (1980) Soil-related constraints to agricultural development in the tropics. International Rice Research Institute, Los Banos

Easterling WE, Aggarwal PK, Batima P, Brander KM, Erda L, Howden SM, Kirilenko A, Morton J, Soussana J-F, Schmidhuber J, Tubiello FN (2007) Food, fibre and forest products. In: Parry ML et al (eds) Climate change 2007: Impacts, adaptation and vulnerability. The fourth assessment report of the intergovernmental panel on climate change. Cambridge University Press, Cambridge

FAO (2005) Global agro-ecological assessment for agriculture in the twenty-first century (CD-ROM), FAO Land and Water Digital Media Series. FAO, Rome

FAO/International Institute of Applied Systems Analysis (IIASA) (2012) Global agro-ecological zones (GAEZ v3.0): Model documentation. FAO, Rome

Fischer G, Frohberg K, Keyzer MA, Parikh KS (1988) Linked national models: A tool for international policy analysis. Kluwer Academic, Dordrecht

Fischer G, Shah M, Tubiello FN, van Velhuizen H (2005) Socio-economic and climate change impacts on agriculture: An integrated assessment, 1990–2080. Philos Trans R Soc B 360:2067–2083

Fischlin A, Midgley GF, Price JT, Leemans R, Gopal B, Turley C, Rounsevell MDA, Dube OP, Tarazona J, Velichko AA (2007) Ecosystems, their properties, goods, and services. In: Parry ML et al (eds) Climate Change 2007: Impacts, adaptation and vulnerability. The fourth assessment report of the intergovernmental panel on climate change. Cambridge University Press, Cambridge

Food and Agriculture Organization (FAO) (1978) Report on agro-ecological zones vol. 1: Methodology and results for Africa. FAO, Rome

Gitay H, Brwon S, Easterling W, Jallow B (2001) Ecosystems and their goods and services. In: McCarthy JJ et al (eds) Climate change 2001: Impacts, adaptations, and vulnerabilities. Cambridge University Press, Cambridge

Green Climate Fund (GCF) (2011) Governing instrument for the green climate fund. GCF, Songdo City

Green Climate Fund (GCF) (2018a) Annex III: Investment framework. GCF, Songdo City

Green Climate Fund (GCF) (2018b) Annex IX: Results management framework. GCF, Songdo City

Green Climate Fund (GCF) (2018c) Projects + programmes. GCF, Songdo City. Accessed from https://www.greenclimate.fund/what-we-do/projects-programmes

Groves T (1973) Incentives in teams. Econometrica 41:617–643

Hart O, Holmstrom B (1987) The theory of contracts. In: Bewley T (ed) Advances in economics and econometrics. Cambridge University Press, Cambridge

Hillel D, Rosenzweig C (eds) (2010) Handbook of climate change and agroecosystems: Impacts, adaptation and mitigation. Imperial College Press, London

Houghton RA (2008) Carbon flux to the atmosphere from land-use changes: 1850–2005. In: Trends: A compendium of data on global change. Carbon Dioxide Information Analysis Center, Oak Ridge National Laboratory, US Department of Energy, Oak Ridge

Intergovernmental Panel on Climate Change (IPCC) (1990) Climate change: The IPCC scientific assessment. Cambridge University Press, Cambridge

International Institute of Tropical Agriculture (IITA) (2000) Delivering the African development potential: The IITA vision. IITA, Nigeria

Kurukulasuriya P, Mendelsohn R, Hassan R, Benhin J, Diop M, Eid HM, Fosu KY, Gbetibouo G, Jain S, Mahamadou A, El-Marsafawy S, Ouda S, Ouedraogo M, Sène I, Maddision D, Seo SN, Dinar A (2006) Will African agriculture survive climate change? World Bank Econ Rev 20:367–388

Laffont JJ, Martimort D (2002) The theory of incentives: The principal-agent model. Princeton University Press, Princeton

Mata LJ, Campos M (2001) Latin America. In: McCarthy JJ, Canziani OF, Leary NA, Dokken DJ, White KS (eds) Climate change 2001: Impacts, adaptation, and vulnerability. The third assessment report of the IPCC. Cambridge University Press, Cambridge

Matthews E (1983) Global vegetation and land use: new high-resolution data bases for climatestudies. J Climatol Appl Meteorol 22(3):474–487

National Aeronautics and Space Administration (NASA) (2018) Missions: Earth Observing System (EOS). NASA, Washington, DC. Accessed from https://eospso.nasa.gov/mission-category/3

National Research Council (NRC) (2013) Abrupt impacts of climate change: Anticipating surprises. Committee on understanding and monitoring abrupt climate change and its impacts. The National Academies Press, Washington, DC

Nordhaus W (2006) Geography and macroeconomics: new data and new findings. Proc Natl Acad Sci U S A 103(10):3510–3517

Nordhaus WD (2008) A question of balance: Weighing the options on global warming policies. Yale University Press, New Haven

Parry ML, Rosenzweig CP, Iglesias A, Livermore M, Fischer G (2004) Effects of climate change on global food production under SRES emissions and socioeconomic scenarios. Glob Environ Chang 14:53–67

Peters CM, Gentry AW, Mendelsohn RO (1989) Valuation of an Amazonian rainforest. Nature 339 (6227):655–656

Rabatel A, Francou B, Soruco A, Gomez J, Caceres B, Ceballos JL, Basantes R, Buille M, Sicart JE, Huggel C, Scheel M, Lejeune Y, Arnaud Y, Collet M, Condom T, Consoli G, Favier V, Jomelli V, Galarraga R, Ginot P, Maisincho L, Mendoza J, Menegoz M, Ramirez E, Ribstein P, Suarez W, Villacis M, Wagnon P (2013) Current state of glaciers in the tropical Andes: A multi-century perspective on glacier evolution and climate change. Cryosphere 7:81–102

Rosenzweig C, Parry M (1994) Potential impact of climate change on world food supply. Nature 367:133–138

Rosenzweig C, Allen LH, Harper LA, Hollinger SE, Jones JW (eds) (1995) Climate change and agriculture: Analysis of potential international impacts. American Society of Agronomy (ASA) Special Publication No. 59. ASA, Madison

Schenkler W, Hanemann M, Fisher A (2006) The impact of global warming on U.S. agriculture: An econometric analysis of optimal growing conditions. Rev Econ Stat 88(1):113–125

Schlenker W, Roberts M (2009) Nonlinear temperature effects indicate severe damages to crop yields under climate change. Proc Natl Acad Sci USA 106(37):15594–15598
Schlesinger WH (1991) Biogeochemistry: An analysis of global change. Academic, San Diego
Schmitz OJ, Wilmers CC, Leroux SJ, Doughty CE, Atwood TB, Galetti M, Davies AB, Goetz SJ (2018) Animals and the zoogeochemistry of the carbon cycle. Science 362. https://doi.org/10.1126/science.aar3213
Seo SN (2010a) A microeconometric analysis of adapting portfolios to climate change: Adoption of agricultural systems in Latin America. Appl Econ Perspect Policy 32:489–514
Seo SN (2010b) Managing forests, livestock, and crops under global warming: A micro-econometric analysis of land use changes in Africa. Aust J Agric Resour Econ 54:239–258
Seo SN (2011) A geographically scaled analysis of adaptation to climate change with spatial models using agricultural systems in Africa. J Agric Sci 149:437–449
Seo SN (2012a) Adapting natural resource enterprises under global warming in South America: A mixed logit analysis. Economia: J Lat Am Caribb Econ Assoc 12:111–135
Seo SN (2012b) Adaptation behaviors across ecosystems under global warming: a spatial microeconometric model of the rural economy in South America. Pap Reg Sci 91:849–871
Seo SN (2012c) Decision making under climate risks: An analysis of sub-Saharan farmers' adaptation behaviors. Weather Clim Soc 4:285–299
Seo SN (2013) Refining spatial resolution and spillovers of a microeconometric model of adapting portfolios to climate change. Mitig Adapt Strat Glob Chang 18:1019–1034
Seo SN (2014a) Evaluation of agro-ecological zone methods for the study of climate change with micro farming decisions in sub-Saharan Africa. Eur J Agron 52:157–165
Seo SN (2014b) Coupling climate risks, eco-systems, anthropogenic decisions using South American and Sub-Saharan farming activities. Meteorol Appl 21:848–858
Seo SN (2014c) Adapting sensibly when global warming turns the field brown or blue: A comment on the 2014 IPCC Report. Econ Aff 34:399–401
Seo SN (2015a) Micro-behavioral economics of global warming: Modeling adaptation strategies in agricultural and natural resource enterprises. Springer, Cham
Seo SN (2015b) Adapting to extreme climates: raising animals in hot and arid ecosystems in Australia. Int J Biometeorol 59:541–550
Seo SN (2015c) Helping low-latitude, poor countries with climate change. Regulation. Winter 2015–2016: 6–8
Seo SN (2016a) Microbehavioral econometric methods: Theories, models, and applications for the study of environmental and natural resources. Academic, Amsterdam
Seo SN (2016b) Modeling farmer adaptations to climate change in South America: A micro-behavioral economic perspective. Environ Ecol Stat 23:1–21
Seo SN (2016c) Untold tales of goats in deadly Indian monsoons: Adapt or rain-retreat under global warming? J Extrem Events. https://doi.org/10.1142/S2345737616500019
Seo SN (2016d) The micro-behavioral framework for estimating total damage of global warming on natural resource enterprises with full adaptations. J Agric Biol Environ Stat 21:328–347
Seo SN (2017) The behavioral economics of climate change: adaptation behaviors, global public goods, breakthrough technologies, and policy-making. Academic, Amsterdam
Seo SN, Mendelsohn R (2008a) A Ricardian analysis of the impact of climate change impacts on South American farms. Chilean J Agric Res 68:69–79
Seo SN, Mendelsohn R (2008b) Measuring impacts and adaptations to climate change: A structural Ricardian model of African livestock management. Agric Econ 38:151–165
Seo SN, Mendelsohn R (2008c) An analysis of crop choice: Adapting to climate change in South American farms. Ecol Econ 67:109–116
Seo SN, Mendelsohn R, Dinar A, Hassan R, Kurukulasuriya P (2008) Differential adaptation strategies of African cropland across Agro-Ecological Zones. World Bank Policy Research Working Paper 4601. World Bank, Washington, DC
Seo SN, Mendelsohn R, Dinar A, Hassan R, Kurukulasuriya P (2009) A Ricardian analysis of the distribution of climate change impacts on agriculture across agro-ecological zones in Africa. Environ Resour Econ 43:313–332

Seo SN, McCarl B, Mendelsohn R (2010) From beef cattle to sheep under global warming? An analysis of adaptation by livestock species choice in South America. Ecol Econ 69:2486–2494
Steiger C (2006) Modern beef production in Brazil and Argentina. Choices 21:105–110
Thomas N, Nigam S (2018) Twentieth-century climate change over Africa: Seasonal hydroclimate trends and Sahara Desert expansion. J Clim 39:3349–3370
Titley DW, Hegerl G, Jacobs KL, Mote PW, Paciorek CJ, Shepherd JM, Shepherd TG, Sobel AH, Walsh J, Zwiers FW, Thomas K, Everett L, Purcell A, Gaskins R, Markovich E (2016) Attribution of extreme weather events in the context of climate change. The national academies of sciences, engineering, and medicine. The National Academies Press, Washington, DC
Tsuji GY, Uehara G, Balas S (eds) (1994) DSSAT v3, volumes 1, 2 and 3. International Benchmark Sites Network for Agrotechnology Transfer, University of Hawaii, Hawaii, Honolulu, Hawaii
Tubiello FN, Amthor JS, Boote KJ, Donatelli M, Easterling W, Fischer G, Gifford RM, Howden M, Reilly J, Rosenzweig C (2007) Crop response to elevated CO2 and world food supply: a comment on "Food for Thought …" by Long et al., Science 312:1918–1921, 2006. Eur J Agron 26:215–223
UNESCO (1973) International classification and mapping of vegetation. UNESCO, Paris
Vedeld P, Angelsen A, Bojø J, Sjaastad E, Kobugabe GK (2007) Forest environmental incomes and the rural poor. Forest Policy Econ 9:869–879
World Bank (2008) World development report 2008: agriculture for development. World Bank, Washington, DC
Williams JR, Renard KG, Dyke PT (1984) EPIC—a new model for assessing erosion's effect on soil productivity. J Soil Water Conserv 8:381–383
World Meteorological Organization (WMO) (1985) Report of the international conference on the assessment of the role of carbon dioxide and of other greenhouse gases in climate variations and associated impacts. Villach, Austria
World Resources Institute (WRI) (2005) World resources 2005: The wealth of the poor—managing ecosystems to fight poverty. WRI, Washington, DC

Chapter 7
Economics and Evaluations of the Green Climate Fund

Abstract This chapter lays out an ensemble of economic theories that are pertinent to justifying the Green Climate Fund (GCF)'s activities, based upon which a series of evaluations is conducted of the projects to which GCF-approved grants/loans were allocated since November 2015. The three economic theories are elaborated in the contexts of the GCF's funding decisions: a theory of public goods and expenditures, a theory of economic efficiency in land and resource reallocations, and a theory of a public adaptation to climate change. The evaluations of the GCF-funded projects are conducted with reference to the empirical results and predictions from the four scientific traditions surveyed in this book from Chaps. 3, 4, 5, and 6. The evaluations reveal many intriguing policy features of the GCF. First, a GCF grant is not awarded to the proposed project which would yield the largest amount of the global public good, for example, reduction of carbon dioxide equivalent emissions, from a pool of proposals considered. Second, when a public adaptation project is awarded with a funding, funding recipients or accredited entities are often a public agency which cannot provide the concerned public good efficiently or with the least cost. Third, evaluated from the perspective of the efficient resource reallocations, many GCF grants/loans may fail to induce efficient adaptations of the recipients and even encourage mal-adaptations.

Keywords Green Climate Fund · Investment criteria · Public expenditure · Economic efficiency · Public adaptation · Mal-adaptation

7.1 Introduction

The preceding four chapters have described the theories, models, and major predictions from the four scientific traditions in the climate change literature that are highly pertinent to as well as encompassing of the literature for an analysis and evaluations of the Green Climate Fund (GCF) and its funding decisions. We witnessed that the four traditions have produced the results and predictions that inform us, directly or indirectly, about potential impacts, adaptation and mitigation

S. N. Seo, *The Economics of Global Allocations of the Green Climate Fund*,
https://doi.org/10.1007/978-3-030-18274-8_7

possibilities, and behavioral responses. These results and predictions will turn out to provide a quantitative basis, essential as well as solid, for the evaluations of the GCF-funded projects performed in this chapter.

Coming back to the Green Climate Fund, the present author asks first whether the GCF's investment criteria, which is the organization's funding principles, can be justified and, if so, on what grounds (GCF 2018b, c). This chapter will examine and provide an evaluation of each of the 93+ GCF-approved projects to which a grant or loan has been already allocated since November 2015 (GCF 2018e, f).

Let's start with the basic facts about the GCF and its projects. The GCF was first announced in December 2009 at the final days of the Copenhagen Conference of the United Nations climate convention (UNFCCC 2009). At the 2010 Conference in Cancun, Mexico, a transitional committee for the GCF was created (UNFCCC 2010). At the 2011 Conference in Durban, South Africa, the governing instrument of the GCF prepared by the transitional committee was adopted (UNFCCC 2011). At the 2012 Conference in Doha, Qatar, Songdo city in South Korea was selected as the location of the GCF Secretariat (UNFCCC 2012). As of December 2018, the GCF Secretariat had 172 staff members who arrived from 54+ countries (GCF 2018a).

In 2014, the GCF began its first resource mobilization period that was scheduled to end by the end of 2018. It raised about US$ 10 billion in the first year before the Paris Conference in November 2015, with 80% of the total contribution coming from the five richest countries: the US, Japan, the UK, France, and Germany (See Fig. 2.1 in Chap. 2) (GCF 2018d). The initial fundraising push was made in concerted efforts with a view to making a ground-breaking international deal at the Paris Conference at the end of 2015, with zealous supports of both President Obama and UN Secretary General Ban Ki-moon with the former declaring the climate policies as 'the most consequential legacy of his presidency (NYT 2016).'

The first GCF funding allocations were made in November 2015 timed for the kick-off of the Paris Conference. A batch of eight projects selected 'equitably' from the geopolitical climate regions of the globe were approved by the GCF Board. Two projects were selected from Latin America and the Caribbean countries, one project from South Asia, two projects from the Alliance of Small Island States (AOSIS), and three projects from Africa. As of September 2018, 77 GCF projects were approved with total spending approved for these projects equaling to about US$ 12 billion. The most recent 21st GCF Board meeting on October 2018 approved additional 16 projects with about US$ 1 billion funding in total (GCF 2018f). The GCF funding was mostly in the form of a grant, with a small fraction of which was in the form of a loan to be repaid by recipients (GCF 2018e).

Besides the geopolitical regions of climate change, GCF project proposals were approved by the 'results areas' which are defined by the GCF Board. By the middle of 2018, the largest share of the GCF funding went to the results area of the 'energy generation and access,' with almost US$ 7 billion award. The results area of the 'buildings, cities, industries, and appliances' received US$ 1.8 billion. The results area of the 'health, food security, and water security' was also allocated US$ 1.8 billion. The results area of the 'transport' received US$ 0.6 billion. The results area of the 'livelihoods of people and community' was given US$ 0.68 billion. The results area of the 'forests and land use' was allocated US$ 0.4 billion. The results area of the 'infrastructure and built environment' received US$ 0.2 billion (Refer to Fig. 2.3 in Chap. 2).

With the background information of the institution and activities of the GCF, the central question of this chapter is how we can analyze and evaluate the GCF's various activities and funding decisions in a meaningful manner for climate policy processes. The present author will start with a presentation of an ensemble of economic theories that are particularly pertinent to the economic questions of public resource allocations and public expenditures. These theories would provide a theoretical support for the existence of the GCF and its wide-ranging activities, if they are justifiable. More importantly, the suite of theories may point us to analytical methods or indicators that can be relied upon for evaluating each of the GCF's funding decisions.

The first theory, to be presented shortly, is the theory of public goods and expenditure that can be traced far back to Paul Samuelson (Samuelson 1954, 1955). It provides a fundamental theory on why a governmental expenditure from the fund collected through public taxation is inevitable for a certain project and program, even in a well-functioning market economy (Coase 1960; Friedman 1962). Samuelson asserted that a governmental expenditure would be justified for a special type of goods which is jointly consumed by a large group of consumers. When the good in concern is a public good, there is no free market mechanism to provide this good efficiently and, therefore, the government should intervene and provide the good.

An important economic concept in the theory of public expenditure is economic efficiency, which is the second economic theory that will be explained in this chapter that may lend a rationale for the existence of the Green Climate Fund (Ricardo 1817; Pareto 1906). A Pareto efficient allocation of resources would not be achieved if the concerned resources are a public good. From the efficiency standpoint, it is interpreted that a governmental intervention and expenditure is an attempt by the public sector to restore efficient allocations of the free market economy.

Regarding the Green Climate Fund, questions of efficiency and efficient land uses have been raised (Seo 2015b, 2017b). At the present climate system, an economic agent makes optimal decisions to generate the maximum profit from the land and other resources that she manages (Ricardo 1817; von Thunen 1826; Seo and Mendelsohn 2008; Seo 2010a). Is the GCF a proper instrument for restoring the efficient resource allocations which will be disturbed by climatic shifts? Furthermore, it should be asked whether each GCF-funded project is an efficient public intervention, that is, could pass the test of efficiency such as a benefit-cost ratio.

In the efficiency evaluations of an individual GCF-funded project, the empirical models and predictions that were presented from Chaps. 3, 4, 5, and 6 will provide essential quantitative information, based upon which the efficiency evaluations can be conducted. For example, the empirical impact models of climate change on agriculture are inclined to predicting winners and losers from a climate change scenario (Rosenzweig and Parry 1994; Adams et al. 1999; Mendelsohn and Neumann 1999; Reilly et al. 2007; Seo 2016d). Relying on these predictions, we may be able to show that a GCF grant was awarded to the individuals who are predicted to benefit largely from a warmer climate regime, in which case a clear conclusion about the efficiency effect of the international funding can be drawn. Further, we may be able to judge whether a GCF grant may end up forcing the recipients to take inefficient actions or inducing other behaviorally perverse responses (Hart and Holmstrom 1987; Laffont and Martimort 2002).

The third theory that may provide a justification for the GCF and its activities is the theory of public adaptation (Seo 2015c, 2017b). A public adaptation is defined as an efficient adaptation to climate change that must be undertaken at the level of a community of individuals or a local government (Seo 2011). In providing a public adaptation project, a private adaptation by an individual agent is not sufficient, nor efficient (Mendelsohn 2000; Hanemann 2000). The GCF may exclusively pick the public adaptation projects for awarding a GCF grant.

The public adaptation can be either a community or group (of individuals) adaptation or a governmental adaptation. In the case of a governmental adaptation, the needed activity is carried out on a legal basis through a national political system. In the case of a community/group adaptation, the group in concern should voluntarily coordinate its members for the provision of the community good (Ostrom 1990).

In addition to the three economic theories, additional theories, concepts, or instruments which have the high potential to provide novel insights into the problem of global resource allocations under the GCF scheme will be explained in the next chapter. These will include a novel way for project financing called crowdfunding and an Internet-based global cryptocurrency such as the Bitcoin. The crowdfunding is a new way for financing a commercial or private project through an internet platform (Mollick 2014). The Bitcoin, the first of cryptocurrencies, is a new currency (coin) which is created purely from an internet communication protocol and, as such, is circulated literally globally without interventions from traditional banks and governments (Bohme et al. 2015).

The next chapter will also provide an analysis of the Green Climate Fund with regard to how it relates to other green funds that are internationally circulated. One example of such green funds is the European Union's carbon dioxide permit trading, formally, the Emission Trading Scheme (ETS) (Ellerman et al. 2016). The GCF is also one of the many United Nations specialized funds and programs, including the UN Development Programme (UNDP), UN Environment Programme (UNEP), and World Food Programe (WFP), and the next chapter will highlight distinct and overlapping features of the host of the UN funds with the GCF (UNSCEB 2018).

7.2 A Theory of Public Expenditure

One of the central questions raised in this book is how a global policy-body should allocate the dollars raised by the Green Climate Fund to a bundle of intentionally selected projects. As of the middle of 2018, the fund size stood at US$ 10 billion with the declared target size of US$ 100 billion per annum or even US$ 1 trillion per annum (GCF 2011; GEF 2017). The fund was established by voluntary contributions by the world's leading economies, with about 80% of the total contribution coming from the five richest nations (See Fig. 2.1 in Chap. 2).

An economic theory on how governmental revenue should be raised and then spent for a public project was elucidated long time ago by Paul Samuelson in his seminal paper entitled 'The Pure Theory of Public Expenditure' (Samuelson 1954, 1955). The theory was proposed in the context of a national public expenditure but has been extended, without loss of the central premises and conclusions, to a global public expenditure problem such as a global warming policy (Nordhaus 2006).

The transformational idea Samuelson put forth in the 1954 article is that there is a need for a taxation by the government and public (governmental) expenditure in the case of a special type of goods, initially called by him a jointly (collectively) consumed good, even in the smoothly functioning market economy. A year later, Samuelson started to use the term, a public good. A public good is defined in contrast to a private good as a good that is enjoyed (consumed) by the public once it is provided at all.

Samuelson assumed two types of goods in the economy (Samuelson 1954): 1) private consumption goods denoted by $(X_1, \ldots, X_n)$; 2) collective consumption goods denoted by $(X_{n+1}, \ldots, X_{n+m})$. For the private consumption good, the following condition holds in a competitive market with i being an individual consumer:

$$X_j = \sum_{i=1}^{s} X_j^i. \tag{7.1}$$

Samuelson wrote that the 'collective consumption good' as the good that '*all enjoy in common in the sense that each individual's consumption of such a good leads to no subtraction from any other individual's consumption of that good, so that simultaneously for each and every individual and each collective consumption good,*' the following holds (Samuelson 1954):

$$X_{n+j} = X_{n+j}^i. \tag{7.2}$$

The theory clarified that a public good cannot be provided at the efficient level by the market forces alone because of the publicness of the good. Once the good is provided by someone, it will be available to all citizens for free, that is, without extra cost. The good is non-rivalrous or non-depletable. Therefore, there would be no incentive for a single economic agent to supply the good to the market because she cannot exclude anyone from consuming the good. The good is non-excludable (Mas-Colell et al. 1995). From the perspective of the users of the public good, there is an incentive to free ride, that is, enjoy the public good without paying the cost once someone else provides the good (Buchanan 1965).

In the Samuelson's formulation, the optimality conditions for the private consumption goods and the collective consumption goods in the economy were written, as written below, using the consumer's marginal rate of substitution (left-hand side) and the marginal rate of transformation in the production (right-hand side). With U a social welfare function, U_i a marginal social welfare, and F a production-possibility schedule with $F(X_1, \ldots, X_{n+m}) = 0$ (Samuelson 1954):

$$\frac{u_j^i}{u_r^i} = \frac{F_j}{F_r}; \tag{7.3}$$

$$\sum_{i=1}^{s} \frac{u_{n+j}^i}{u_r^i} = \frac{F_{n+j}}{F_r}; \tag{7.4}$$

$$\frac{U_i u_k^i}{U_q u_k^q} = 1. \tag{7.5}$$

Researchers followed the Samuelson's lead and showed that the problems of non-rivalrousness, non-excludability, and free-riding can be mitigated in certain types of public goods. In a congested public good, some users can be excluded from the enjoyment of the good through congestion or crowding out. In a club good, some potential members can be excluded from the consumption of the good through a requirement of club membership (Buchanan 1965; Brown 1973).

The 'pure' theory provided a theoretical foundation for a wide range of public expenditures by the national government. Governmental interventions and expenditures at varied levels of the government such as national defense, police works, public works, public education, some types of pollution controls can be justified reasonably by the pure theory of public expenditures (Samuelson and Nordhaus 2009).

The fund (revenue) needed for a variety of public intervention programs should be generated through public taxation. A federal taxation is used for a federal public good such as the national defense. A state taxation is used for a state public good such as police works. A municipal taxation is used for a municipal public good such as public works.

Once it is recognized that there is the need for a governmental intervention for a public good project, the thorniest question boils down to how much of the public good in concern should be provided. The answer depends on what the socially optimal amount of the concerned public good is. The answer to the question determines the amount of tax revenue needed to be raised from the citizens.

The needed tax revenue should be collected from the citizens according to a rule or a set of rules. For example, the total tax burden may be divided equally across the citizens or may be assigned in accordance to a certain principle. To give an example, a polluter-pay principle would state that the amount of damage caused by a polluting firm should be set as the company's tax burden in a pollution control policy (Baumol and Oates 1988). In the public education, a progressive taxation rule may be relied upon for generating tax revenue needed for financing public schools.

The taxation principles are not set in stones in the economics literature. There are more than one principle or rule that are justified and used historically by the governments even in the United States (Mankiw 2014). For a global public good policy, such a taxation principle is not found and the above-mentioned principles applied to a national public good policy cannot be extended to the global policy issue.

Is there a principle or a rule with regard to the contributions of member countries to the GCF on who should pay and how much should each party pay? There is no rigorously defined principle on this issue in the GCF operations, but the climate change negotiations led by the United Nations convention rely loosely on the principle of 'a common but differentiated responsibility and respective capabilities' among the parties of the convention (UNFCCC 1992). This principle of burden sharing can be traced back to the very first United Nations conference on human environmental issues held in Stockholm, Sweden in 1972 (UN 1972).

The government is obligated to spend the tax revenue, the total of it, to provide the good to the public for which the tax revenue was raised. In the case of national defense, it means that the government should allocate the tax revenue into various activities such as developing and purchasing weapons, building facilities, employing and training soldiers, and other activities, for the purpose of providing national security.

What is the socially optimal level of the public good, e.g., national defense or environmental protection? The socially optimal level of the public good is determined by the social marginal benefit function and the social marginal cost function of the national defense (Samuelson 1955). The former is derived from the social benefit function and the latter is derived from the social cost function of providing national defense. At the point where the social marginal benefit and the social marginal cost are equated, the social welfare is maximized (Baumol and Oates 1988).

The social marginal cost of the national defense system is determined by engineering constraints as well as market factors. The engineering constraints determine which weapons, facilities, soldiers, and researches should be employed for national defense, whose prices are determined in the market.

It is harder to measure the social marginal benefit function of a public good and, for a certain class of a public good, for example, an environmental amenity or clean air, it is much harder (Mendelsohn and Olmstead 2009). At a fixed level of the public good that is concerned, individuals' marginal benefits for the good should be vertically summed up to produce the social marginal benefit at that level (as in Eq. 7.4 above), which should be repeated for the entire range of the public good. However, because an individual's marginal benefit function for the public good is not directly observable in the market, it must be derived, through statistical and experimental methods, from associated market behaviors of individuals which reveal consumers' preferences for the public good embedded in the marketed goods as an attribute (Freeman et al. 2014).

Economists have developed a family of revealed preference methods in order to derive an individual's willingness to pay (WTP) for the public good, including an averting behavior method, a hedonic price method, and a travel cost method. An averting behavior method estimates an individual's willingness to pay for a public good by examining individuals' behaviors that are undertaken to avoid a certain event. For example, an individual may purchase a bottled water in order to avoid polluted tap water.

Another is the hedonic method which singles out an individual's willingness to pay for one attribute out of the many attributes of the good purchased by the individual (Ridker and Henning 1967; Thaler and Rosen 1976). For example, an individual's purchase of a residential property at a certain price is determined by many characteristics of the property including structural characteristics (that is, the number of rooms, floors, etc.), safety of the neighborhood, public schools, pollution levels, visibility, among other things. With an application of a host of econometric techniques, the individual's willingness to pay for the specific characteristic of the property, say, a lower pollution level, is derived.

When any of the revealed preference methods cannot be applied to extract an individual's WTP for a certain public good, researchers may alternatively proceed to ask the individual directly to state her willingness to pay for a particular service. Such a situation may arise in the estimation of existence value of an endangered species such as bald eagle or monarch butterfly. A contingent valuation (CV) method is one of such valuation methods referred to as the stated preference method (Hanemann 1994). The CV method is, however, vulnerable to biases such as ignorance, framing, and strategic bias (Diamond and Hausman 1994).

7.3 A Theory of Ricardo-Pareto Efficiency in Resource Uses

The determination of the size of public expenditure is inextricably associated with an individual's motivation to allocate her resources optimally or efficiently. An optimal decision for an economic agent is one that maximizes the profit she can earn taking into account numerous constraints that she is faced with. For the society's decision, an optimal resource allocation by the society can be understood in the same way.

The concept of optimality was expounded most clearly by Italian economist Vilfred Pareto in the national economy context or, more precisely, macro economy (Pareto 1906). The efficiency in the context of an individual's resource allocation decision can be traced back to the David Ricardo's marginal rent theory as well as the von Thunen's spatial land use theory (Ricardo 1817; von Thunen 1826).

The theory of optimality or efficiency is one of the most critical concepts in economics and policy evaluations (Baumol and Oates 1988; Hahn and Dudley 2007; Muller and Mendelsohn 2009). As such, it will certainly be proven to be a foundational tool for the evaluations of the Green Climate Fund's allocation decisions.

Let's imagine an individual resource manager with 10 acres of land in Des Moines, Iowa, a major agricultural region in the United States. She can invest the land for a single crop or other activities in order to earn the highest income from the land (Ricardo 1817). She may choose a single crop from a large pool of available crops such as corn, wheat, soybeans, canola, and alfalfa. Or, she may choose to raise cattle, sheep, turkeys, or goats on the land. Or, she may opt for tree crops such as citrus, date, persimmons, mushrooms, *etc* (Seo 2016b, d).

More formally, let's say that there are J possible portfolios that she can choose from. The manager's decision problem can be written as the following maximization problem:

$$Max_j\{\pi_1, \pi_2, \ldots, \pi_J\}. \tag{7.6}$$

The set of J portfolios can be defined to include both a portfolio composed of a single agricultural product and a portfolio composed of multiple agricultural products. Similarly, the set can include both a portfolio composed of a single system of agriculture and a portfolio composed of multiple systems of agriculture. In other words, both a specialized portfolio and a diversified portfolio can be accounted for by the set in Eq. 7.6.

The π_j is the maximum profit earned by the manager from choosing and managing portfolio j in the long-term. The profit must be defined to include both all components of revenues and all components of costs from the selected portfolio. For instance, it can be defined to include both the revenue from market sales of farm-produced products and the value of household consumptions of self-produced products. Similarly, on the cost side, the cost of family labor hours should be counted, in addition to the cost of hired labor hours (Seo and Mendelsohn 2008).

An important scientific advancement made through the climate change economics research, sadly often overlooked though, is that the natural resource manager's profit is dependent on, among other conventional factors of production (X_j), a climate regime (C_j) faced by the resource manager (i). That is,

$$\pi_{ij} = f\left(C_i | X_j\right). \tag{7.7}$$

Owing to the influence of the present climate regime on the profits of the available agricultural portfolios, a shift of the climate regime from one regime to another in the future date would cause a change in the maximum profit that can be earned on the land from each of these portfolios. For each portfolio defined as above, let the profit that can be earned from the portfolio under the new climate regime be written as $\tilde{\pi}_j$.

Let's assume that the j_{th} portfolio is the profit maximizing choice under the current climate regime (C) and the k_{th} portfolio is the profit maximizing choice under the new climate regime ($\tilde{C}$):

$$\begin{aligned} \pi_j(C_i) &= Max\{\pi_1, \pi_{2,\ldots,} \pi_J\} \\ \tilde{\pi}_k\left(\widehat{C_i}\right) &= Max\{\tilde{\pi}_1, \tilde{\pi}_{2,\ldots,} \tilde{\pi}_J\}. \end{aligned} \tag{7.8}$$

An adaptation behavior (AD) by the manager to the change in the climate regime can be defined as the switch in the farm portfolio from one to another caused by the climatic change, with the double vertical bars implying the change from the post-bar state to the pre-bar state:

$$AD_i\left(\tilde{C} \| C\right) = k \| j. \tag{7.9}$$

This is a Pareto optimal or economically efficient adaptation to the external change (Mendelsohn 2000; Hanemann 2000). The impact of the climatic change is then measured as the change in the maximum profits in the two climate states (Seo 2010a, 2016a):

$$\Delta\pi = \tilde{\pi}_k - \pi_j. \tag{7.10}$$

If the differential in Eq. 7.10 is positive, then the climatic change is beneficial to the manager. If it is negative, the climatic change is harmful to the manager. The adaptation behavior described in Eq. 7.9 and the economic impact described in Eq. 7.10 would not be meaningful to another manager. That is, each manager has to make her own adaptation decision and face its own economic consequences.

The impact of the climatic shift to the individual will be quite different if she would not take an efficient adaptation and instead stick to the current portfolio, which can be measured as follows (Seo 2010a, 2016a):

$$\overline{\Delta}\pi = \tilde{\pi}_j - \pi_j. \tag{7.11}$$

Another scenario for the individual's response to the climatic shift is that she is forced to or voluntarily adopt an inefficient portfolio. This is a scenario of mal-adaptation (Seo 2014b, 2015b). The impact can be measured as follows, with m being the inefficient portfolio:

$$\overline{\overline{\Delta}}\pi = \tilde{\pi}_m - \pi_j. \tag{7.12}$$

The economic theory dictates the following relationship, which states that the damage from the climatic shift will be smaller when an efficient adaptation is undertaken than when either no-adaptation or a mal-adaptation is undertaken:

$$|\Delta\pi| < Min\left\{\left|\overline{\Delta}\pi\right|, \left|\overline{\overline{\Delta}}\pi\right|\right\}. \tag{7.13}$$

The Eqs. 7.9, 7.10, 7.11, 7.12, and 7.13 are critical numerical quantities for the series of evaluations of the Green Climate Fund and its funding allocation decisions, the chief subject of this book. What are their critical implications? Let me clarify some of them here and discuss more thoroughly later in the empirical results sections. First of all, the equations tell that policy-makers should be aware that even if a particular crop or a natural resource product were to be harmed greatly by a forecasted climate change, it does not necessarily mean that the farm with the crop portfolio will be harmed severely. The farm may be able to adapt to a new crop portfolio and, as a result, even gain economically from the new climate regime, in which case giving a grant to the farm cannot be justified solidly (Seo 2014a, b).

Second, the series of equations imply that a GCF grant given today to a particular crop which is presently dominant in a particular region may turn out to hinder a farmer's efficient switch to a different crop which is more profitable in the new climate regime. The GCF grant may give the crop manager a perverse incentive to stick to the currently-dominant crop with the intention of receiving the GCF subsidy (grant), despite a significant predicted decrease in the crop's profitability owing to the climatic shift (Seo 2015b).

This is one of the most policy-important findings on behaviorally efficient adaptations to climate change. Any grant, foreign aid, or subsidy program in agriculture at the present time is predominantly given to the most important crop in the recipient farm and the region. For example, the U.S. crop insurance and subsidy program is dominated by four major crops in its subsidy (premium support) allocations: Corn, wheat, cotton, and soybeans (Sumner and Zulauf 2012). International development aids to Sub-Saharan agriculture have been concentrated on the major staple crops such as maize (Byerlee and Eicher 1997; Seo 2010b; Kafle et al. 2016). This informal rule of aid distributions is politically convenient, but may fail to encourage efficient behavioral adjustments by the concerned parties when adopted by the GCF Board.

Third, a behavioral adaptation in Eq. 7.9 is presented to take the form of a portfolio switch, i.e., not a product switch. A portfolio selection incorporates, in addition to a crop switch, a selection of a farming system. For example, a rural farm can switch its farming system from a crops-only system to a mixed crops-livestock system of agriculture (Seo 2010a, b). A farmer will have many farming system options to choose from, including a crops-only, a livestock-only, a forests-only, a mixed crops-livestock, a mixed crops-forests, a mixed livestock-forests, and a fully integrated crops-livestock-forests system (Seo 2010c, 2012a).

In addition, a portfolio selection decision incorporates changes in numerous farming practices that pertain to each system of agriculture. More straightforwardly, a farmer would adopt or not adopt a particular farming practice, given the portfolio of her choice, to maximize the return from managing it. What are some of the notable farming practices? A crop farming with an inter-cropping technique is one example of farm practices, which is a particular method of crop agriculture. An irrigated crop farming is another example, which can take the form of rainfed irrigation, surface water irrigation, ground water irrigation, gravity irrigation, drip irrigation, or other precision irrigation techniques (Dinar et al. 1992; Schlenker et al. 2005; Kurukulasuriya et al. 2011). Another example of the cropping techniques is a soil-carbon conservation agriculture such as no-tillage agriculture (Smith et al. 2008). A fourth example is a methane emission reduction technique such as feed additives and dietary changes for animal husbandry (Hristov et al. 2015; CSIRO 2018).

A Pareto efficiency is defined formally as the state of allocation of resources in the society from which one's welfare cannot be improved without having another's welfare decreased (Mas-Colell et al. 1995). Stated differently, at the state of Pareto efficiency, the society cannot improve the welfare of someone without hurting the welfare of another. This concept holds well in the farmer's adaptation decisions in response to a climatic change which are economically efficient (Ricardo 1817; von Thunen 1826). Efficient adaptations of individuals, as described in the above model, will guarantee the Pareto efficiency under the new climate regime. However, this would not be the case in the exceptional situation that is to be explained in the next section: a public adaptation.

7.4 A Theory of Public Adaptation

Of the whole range of adaptation measures and strategies that the society must implement in order for the society to adapt efficiently to the new climate system, some measures and strategies may not be taken by an individual manager. A situation can arise in which it is less expensive to take the adaptation action together as a group than as an individual.

In such a situation, an adaptation measure of this type is a public good in the Samuelsonian sense (Samuelson 1954). A supply of the adaptation measure of this kind leads to a joint consumption of the good by multiple or a large number of people. An individual's consumption of the public adaptation good does not diminish the amount of consumption available to the others in the group. Further, it is very costly, if not impossible, to exclude an individual agent in the group from the consumption of the public good.

The publicness of the adaptation measure of this kind, however, can be manifested at varied geo-political scales: at the level of a village community, municipality, state, or country. At one of these levels, a public adaptation measure can be efficiently provided. That is, a village-level adaptation measure cannot be efficiently provided by a national government. A national-level public adaptation measure cannot be provided efficiently by a municipality government.

Is there really a public adaptation measure that cannot be provided efficiently in the market, that is, by actions of individuals? A purely public adaptation at a national level may not exist, but a public adaptation measure at a lower level may exist, e.g., at a village or district level, and has been reported. Several examples introduced presently would clarify this point.

Building a coastal wall for protection against a continued sea level rise owing to a globally warming world is often cited as a public adaptation (Yohe and Schlesinger 1998; Church et al. 2013). It would be meaningless for an individual coastal land owner to build a wall on her land that adjoins the sea. A private wall that surrounds his house, which is a private adaptation, would not be effective in stopping inundation of a rising sea. For the coastal wall to be effective, it should be built along the entire coastline, that is, without the weakest point where sea water can flood in. The coastal wall therefore should be provided by a coastal municipality, that is, a city or a town, or a community group who would coordinate the effort.

Another reported example of a public adaptation measure is a certain type of agricultural irrigation (Seo 2011). Most farmers irrigate their farm lands privately or rely on rainfed agriculture. A private irrigation can be costly, for example, groundwater irrigation by pumping groundwater from undergrounds, building a precision irrigation system such as a sprinkler system (Schoengold and Zilberman 2007). Other private irrigation practices are less costly, to mention a few, making irrigation channels around the farmland plots, building water storage on the farmland plots.

A government-provided irrigation is also very common in major agricultural regions where severe drought events are recurring disasters, such as California and Arizona (Schlenker et al. 2005). A public irrigation scheme includes a dam, a reservoir, an aqua duct built by the public agency. It also includes a water allocation market at which irrigation water is distributed to individual farmlands.

In a drought-prone region, it would be more efficient for the region and individual farmers to establish the public irrigation scheme than to irrigate their lands privately (Seo 2011). In a moderately drought year, farmers may be able to rely on a host of private irrigation measures to supply water to their farmlands. But, in a severe drought year, farmers can resort to the public irrigation scheme to acquire water for their farms. This means that the private irrigation and the public irrigation could coexist, both schemes complementing each other (Schoengold and Zilberman 2007).

The concept of a public adaptation is important and inevitable in this book because the concept necessitates the need to raise a public money by the public authority and redistribute it to the public adaptation project, the primary transaction that this book has taken to elucidate. Stated another way, the Green Climate Fund and its numerous funding projects may be justified on the grounds of the ineluctable need for a variety of public adaptation projects.

Notwithstanding, there are two conflicting issues in justifying the GCF scheme on the grounds of the concept of public adaptation, although the above-described theory of public adaptation can rightly be a basis for a climate change adaptation fund which is coordinated and implemented by a public agency.

The first conflict is a right scale of a policy intervention. As is evident in the examples of the coastal wall and the agricultural irrigation, a public adaptation effort is called for at a local level, that is, at a city level or a village level. It is very rare that a public adaptation measure is needed at a national level or at a multi-national level.

Therefore, a global revenue generation for the establishment of the GCF and a global redistribution of the fund to numerous local public projects cannot be justified either on the efficiency grounds or on the grounds of the theory of public adaptation. That is, a village-level public adaptation project should resort to the fund generated at the village level and redistributed at the village level, if it is intended to be an efficient intervention. A matching of a public adaptation project with a public agency at the appropriate level would ensure efficiency in the reallocations of the fund.

At present, coastal protection is engineered and of high priority to the low-lying cities and countries of the world. Take the Netherlands for example where 27% of the territory is below the mean sea level. Local coastal governments are in charge of protecting the coastal lines, more specifically, local water boards which have been historically formed (Raadschelders 2005). Similarly, a public irrigation scheme for agriculture is most often managed at a municipality level, for example, a county, unless irrigation water should be drawn from a neighboring county.

The second conflicting issue is a possible collision between a private adaptation and a public adaptation. A public adaptation project can collide with a private adaptation project. In the agricultural irrigation example, farmers can and do irrigate their lands privately. A private adaptation measure can be taken efficiently by an individual farmer, considering the complexity of conditions of her farmland, since she has the full knowledge of the land.

A large portion of the farmers in a certain agricultural region may not need a public irrigation scheme at all if they are located in the areas that have easy access to surface water as well as sufficient rainfall. In this situation, it is not an easy decision to determine whether a public agency should intervene to provide an irrigation scheme or not, considering the complexities in incentives of the involved parties (Hart and Holmstrom 1987).

From another angle, it is the case in many farming areas that farmers can irrigate their farmlands privately without the public support most of the time, but a public irrigation scheme can be helpful to them in a severe drought event which occurs infrequently, say, once per a couple of decades. At what level of frequency of the drought events should the public agency intervene to provide a public irrigation scheme in this situation?

Pursuing this line of reasoning further, we can arrive at a situation in which a public irrigation scheme turns out to induce the problem of adverse selection similar to that in the insurance markets (Laffont and Martimort 2002). More concretely, a publicly-supported reservoir built for agricultural irrigation would force farmers in the region to adopt a certain type of agriculture which would not be adopted without the public reservoir system due to frequent profit losses. The public project induces adverse selection.

Further, even if a public irrigation scheme is justified on some grounds, a public irrigation scheme can be over-supplied or under-supplied since the public agency does not have the full knowledge and flexibility with regards to the complicated conditions and factors of the farmlands. For example, the number of public water tanks or the size of the dam/reservoir provided by the public sector may turn out to be too large or too small against the socially optimal level.

A recent example would make this point clear. A technologically novel way to be prepared against a severe drought event that may last several years is a public construction of a desalination plant, i.e., a plant that removes salinity from sea water to make it suitable for uses in agricultural fields. From an engineering perspective, it is a highly promising technology that can be utilized in the areas where a severe drought tends to occur regularly and last multiple years.

There are several successfully operating desalination plants in Israel and California. The California desalination plant was built as a response to a severe drought in the State during the 1990s. Because of the technological difficulties, the cost of a desalination plant is very high. On top of this, the plant became little utilized since its construction because the severe drought season was followed by two decades of high rainfall years and only minor drought years (Carlsbad 2016; USDA 2016).

7.5 Evaluation: A Public Adaptation Theory Test

Having explained the three economic theories that could provide a foundation for the establishment and activities of the GCF, the present author will henceforth conduct a set of evaluations of the GCF-funded projects. The evaluations are conducted, on top of the theoretical foundations described in this chapter, by referring to the results and predictions from the empirical models presented in the preceding chapters.

Table 7.1 Summary evaluation statistics for the GCF projects on forests and land use

Recipient countries	Project title	Accredited entity	Expected contribution (tons CO2 avoided)	Funding size (US $)	Duration (years)
Ecuador	Priming Financial and Land-Use Planning Instruments to Reduce Emissions from Deforestation	United Nations Development Programme (UNDP)	15 m tons	$84 m	5
Morocco	Development of Argan orchards in Degraded Environment – DARED	Agency for Agricultural Development of Morocco (ADA)	604.2 k tons	$49.2 m	5
Guatemala, Mexico	Low-Emission Climate Resilient Agriculture Risk Sharing Facility for MSMEs (Micro, Small, Medium-sized Enterprises)	Inter-American Development Bank (IDB)	9.2 m tons	$158.0 m	15
Cambodia	Climate-Friendly Agribusiness Value Chains Sector Project	Asian Development Bank (ADB)	240.0 k tons	$141.4 m	6
Peru	Building the Resilience of Wetlands in the Province of Datem del Marañón, Peru	Peruvian Trust Fund for National Parks and Protected Areas (Profonanpe)	2.6 m tons	$9.1 m	5

Note: *k* thousand, *m* million

We begin with the test of the public adaptation theory. In Table 7.1, five projects to which the GCF grants were allocated are summarized with pertinent quantitative data made available in the GCF project documents such as the expected tons of CO2-equivalent emissions removed, funding size, and a project duration. The projects in Table 7.1 are classified by the GCF as the results area of the forests and land use.

We can assess the projects in Table 7.1 from the viewpoint of who received the grants. The funding recipients, called accredited entities or executing entities by the GCF, are mostly international policy organizations. The three organizations in Table 7.1 are such international organizations as the United Nations Development Programme (UNDP), the Inter-American Development Bank (IDB), and the Asian Development Bank (ADB). The two others are national policy organizations: The Agency for Agricultural Development of Morocco (ADA), the Peruvian Trust Fund for National Parks and Protected Areas (Profonanpe).

Let's take a closer look at the GCF project in Guatemala and Mexico: The Low-Emission Climate Resilient Agriculture Risk Sharing Facility for MSMEs (Micro, Small, Medium-sized Enterprises). The recipient of the funding—the accredited entity—is the IDB. The risk sharing facility in Guatemala and Mexico should be operational at a local or a village level, for which reason it can be said that the IDB may not be an appropriate organization to provide a public adaptation of the low-emissions and climate resilient agriculture's risk sharing facility for the MSMEs.

A low-emissions agriculture as well as a climate resilient agriculture must be tailored for a local situation. Many types of the risk sharing 'facility' are also locally designed and operational, to name but a few, a community-wide irrigation scheme, a village-level diversification of portfolios, a village-level risk pooling, and locally-tailored extension services (Seo 2016a, c, d).

Even in an economy-wide risk sharing mechanism such as banking and other credit services that would target vulnerable populations in developing countries to future climatic changes, local knowledge and expertise are inevitable for designing a locally appropriate service or product (Conning and Udry 2007). A large number of agricultural research and extension centers, locally-based agricultural cooperatives, product-specific agricultural credit services that we observe today in Mexico and Guatemala are some examples of such services and products. The IDB is not an 'efficient' institution to take on the proposed risk sharing activities.

The same line of reasoning can be extended to the "Climate-Friendly Agribusiness Value Chains Sector Project" submitted by Cambodia, whose recipient of the funding was the Asian Development Bank (ADB), a comparable organization in Asia to the IDB in Latin America.

An even clearer example is the project entitled "Development of Argan orchards in Degraded Environment" in Morocco. The recipient of the funding was the Agency for Agricultural Development of Morocco (ADA). At first, it is not apparent whether the development of Argan orchards in degraded environment can be classified as a public good. Even if it is, it would not be a national-scale public good for which a national level intervention through the ADA of Morocco must be made (Parry et al. 2004; Butt et al. 2005; Seo et al. 2009).

A similar evaluation on the basis of the public adaptation theory of the GCF grants can be performed on the funded projects in other results areas than the forests and land use area shown in the table. The projects in Table 7.1 are singled out by the present author for illustrative purposes, so are not the exclusive list of projects to which a public adaptation test can be applied. Readers will, I believe, be able to conduct a similar examination of the GCF projects not included in Table 7.1.

7.6 Evaluation: A Samuelson Public Expenditure Test

The theory of public goods and expenditure can also be applied to the evaluation of the GCF funding decisions. In Table 7.2, the present author tabulates the pertinent quantitative data of the GCF-funded projects which are classified by the GCF Board as the results area of the energy generation and access. The key statistics for our evaluations in this section are the expected contribution of a project in terms of tons of CO2-equivalent emissions avoided and the grant size that the project received.

Table 7.2 Summary evaluation statistics for the GCF projects on energy generation and access

Recipient countries	Project title	Expected contribution (tons CO2 avoided)	Funding size (US $)	Duration (years)
Rwanda, Kenya	KawiSafi Ventures Fund in East Africa	1.5 m tons	$110.0 m	5
Chile	Climate Action and Solar Energy Development Programme in the Tarapacá Region in Chile	3.7 m tons	$265.0 m	20
Dominica, etc.	Sustainable Energy Facility for the Eastern Caribbean	9.4 m tons	$190.5 m	8
Armenia, etc.	GCF-EBRD Sustainable Energy Financing Facilities	27.5 m tons	$1.4b	15
Madagascar	Sustainable Landscapes in Eastern Madagascar	10.0 m tons	$69.8 m	10
Benin, etc.	Universal Green Energy Access Programme	50.6 m tons	$301.6 m	15
Mongolia	Business loan programme for GHG emissions reduction	1.2 m tons	$60.0 m	5
South Africa	SCF Capital Solutions	33.0 m tons	$34.1 m	10
Argentina	Catalyzing private investment in sustainable energy in Argentina – Part 1	15.3 m tons	$653.0 m	4
Mauritius	Accelerating the Transformational Shift to a Low-Carbon Economy in the Republic of Mauritius	4.3 m tons	$191.4 m	6
Cook Islands, etc.	Pacific Islands Renewable Energy Investment Program	3.0 m tons	$26.0 m	7
Bahamas, etc.	Geeref Next	769.0 m tons	$765.0 m	5
Egypt	Egypt Renewable Energy Financing Framework	18.9 m tons	$1.0b	5
Tajikistan	Tajikistan: Scaling Up Hydropower Sector Climate Resilience	7.5 m tons	$133.0 m	5
Solomon Islands	Tina River Hydropower Development Project	2.5 m tons	$234.0 m	5
Mongolia	Renewable Energy Program #1 – Solar	306.7 k tons	$17.6 m	10
Kazakhstan	GCF-EBRD Kazakhstan Renewables Framework	12.9 m tons	$557.0 m	5
Bhutan	Bhutan for Life	35.1 m tons	$118.3 m	14
Barbados	Water Sector Resilience Nexus for Sustainability in Barbados (WSRN S-Barbados)	220.2 k tons	$45.2 m	5.25
Paraguay	Poverty, Reforestation, Energy and Climate Change Project (PROEZA)	7.9 m tons	$90.3 m	5
Argentina	Promoting risk mitigation instruments and finance for renewable energy and energy efficiency investments	9.1 m tons	$163.9 m	5

(continued)

Table 7.2 (continued)

Recipient countries	Project title	Expected contribution (tons CO2 avoided)	Funding size (US $)	Duration (years)
Bangladesh	Global Clean Cooking Program – Bangladesh	2.9 m tons	$82.2 m	3.5
Rwanda	Strengthening climate resilience of rural communities in Northern Rwanda (SCRNRP)	273.7 k beneficiaries	$33.2 m	6
Zambia	Zambia Renewable Energy Financing Framework	4.0 m tons	$154.0 m	5
India	Line of Credit for Solar rooftop segment for Commercial, Industrial and Residential Housing sectors	5.2 m tons	$250.0 m	5

Note: *k* thousand, *m* million, *b* billion

As can be verified from the project titles in the table, the energy generation and access projects are directed toward a variety of energy programs that pertains to carbon dioxide emissions or renewable energy sources. Readers will easily come across in Table 7.2 a popular energy nomenclature such as solar energy, hydropower, green energy, renewable energy, sustainable energy, low-carbon energy, and clean cooking.

The funding size is most often in the range of several hundred million US dollars, with the range from US$ 26 million allocated to the Pacific Islands to US$ 1.4 billion allocated to multiple countries of Armenia, Egypt, Georgia, Jordan, Republic of Moldova, Mongolia, Morocco, Serbia, Tajikistan, and Tunisia.

A justification of a governmental taxation and public expenditure for the purpose of a climate change policy comes only from the predicted reduction of damage from a potential climatic change on the human society and ecosystems. In other words, a taxation and public expenditure must be justified on the grounds of the avoided damage that can be expected from implementing the GCF programs, that is, from spending the GCF grants and loans (Nordhaus 1994; Mendelsohn and Neumann 1999; Hahn and Dudley 2007).

Nonetheless, as can be seen from Table 7.2, there is no quantitative information on climate change-induced damage, that is, the reduced climate change-induced damage by implementing a GCF project. The further details of the projects shown in Table 7.2 can also be verified from the GCF projects' documentations made available on the organization's website (GCF 2018e). Even in the detailed documentations, there is no assessment expressed in terms of avoided damage in US dollars in comparison against the cost of the proposed program.

Neither is there an assessment of avoided damage which is quantified in terms of the degree of global temperature increase avoided or the degree of global warming mitigated by the implemented program, which is an alternative physically-based indictor of climate change-induced damage widely appropriated by the parties, both scientists and negotiators. If you are not clear on this point, refer to the United Nations emissions gap report in which the total commitment of the nations' emissions reductions is gauged against the 2-degree Celsius target (UNEP 2017). The spirit of the UNEP emissions gap exercise can be extended to each GCF project.

A proponent of the GCF may argue that Table 7.2 does indeed present the quantitative information in terms of avoided CO2-equivalent emissions. However, it should be noted that the carbon dioxide is itself not harmful to either humanity or ecosystems. What is harmful is the consequences of increased carbon dioxide, say, a hotter and/or more unpredictable climate system beyond a certain threshold which causes ecosystem and anthropogenic damages (Mendelsohn and Neumann 1999; Tol 2009; Seo 2017b). An increase in carbon dioxide in the atmosphere, on the contrary, helps plant and ecosystem growth by enhancing photosynthesis processes (Ainsworth and Long 2005; Reich et al. 2018).

A more critical flaw of the GCF funding decisions from the viewpoint of the theory of public expenditure, which is prominently featured in Table 7.2, is that the GCF grants and loans on the results area of the energy generation and access are not directed to the recipient countries where such grants could achieve the largest impact, that is, the largest emissions reductions. Put differently, it is evident in the table that the GCF grants are directed to under-developed poor countries in Africa, Latin America, South Asia, and Pacific Islands. The recipient countries in these regions are developing at a slow growth rate, therefore, the increase in the CO2-equivalent emissions is not very large in comparison with those from fast-growing or large economies (GCP 2017).

On the other hand, the fast-growing economies such as China, India, and the US are excluded from the GCF energy-generation-access project grants. As shown in Table 7.2, there is only one project in India which received a grant, whose proposal is, however, limited to rooftop solar installations. Nonetheless, the CO2 emissions growth at the global level is highly concentrated in the three countries and is expected to be so throughout the twenty-first century (GCP 2017). This means that the GCF funding programs will do little to reduce the growth rate of carbon dioxide emissions during this century at a global level, despite the trillions of dollars to be disbursed by the GCF Board in this century.

In Fig. 7.1, percentage allocations of the GCF's energy-generation-access area funding in Table 7.2 to China, India, and the US are drawn next to percentage emissions of the total global CO2 emissions of the three countries. The emission estimates are carbon dioxide equivalent emissions in 2016 from fossil fuels and cement production compiled by the Carbon Dioxide Information and Analysis Center (CDIAC) in the United States (Boden et al. 2017).

The China's emissions amounts to 28% of the total global emissions, but the GCF funding was not awarded. The US emissions amounts to 15% of the total global emissions, but there is no GCF funding directed to the country. The Indian emissions accounts for 7% of the global emissions of CO2 while the percentage of the total GCF funding was about 3.5% of the total funding directed to the results area of the energy generation and access. In other words, 97% of the GCF's energy area funding was awarded, nearly excluding the 50% of the global CO2 emissions from fossil fuels and cement production.

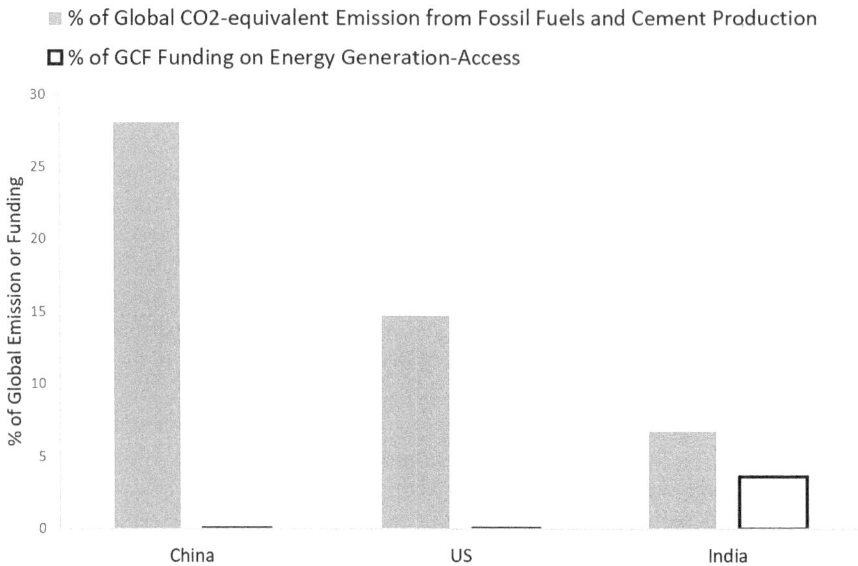

Fig. 7.1 Percentage allocations of the GCF energy funding *vis-a-vis* percentage global emissions

An astute reader may find that the analysis in Fig. 7.1 is another way to express the emissions gap calculated by the United Nations Environment Programme as well as the 1.5 °C report by the Intergovernmental Panel on Climate change, both widely hailed by the global media (UNEP 2017; IPCC 2018). The difference between the GCF funding percentage and the emissions percentage for each country shown in Fig. 7.1 is expected to become even larger in the coming decades.

A similar analysis can be conducted on the GCF-funded projects on the results area of the buildings, cities, industries, and appliances. As summarized in Table 7.3, the projects in this policy area are assorted programs that propose to construct or retrofit old ones into energy-saving or energy-efficient buildings, cities, and industries. Examples of the GCF-funded building projects include the energy efficient building retrofits in Armenia, Brazilian energy efficient cities, Ulaanbaatar green affordable housing and resilient urban renewal project in Mongolia, a scaling-up project of energy efficiency for industrial enterprises in Vietnam, and a sustainable and climate resilient connectivity project for an island of Nauru in Micronesia.

For the buildings and homes, an improvement in energy efficiency takes many forms depending on how energy is consumed there: a change in lighting methods from incandescent lightbulbs to Compact Fluorescent Lamps (CFL) or Light Emitting Diodes (LED), good insulation, heat-reflecting roof tiles, rooftop solar panels, energy efficient air conditioners and space heaters, energy-efficient electrical machines, and trees for natural cooling (Akasaki et al. 2014; Graziano and Gillingham 2015).

Again, the GCF grants in the results area of the buildings and cities are directed to under-developed countries and leave out fast-growing or large economies. The GCF-funded projects do not specify their contributions in terms of either avoided

Table 7.3 Summary evaluation statistics for the GCF projects on buildings, cities, industries, and appliances

Recipient countries	Project title	Expected contribution (tons CO2-equivalent avoided)	Funding size (US $)	Duration (years)
Mexico	Energy Efficiency Green Bonds in Latin America and the Caribbean	2.6 m tons	$184.5 m	5
El Salvador	Energy savings insurance for private energy efficiency investments by Small and Medium-Sized Enterprises	562.0 k tons	$41.7 m	5
Armenia	De-risking and scaling-up investment in energy efficient building retrofits in Armenia	1.4 m tons	$29.8 m	6
Bosnia and Herzegovina	Scaling-up Investment in Low-Carbon Public Buildings	2.0 m tons	$122.6 m	8
Paraguay	Promoting private sector investments in energy efficiency in the industrial sector in Paraguay	4.0 m tons	$43.0 m	5
Brazil	Financial Instruments for Brazil Energy Efficient Cities	17.4 m tons	$1.3b	7
Vietnam	Scaling Up Energy Efficiency for Industrial Enterprises in Vietnam	120.0 m tons	$88.0 m	5
Nauru	Sustainable and Climate Resilient Connectivity for Nauru	535.4 k tons	$65.2 m	5
Mongolia	Ulaanbaatar Green Affordable Housing and Resilient Urban Renewal Project (AHURP)	7.9 m tons	$544.0 m	8.5

Note: *k* thousand, *m* million, *b* billion

economic damage or reduced temperature increase that can be achieved by each of the awarded projects. Nor do many of these projects specify the degree to which energy efficiency is improved by implementing the awarded project.

The benefit from a dollar spent on the energy efficient buildings and cities would be larger where the city is developing at a faster rate with new buildings and houses constructed at a faster rate than comparatively stagnant cities (Hsu et al. 2017). In such a city, annual total energy produced and used would increase at a faster rate, therefore, an improvement in energy efficiency in the buildings and facilities is expected to have a larger impact on both climatic changes and economic damages.

When it comes to the GCF building grants to under-developed countries, the awards would be more likely perceived as a foreign development aid than a climate change policy measure. Although puzzling at first, it will take only a few moments for the readers to comprehend why. The foreign grants to new buildings and city facilities with some energy-saving features in them will certainly encourage property builders and local politicians to increase the supply of new buildings and houses in the recipient city, which is then certain to increase the emissions of carbon dioxide equivalent emissions in the city through increased energy uses. In other worlds, the grants would not decrease the emissions of CO2-equivalent gases.

Then, readers might wonder what the expected contribution shown in the third column of Table 7.3 in terms of the avoided carbon dioxide emissions may reflect. The figure can be best understood as the reduced amount of carbon dioxide-equivalent emissions from the project implementation from the baseline hypothetical emissions level emitted by the newly built hypothetical buildings with no proposed energy-efficiency improvements, that is, old technologies.

7.7 Evaluation: A Ricardo-Pareto Efficiency Test

The third evaluation of the GCF-funded projects can be conducted on the grounds of the Ricardo-Pareto efficiency, more specifically, efficiency in resource re-allocations. As described in the theory section of this chapter, we can ask for this purpose whether an activity which is funded, therefore forced, by the GCF is the most profitable enterprise of the recipient of all possible enterprises given the future climate change regime. This is one of the ways to test the Ricardo-Pareto efficiency of the GCF projects, but the most fundamental test of all.

This evaluation, however, would not be a handy task at all. Fundamentally, an evaluator should be certain about which enterprise is the most profitable one of all possible enterprises in a certain climate system and, further, the identification should be done for each individual farm. Second, the future climate system cannot be determined with certainty by either farmers or an evaluator. Climate scientists can only offer a range of predictions based on an ensemble of future storylines (scenarios) (Nakicenovic et al. 2000; IPCC 2014).

Notwithstanding these complications, it can be shown that the efficiency test of an individual GCF-funded project can be conducted meaningfully for the reasons to be described shortly. For this evaluation, the present author summarizes in Table 7.4 the GCF-granted projects on food security which are separated by the author from the results area of the health, water security, and food security defined by the GCF.

A salient feature noticeable at the first sight of Table 7.4 is that 12 out of 14 total GCF-funded projects on food security has expressed the goal of the project as 'resilience.' In Table 7.4, the present author highlighted the word using a bolder font. It is remarkable that nearly all GCF-awarded projects agree on the single indicator/concept when it comes to food security under a steadily warming world.

What it means for agriculture or food security to be resilient under climatic shifts is not defined in any of the project documents of the funded projects. Further, the resilience expressed in these projects may implicitly have a different meaning from one project to another. Let me give several interpretations of the resilience that are frequently appropriated by researchers. Resilience often means in agricultural insurances that farmers can survive a threshold weather event such as an extremely severe drought or an extreme heat wave (Udry 1995). Alternatively, in the climate change economics literature, resilience can mean that one system of agriculture is less

Table 7.4 Summary evaluation statistics of the GCF projects on food security

Recipient countries	Project title	Expected contribution (number beneficiaries)	Funding size (US $)	Notes
Gambia	Large-scale Ecosystem-based Adaptation in the Gambia River Basin: developing a climate **resilient**, natural resource based economy	57,750	$25.5 m	
Sri Lanka	Strengthening the **resilience** of smallholder farmers in the Dry Zone to climate variability and extreme events through an integrated approach to water management	2.0 m	$52.1 m	
Namibia	Climate **Resilient** Agriculture in three of the Vulnerable Extreme northern crop-growing regions	21.0 k	$10.0 m	
Namibia	Empower to Adapt: Creating Climate-Change **Resilient** Livelihoods through Community-Based Natural Resource Management in Namibia	76.5 k	$10.0 m	
Vanuatu	Climate Information Services for **Resilient** Development in Vanuatu	260.7 k	$21.8 m	
Tanzania	Simiyu Climate **Resilient** Development Programme	3.0 m	$176.0 m	
Morocco	Irrigation development and adaptation of irrigated agriculture to climate change in semi-arid Morocco	15.5 k	$93.3 m	
India	Ground Water Recharge and Solar Micro Irrigation to Ensure Food Security and Enhance **Resilience** in Vulnerable Tribal Areas of Odisha	16.0 m	$166.3 m	Groundwater recharge shafts in 10,000 tanks; solar pumps for irrigation
Senegal	Building the Climate **Resilience** of Food-insecure Smallholder Farmers through Integrated Management of Climate Risks	526.5 k	$10.0 m	A weather index insurance programme

(continued)

Table 7.4 (continued)

Recipient countries	Project title	Expected contribution (number beneficiaries)	Funding size (US $)	Notes
Tajikistan	Building climate **resilience** of vulnerable and food insecure communities through capacity strengthening and livelihood diversification in mountainous regions of Tajikistan	120.0 k	$10.0 m	Integrated: climate information, capacity building, water management, agriculture and forestry.
Zambia	Strengthening climate **resilience** of agricultural livelihoods in Agro-Ecological Regions I and II in Zambia	3.9 m tons of CO2 Avoided	$137.3 m	
Uganda, Ghana, Nigeria	Acumen **Resilient** Agriculture Fund	10.0 m	$56.0 m	
Namibia	Improving rangeland and ecosystem management practices of smallholder farmers under conditions of climate change in Sesfontein, Fransfontein, and Warmquelle areas	44.4 k	$10.0 m	
Mali	Africa Hydromet Program – Strengthening Climate Resilience in Sub-Saharan Africa: Mali Country Project	5.3 m	$27.3 m	
Burkina Faso	Africa Hydromet Program – Strengthening Climate Resilience in Sub-Saharan Africa: Burkina Faso Country Project	7.0 m	$25.0 m	
Morocco	Saïss Water Conservation Project	2.2 m	$253.6 m	Sustainable surface water; A bulk water transfer scheme

Note: *m* million, *k* thousand

vulnerable to the range of climate regimes, that is, to the changes in the climate regime from one to another (Seo 2010a, b, 2012b). More drastically, a resilience system can mean the opposite of a catastrophe system, that is, the system that can be recovered to a pre-catastrophe state after a catastrophic event (Thom 1975; Mandelbrot 2001; Weitzman 2009; Seo 2018a).

What does each project in Table 7.4 propose to do to achieve its own definition of resilience? The table indicates that each project proposes a different strategy or technique for achieving resilience. Proposed strategies are highlighted by an underline by the present author in the table. Two projects identify an 'integrated' management as a needed strategy, another identifies 'diversification', and another identifies a 'community-based' management. Two projects identify an 'ecosystem-based' management and the other two identify 'irrigation' as the primary strategy for attaining the resilience. Other projects pinpoint climate 'information services' or a creation of a 'resilience fund' as a necessary intervention.

It is conspicuous that none of the 14 food security projects is proposed on the basis of 'efficiency', or 'optimality', or 'efficient adaptation', or 'most profitable or best practices' under a new climate regime (Mendelsohn et al. 1994; Seo 2010a, 2016a, c). Therefore, it can be said that the GCF-funded food security projects most likely fail to satisfy the Ricardo-Pareto efficiency test in the sense of Eqs. 7.8 and 7.9 in this chapter. In other words, an individual GCF project fails to adapt—more correctly, propose to adapt—efficiently to a new climate regime by adopting the most profitable practice or enterprise after a thoughtful deliberation of the new reality.

Let me elaborate this point further. There is no indication in the documents of the GCF projects that the GCF grants on the food security projects are allocated in a way to force the grant recipients to adopt the best, that is, optimal, system or practice of agriculture, leaving aside for the moment a critical question of whether such a grant from an international organization is needed at all for such an optimal individual decision.

The second point is that there is also evidence to show that many of these projects are not efficient or optimal strategies, but rather likely to create an incentive for a grant recipient to take inefficient actions. More specifically, the funding is often selectively given to the most vulnerable enterprise or the most vulnerable/impoverished populations, which can force inefficient actions.

Let's take for consideration the three GCF-funded projects in Table 7.4 that went to Namibia, an arid and desert-zone southwestern African nation known for the Namib desert. The first project entitled "Climate Resilient Agriculture in Three of the Vulnerable Extreme Northern Crop-growing Regions (CRAVE)," the second project entitled "Empower to Adapt: Creating Climate-Change Resilient Livelihoods through Community-Based Natural Resource Management in Namibia," and the third project entitled "Improving Rangeland and Ecosystem Management Practices of Smallholder Farmers under Conditions of Climate Change in Sesfontein, Fransfontein, and Warmquelle areas of the Republic of Namibia."

The three projects share in common the characteristic that the GCF funding is directed to the most vulnerable region in the country as well as to the most vulnerable agricultural system of possible agricultural enterprises. To be more concrete, the funding was awarded to the most vulnerable northern regions of the country where climate is arid and semi-arid (Butt et al. 2005; Fischer et al. 2005; Seo et al. 2009). Also, the funding was awarded to the smallholders with the crop system of agriculture and the rangeland management that could be the most vulnerable agricultural enterprises to climate change (Reilly et al. 1996; Schlenker and Roberts 2009; Kurukulasuriya et al. 2011).

The Sri Lankan project entitled "Strengthening the Resilience of Smallholder Farmers in the Dry Zone to Climate Variability and Extreme Events through an Integrated Approach to Water Management" can be similarly analyzed. The GCF monetary transfer is directed to the dry zones—the most vulnerable locations— as well as to the smallholders—the most vulnerable populations, but the targeted groups are in fact projected to suffer the most from climatic shifts (Seo et al. 2005). The grant may end up with unintended inefficiency consequences.

In this particular aspect, we may say that the GCF funding principle is the principle of an agricultural insurance policy such as that of the US crop insurance and subsidy program. The purpose of the US governmental subsidy is to protect the vulnerable populations of the country from natural disasters that are 'beyond' human forces (Sumner and Zulauf 2012; Goodwin and Smith 2014). The principle of insurance, however, would not be recommended for a climate change fund which must induce vulnerable populations to take best actions given the inevitable changes in the current climate.

The third point is that without identifications of the best or efficient practices/ options under a future climate system either by the GCF Board or the grant applicants, the GCF fund transfer is most likely to be appropriated on non-climate-related purposes. For instance, the funding would only serve recipients as a development aid or an income redistribution to the poor by an international organization. As such, there would be no climate policy purposes to be served.

One of the evidences for such a distortion occurring was given in Chap. 2 where it was pointed out that all GCF projects on the results area of the health, food, and water security are classified as a 'sustainable development' project in the sub-category classification. Some of the GCF projects on food security shown in Table 7.4 are also explicit on their ultimate pursuit of the sustainable development goal: To mention a few, Vanuatu's "Climate Information Services for Resilient Development in Vanuatu" and the United Republic of Tanzania's "Simiyu Climate Resilient Development Programme."

If the funding recipient were to perceive the monetary transfer as a foreign aid— or a development aid— to the poor or a justified income redistribution to the poor, it would drive her/him to stick to the existing form of agriculture, in the fear of losing the grants in the future, even if s/he foresees it to be the most vulnerable form of agriculture in the future climate (Seo 2014b, 2015b).

It should be noted that the GCF funding is only one of the many foreign or development aids with which farmers in developing countries are deeply engaged. Just for the United Nations' level, there are two dozen UN specialized funds and programs, many of which offer similar monetary aids to the poor to that of the GCF (UNSCEB 2018). These funds include, which will be explained in depth in Chap. 8, the United Nations Development Programme (UNDP), World Food Programme (WFP), United Nations Environment Programme (UNEP), United Nations Children's Emergency Fund (UNICEF), United Nations International Strategy for Disaster Reduction (UNISDR), International Fund for Agricultural Development (IFAD) (UNDP 2018; WFP 2018). Accustomed to the multiplicity of similar aids and even ingrained, poor farmers may simply react to the funding from the GCF in the same way as they react to the other aid programs that promote no climate-change policy goals.

The fourth critical point with regard to the GCF's efficiency effects on recipients' decisions is that the GCF monetary awards may force mal-adaptation of grant recipients. Let me point to and explain several projects on this regard. A grant is allocated to the project in Morocco entitled "Irrigation Development and Adaptation of Irrigated Agriculture to Climate Change in Semi-arid Morocco" (Table 7.4). The grant to establish the irrigation in the semi-arid regions of Morocco would certainly encourage the Moroccan farmers to stick to or switch even from another production system to an irrigated crop agriculture.

However, the semi-arid climate zone in Africa has the ecosystem of rangelands, grasslands and deserts, which makes it more favorable to livestock systems of agriculture than crop systems (Kazianga and Udry 2006; Seo 2010c). The international aid for the irrigation development in the semi-arid climate zone would even compel the recipient farmers to reduce or remove the livestock portfolio to increase irrigation-based crop portfolios. The latter, however, will be harmed more severely by a future climate regime with a hotter temperature than the former (Byerlee and Eicher 1997; World Bank 2009; Seo 2010b, 2015a).

Let's take another project for the analysis: the Senegal's project entitled "Building the Climate Resilience of Food-insecure Smallholder Farmers through Integrated Management of Climate Risks." The details of the project are described in the project document as follows (GCF 2018e):

> Risk-reduction activities such as water and soil conservation measures, increased water availability, livelihood diversification and training on climate-resilient practices will be undertaken. These activities will be complemented by risk transfer through a weather index insurance programme that will transfer risk to the international market and provide farmers with compensation in case of climate shocks. The government of Senegal will contribute half the cost of the insurance premium to enrolling households.

The "transfer risk to the international market" means that the cost of the weather index insurance programme will be borne by the Green Climate Fund and ultimately by the rich nations and donors that supply money to the GCF. According to the document, half the cost of the insurance premium will be paid by the Senegal government. However, the weather index insurance programme will with little doubt result in mal-adaptation of Senegalese farmers, which may occur in tandem with water and soil conservation measures, increased water availability, and livelihood diversification.

Why would it be so? The weather index insurance programme may force rational adaptation behaviors of individual farmers if all index design issues were resolved, but even in such cases only when the farmers are required to bear the consequences of their farming decisions (Carter et al. 2015). If the cost would be borne by an international entity, there would be no need for the Senegalese farmers to make efficient decisions.

For many readers of this book, it would be easier to gauge the efficiency consequences of the weather index insurance by considering it in the same line with the efficiency effects of the crop insurance program in the United States which is largely supported by the government subsidy (Halcrow 1949; Sumner and Zulauf 2012). There is one major difference between them. In the proposed weather index insurance, the financial supporter of the program is not even a national government but is an opaque international agency. This adds to the well-perceived vulnerability of the proposed weather index insurance program, that is, additional risk arises from potential rapid changes in international climate negotiations and geopolitics.

Let's take another project for examination: India's project entitled "Ground Water Recharge and Solar Micro Irrigation to Ensure Food Security and Enhance Resilience in Vulnerable Tribal Areas of Odisha." The project document describes its proposed activities as follows (GCF 2018e):

> The Indian state of Odisha is highly vulnerable to climate change due to high monsoon variability.....These climate related challenges will be addressed by enhancing ground water recharge in the community ponds through structural adaptation measures, and the use of solar pumps for micro irrigation. ...... through the installation of groundwater recharge shafts in 10,000 tanks. ...... Ground water recharge shafts can be transformative by demonstrating their ability to recharge the underlying aquifier system, developing long-term groundwater reserves. At the same time, resilient crop planning through irrigation will improve food security in the region, whilst the use of solar pumps for irrigation will increase energy access and contribute to the state's climate-resilient, low emission crop planning.

The Indian State of Odisha is located in one of the severest monsoon climate regimes in India and perhaps in the world. The monsoon season precipitation reaches as high as 2855 mm per month, which is an extremely large amount of rainfall. In the non-monsoon season, the monthly rainfall reaches 114 mm per month. Deadlier is the fact that the ratio of monsoon season rainfall over non-monsoon season rainfall fluctuates year by year by as much as 250%, one of the largest in India (IITM 2012; Seo 2016b). To make things even worse, the State is a highly mountainous tribal region.

Again, because the cost of installing the various structural components and implementing the program is borne by the foreign entity, the Indian State may have little motivation to make efficient, cost-effective decisions when the proposed structural measures are in place. If such is the case, many measures described in the project document are likely to increase the risk from the monsoon climate, that is, rather than to decrease it.

The installations of a community pond and water tanks by a foreign policy agency would drive Odisha farmers to grow more crops and increase other agricultural activities during the monsoon season as well as the non-monsoon season. This will certainly increase the risk of farming in the State since an extremely intense rainfall is more often unpredictable in advance especially in a monsoon climate regime. Further, this unpredictability is predicted by many climate scientists to increase in the Odisha state in the future owing to a global climate change (Meehl and Hu 2006; Goswami et al. 2006).

Second, the construction of the water tanks and recharge shafts from a community pond may increase the capacity of groundwater recharge which can then be drawn for agricultural activities in the non-monsoon season. Although this has yet to be proven by the project, i.e., by overcoming the soil permeability problem, we can foresee the increase in the risk that arises from the new structures. For example, an extremely intense rainfall, quite common in the severe monsoon regime, may overwhelm the water tanks and shafts.

Besides the structural problems, an economic incentive problem may even be severer. The cost of installations and remedies of potential structural failures is likely very large, which is paid by the international entity, that is, the GCF. Will agricultural activities in the Odisha State increase as much as to justify the cost for the project? The answer would be no if we would consider the monsoon climate regime in Odisha as well as the future climate change which would increase monsoon variability even further, even if we assume many sensible changes in farming practices (Kala 2015). Even if we assumed the project's technical success, it would not increase agriculture in the non-monsoon season as much owing to an intensified monsoon climate regime as to justify the investment.

A third point can be made on other technical components of the project, more specifically, solar pumps and irrigation. A solar energy generation is only feasible when the Sun is out of clouds and in non-rainy days. It is also well established that solar energy is practically not storable. In a monsoon climate regime, solar pumps therefore are less effective in generating energy than in a non-monsoon climate regime owing to more frequent cloud-covered skies and rainy days. Further, an extremely intense rainfall, which occurs quite often in the region, could flood the entire local area, leading to landslides, earth breakups, trees broken and uprooted, and washed-away plains due to flooding, all of which would make installed solar panels easy to break or malfunction. These accidents are quite frequent and pose big challenges to the solar installations in many places in the world. Therefore, local farmers who received the GCF grant should make sure that the solar panels/pumps should be stored away in storage spaces during the monsoon and reinstalled every time during the non-monsoon season, which is costly.

The Indian project can be evaluated for its efficiency effects with reference to the Indian farmers' behaviors in response to the monsoon risk that are actually observed (Seo 2018b). A study of Indian farmers' behaviors reveals that Indian farmers have adapted to a high monsoon risk by switching to an alternative asset such as goats. The study finds that goats are raised in larger numbers, the larger the monsoon climate risk is (Seo 2016b). A similar behavioral response of choosing goats was also reported in the humid climate zones of Sub-Saharan Africa (Seo and Mendelsohn 2008).

Let's move on to another food security project for evaluation: The Tajikistan's project entitled "Building Climate Resilience of Vulnerable and Food Insecure Communities through Capacity Strengthening and Livelihood Diversification in Mountainous Regions of Tajikistan." The project document describes its proposed activities as follows (GCF 2018e):

> Tajikistan is experiencing increasing temperature and rainfall variability and recurrent natural disasters, particularly droughts and floods. Local communities in mountainous areas have low adaptive capacity to cope with these impacts, due to a lack of adequate climate information to plan their agricultural production while they rely heavily on climate-sensitive sources of income.
>
> This initiative will introduce adaptation measures to address climate change effects leading to declines in agricultural yields, increases in food prices and reduced agricultural wages. It will focus on the most vulnerable and food insecure communities in the Rasht valley, Khatlon and Gorno-Badakhshan Autonomous Region (GBAO) regions. It will include an integrated approach to provide climate information services, capacity building, sustainable water management and resilient agriculture and forestry.

To give readers a brief background, Tajikistan is a landlocked, mountainous country in Central Asia in the north of Afghanistan and Pakistan. The country is covered by high mountains in the Pamir range and more than 50% of the country is located above 3000 meters in altitude above the sea level (CIA 2018). As such, it is expected that forestry is an important part of the natural resource sector of Tajikistan and so the forestry sector is mentioned in the document.

Evidently, the project focuses on "declines in agricultural yields, increases in food prices" that may occur due to projected climate changes. In all likelihood, this means that the GCF funding will be used to increase crop productions and crop yields in the selected region, that is, "the most vulnerable and food insecure communities in the Rasht valley."

This would be a very inefficient way to cope with a future climate change in Tajikistan. Owing to the geographical factors described above, the country is already low in crop investments and high in non-crop investments such as livestock management, forestry, natural resources, and specialty crops. Only 7% of the land is arable land and the country imports 70% of its food consumption (CIA 2018). As such, cotton is a predominant crop, accounting for 60% of agricultural production in the country and supporting 75% of rural population. For another, aluminum production is the primary source of income for the country. A GCF funding-induced push away from these more efficient sectors would make the country more insecure, that is, not less insecure, in terms of food production. Imagine how inefficient it will be if an international aid were to be given to the farmers in the Andes Mountain Ranges and the Amazon Basin who would propose to slash the forests and dam the rivers in order to create a vast maize growing area (WRI 2005, Seo 2012a, 2016c)!

So far, our evaluations of efficiency consequences of the GCF intervention have been on the food security projects funded by the international organization. Another set of the GCF-funded projects is summarized in Table 7.5, which can be categorized as water security projects. The listed projects in Table 7.5 are originated from small island nations such as Maldives, Fiji, and Grenada, as well as local areas from the countries with well-recognized water-supply problems such as Bangladesh.

The water security projects in Table 7.5 have in common that they have a clear objective of improving water security through enhancing access to drinking water, increased water supply, and improved sewage systems. These goals are suggested to be achieved through various technologies and measures such as harnessing rainwater and groundwater, a desalination plant, a dam, access channels to a dam, a waste water treatment plant, solar pumps, and a water tariff.

To evaluate the efficiency effects, let's first take a closer look at the Maldives' project entitled "Support of Vulnerable Communities in Maldives to Manage Climate Change-Induced Water Shortages." Maldives is a small island nation located in South Indian Ocean. The country is an archipelago consisting of 1190 small, low-lying coral islands spread over 90,000 square kilometers and one of the most popular tourist destinations in the world. The project documentation states the following (GCF 2018e):

Table 7.5 Summary evaluation statistics of the GCF projects on water security

Recipient countries	Project title	Expected contribution (number beneficiaries)	Funding size (US $)	Details
Maldives	Support of Vulnerable Communities in Maldives to Manage Climate Change-Induced Water Shortages	105,000	$28.2 m (5 years)	Scale up an integrated water supply system based on rainwater, groundwater, and desalinated water into a low-cost delivery system for vulnerable households.
Fiji	Fiji Urban Water Supply and Wastewater Management Project	290,854	$405.1 m (7 years)	Building and renovating infrastructure to improve access to safe water and sewerage systems in the greater Suva area of Fiji. Creating a new river water intake station on the River Rewa and improving the Kinoya wastewater treatment plant and associated sewer coverage.
Colombia	Scaling Up Climate Resilient Water Management Practices for Vulnerable Communities in La Mojana	405.6 k	$117.2 m (8 years)	The region of La Mojana, located in an inland deltaic plain with wetland systems. Scaling up climate resilient integrated water resource management practices will tackle barriers derived from climate change through a number of activities.
Ethiopia	Responding to the Increasing Risk of Drought: Building Gender-responsive Resilience of the Most Vulnerable Communities	1.3 m	$50.0 m (5 years)	Providing rural communities with critical water supplies for year-round drinking water and small-scale irrigation to address risks of drought and other climate impacts.
				The three main activities will be introducing solar-powered water pumping and small-scale irrigation.

(continued)

Table 7.5 (continued)

Recipient countries	Project title	Expected contribution (number beneficiaries)	Funding size (US $)	Details
Grenada	Climate-Resilient Water Sector in Grenada	107.3 k	$51.6 m (6 years)	Creating a climate resilient water sector; A Water Resource Management Unit (WRMU) will be established, and a water tariff will create climate-responsive price signals for users, whilst providing resources to upgrade infrastructure.
Bangladesh	Enhancing adaptive capacities of coastal communities, especially women, to cope with climate change induced salinity	719.2 k	$33.0 m (6 years)	Strengthening the adaptive capacity of coastal communities, especially women, to cope with the impacts of climate-induced salinity on their livelihoods and water security;
				To reduce the adverse impacts to agricultural livelihoods.

Note: *m* million, *k* thousand

> There are high levels of poverty on the outer islands, which experience drinking water shortages during the dry season causing significant human, environmental, and social impacts. Groundwater becomes increasingly saline as a result of climate change-induced sea level rise (3.1 mm/year) and variable rainfall patterns. Responses are constrained by remoteness and limitations on land space.
>
> The project will scale up an integrated water supply system based on rainwater, groundwater, and desalinated water into a low-cost delivery system for vulnerable households. This will provide uninterrupted supply to 49 islands that currently rely on emergency water deliveries for three months of each year. Decentralized and cost-effective dry season water supply systems will also be introduced. Water desalination production plants will be built on four larger islands that will contribute to this improved dry season water distribution network to outer atolls and local supply systems.

According to the document, drinking water shortages occur during the dry season on the outer islands, lasting 3 months each year. The proposal asserts that increased salinity in groundwater will become a more serious problem owing to climate change-induced sea level rise, quoting a 3.1 mm rise annually, and variable rainfall patterns (Church et al. 2013).

However, it seems evident that because of the natural conditions of Maldives comprising thousands of small atolls as well as being the world's lowest country with an average elevation of 1.8 meters above the sea level (CIA 2018), the country's priority should be protecting its citizens from the sea level rise by building loops of coastal walls, which is more urgent than solving drinking water shortages induced by future climate changes. The construction, however, of a multitude of coastal walls—a wall for each island—would be almost unrealistic for Maldives because tourist beaches of serendipity will be gone and replaced by ugly walls while the large number of small islands makes it very costly to build even such unattractive walls (Ng and Mendelsohn 2006). Notwithstanding, for the long-term horizon by which the sea level is projected to have risen by more than a meter and even as high as seven meters (Kopp et al. 2014), the Maldives water security project for drinking water access would be highly inefficient without the coastal protection firsthand against the rising seas.

Can the drinking water shortages be solved in the long-term without building costal walls, that is, not destroying the serene beaches which are critical for tourist incomes for the country? The four desalination plants proposed in the document may provide an alternative to rely on since it offers a way to utilize the salt water for supplying drinking water (Carlsbad 2016). However, the efficiency of allocating the GCF money to the desalination plants is highly doubtful. First, it would be much cheaper to build the loops of coastal walls in selected islands to prevent increased salinity of groundwater than to construct the desalination plants. The Carlsbad desalination plant in San Diego, California took 14 years to construct with the total cost estimate of US$ 1 billion for a single plant. The operational cost of a desalination plant is relatively high owing to heavy energy uses in its processes such as water conveyance, water treatment, water reuse, brackish water desalination, sea-water desalination, which may use approximately 10 kilowatt hours of electricity per cubic meter of water produced (NRC 2008).

Further, unlike the sea walls which would protect people's lives against the sea surges and inundation caused by climate change-induced hurricane activities in Indian Ocean, earthquake-induced tsunamis, and ever higher sea levels, even though the walls could spoil serene beaches, the desalination plants do not offer such most critical protections (Seo and Bakkensen 2017; Seo 2017c, 2018c).

Be reminded that our evaluations in this section and of this project are only concerned about the efficiency aspects of the GCF funding projects. More concretely, we ask whether the proposed and funded project yields the benefit that justifies the cost or whether the project is the most beneficial project for the recipient population and country. The three paragraphs about the Maldives project can be interpreted, I hope, as an intellectual exercise of comparing multiple options and finding the best, which is the essence of the economic concept of efficiency (Ricardo 1817).

Some of the water security projects in Table 7.5 provide too little information on the specific actions to be performed, making a precise evaluation of efficiency effects impossible. The Bangladesh project entitled "Enhancing Adaptive Capacities of Coastal Communities, Especially Women, to Cope With Climate Change Induced Salinity" proposes 'a community-based approach' but lacks the description of the specific activities and strategies that define the approach.

The Colombian project is another such example, entitled "Scaling Up Climate Resilient Water Management Practices for Vulnerable Communities in La Mojana." Although it proposes 'climate resilient integrated water resource management practices,' such practices are not described concretely or consistently in the project document, making it impossible to provide an evaluation of efficiency effects. For example, the project refers to 'systemized knowledge management', 'ecosystem restoration', and others which seem vague and only loosely tied to the project's main objective.

Let's take another water security project for efficiency evaluations: The project entitled "Climate-Resilient Water Sector in Grenada (G-CREWS)" submitted by a team from Grenada. The project has two components. On the one hand, it aims to increase fresh water availability by upgrading the present infrastructure. On the other hand, it aims to reduce water demand by creating a water tariff. The goal of the project is stated to be 'creating a climate resilient water sector.'

Grenada is a country located in the Southeastern Caribbean Sea, whose land size is about 1/tenth of Rhode Island, the smallest State in the United States. The problem of water scarcity is described as follows in the project documentation (GCF 2018e):

> Grenada relies heavily upon surface water and rainwater catchment for its water supply. Climate change is aggravating water scarcity problems with increasing average temperatures, more erratic rainfall, more frequent heavy rainfall events, and saltwater intrusion in groundwater due to sea level rise. G-CREWS will support the water sector by reducing water demand and increasing water availability.

Although the above description argues that more erratic rainfall and more frequent heavy rainfall events are worsening the water scarcity problems in the country, more rainfall induced by a warmer climate could, on the contrary, increase water availability. Excluding these perhaps offsetting changes in rainfall patterns, we may narrow the project focus to the 'saltwater intrusion in groundwater caused by sea level rise,' the same problem faced by Maldives which was analyzed above.

The problem of saltwater intrusion in groundwater cannot be addressed successfully by either upgrading the present infrastructure or water tariff. The water tariff creates an incentive to conserve water, but does not address water shortages due to increased water salinity. The present infrastructure is certainly not adequate to protect the coastal cities and farms from a sea level rise as high as several meters (Kopp et al. 2014).

The salinization problem can be addressed, as already analyzed for the Maldives project, by either constructing a desalination plant or a set of protective structures around groundwater sources. These technologies were already explained above. The point I am making here is that, taking for granted the climate change projections and sea level rise projections that the proposal applicant relies on, the proposed measures and actions will not be pertinent nor sufficient to resolve the problem the proposal applicant expressed to seek to address. Stated in the economics terminology, the project will incur significant efficiency losses from the allocation of the GCF grant to the project.

7.8 Additional Verdicts

This chapter has provided an ensemble of economic evaluations of the Green Climate Fund (GCF)'s activities, especially of their funding decisions on the projects to which grants/loans have been allocated. The evaluation of each project was conducted on the theoretical grounds of the three economic theories: a theory of public goods and expenditure, a theory of efficient land and resource allocations, and a theory of public adaptation. In evaluating each project on the criteria of the three economic theories, the empirical results and predictions gathered from the four scientific literatures surveyed in the preceding chapters were presented as supporting evidence.

What can we learn from the series of evaluations conducted in this chapter of the GCF funding decisions and respective projects? Above all, we can conclude that the GCF allocation decisions are made based on a selective indicator or multiple indicators which cannot be justified in an economic sense, more specifically, in the sense of the three essential economic theories. The indicator relied upon is either the expected contribution in terms of CO2-equivalent emissions avoided or the expected contribution in terms of the number of beneficiaries from implementing a funded project (GCF 2018e). These indicators are not in themselves constructed as a scientific measure that embraces the possibility of alternative options as well as the costs of the range of options.

Second, the GCF Board developed the GCF investment guideline, called the investment framework, to justify a variety of funding allocation decisions (GCF 2018b). The guideline comprises the following criteria, in the order of importance listed by the GCF Board, (i) impact potential (mitigation impact or adaptation impact), (ii) paradigm shift potential, (iii) sustainable development potential, iv) needs of the recipient, v) country ownership, and (vi) efficiency and effectiveness.

The investment framework does not tackle directly the fundamental question of how the global community should provide the global public good, that is, a 'favorable' global climate system in an optimal way. Or put differently, it is not concerned about how the global community should allocate its resources to manage the global climate system in the best way (Koopmans 1965; Nordhaus 1994).

More importantly, the investment criteria are not rigorously defined, nor are concrete indicators of each criterion provided by the decision Board. It is difficult for a researcher to assign a quantitative value (weight) on either the paradigm shift potential or sustainable development potential. It may be even more difficult to justify some guidelines such as needs of the recipient or country ownership in the economic, non-political sense. The investment criterion of 'efficiency' which is listed at the end of the list is not comparably defined by the GCF to the aforementioned essential economic theories.

Because of these shortcomings in carving out the investment framework, the present author is able to lay a series of critical verdicts on the current status of the GCF program. To begin, the GCF funding programs can be bettered viewed as a foreign aid program implemented at a global scale than a climate change policy program. The international programs of foreign aids have a much longer history, some of which are a century-old, than the short history of climate change-concerned funding programs.

Some of the most active international aid programs include the UNICEF's emergency and disaster relief, the Save the Children Fund, the World Food Programme (WFP), and the Tzu Chi Foundation's international relief. The UNICEF relief funds encompass many climate-related programs such as the earthquake relief, children in conflict, food crises, flood relief, hurricane relief, tsunami relief, child refugee and migrant crisis (UNICEF 2018). The World Food Programme is the humanitarian food-assistance program that annually provides food assistance to about 80 million people in about 80 countries (WFP 2018). The aid programs bear a strong resemblance to the GCF funding programs that are discussed in this chapter with regard to their assistance areas, institutional backgrounds, among other things. A full description of the United Nations' level specialized funds is one of the key topics of the next chapter where the present author will deliver a more thorough analysis on the relationships between the GCF and the UN specialized funds.

Second, a further examination of the international aid programs reveals that the GCF funding decisions can be better rationalized as a sustainable development program than a climate change policy program. As observed in Chap. 2, nearly all GCF-funded projects are classified by the GCF Board as a sustainable development program in its sub-category. It is not rare to see that the GCF policy-makers and UN climate change negotiators justify the GCF program as one of the ways to accomplish the UN Millennium Development Goals (MDG) from the United Nations Millennium Declaration. In the UN Declaration, although addressing climate change was recognized as one of the MDGs, it was treated only as a slight component of the sustainable development grand project and it was only touched with reference to the now-defunct Kyoto Protocol (UN 2000). Notwithstanding, it may not be an exaggeration to say that the monetary size of the international climate change funding and other programs outweigh, as of the end of 2018, that of all the programs implemented under the other MDG goals (GEF 2017).

However, it should be recognized by the GCF decision-makers that tackling climate change challenges takes a very different set of actions and priorities from the actions that are called for for solving the problems of sustainable development (Solow 1993; World Bank 2008). As demonstrated plentifully in this chapter, it should be understood that a sustainable development aid program will be ineffective in solving climate change problems at both a local scale and a global scale. In this chapter, the present author clarified many situations in which a sustainable development aid program can unexpectedly result in a higher CO2-equivalent emissions level as well as amplified vulnerability of aid recipients.

From a positive angle, it was repeatedly stated by, e.g., the GCF governing instrument and the Paris Agreement, that the GCF funding programs should serve as a primary financial instrument to fulfill a pivotal agreement reached by the parties at the establishment of the United Nations Framework Convention on Climate Change (UNFCCC) declared at the very beginning of international climate change negotiations at the Rio Earth Summit in 1992 (GCF 2011; UNFCCC 2015). That is the Article 3 of the UNFCCC document which declared 'common but differentiated responsibilities' as one of the fundamental principles of the UN climate convention (UNFCCC 1992). The Article 3 of the UNFCCC enentitled "Principles" states the first of the principles as follows:

> The Parties should protect the climate system for the benefit of present and future generations of humankind, on the basis of equity and in accordance with their common but differentiated responsibilities and respective capabilities. Accordingly, the developed country Parties should take the lead in combating climate change and the adverse effects thereof.

This principle has evolved over the course of three decades to become the thorniest issue of all in any climate negotiations since then. Negotiators from developing nations argued that a GCF-like monetary aid should be preceded before any agreement is reached on a global climate policy, which in effect derailed many ambitious UN negotiations including the Copenhagen Conference in 2009 and the Paris Conference in 2015 (UNFCCC 2009, 2015; Nordhaus 2010; Seo 2017a).

Having stated that the first principle of the UNFCCC lies at the heart of the GCF funding decisions, the root of the first principle goes way further back to the initial United Nations conference on the human environment held in Stockholm, Sweden in 1972 (UN 1972). The Stockholm Conference declared 20 principles of the UN actions on the environment, of which principles 11 and 12 defined the development and environment issue in a particular way:

> The environmental policies of all States should enhance and not adversely affect the present or future development potential of developing countries, nor should they hamper the attainment of better living conditions for all, and appropriate steps should be taken by States and international organizations with a view to reaching agreement on meeting the possible national and international economic consequences resulting from the application of environmental measures (Principle 11).

> Resources should be made available to preserve and improve the environment, taking into account the circumstances and particular requirements of developing countries and any costs which may emanate from their incorporating environmental safeguards into their development planning and the need for making available to them, upon their request, additional international technical and financial assistance for this purpose (Principle 12).

The Stockholm principles on the development and environment nexus have dominated all subsequent conferences held by the United Nations on environment, sustainability, and climate change (WMO 1985; UNFCCC 1992, 2015). At the same course of history, they have become the greatest barrier to climate policy negotiations and the GCF decisions.

We can end this chapter conclusively with a favorable verdict that the GCF's funding allocation decisions are faithful to the first principle of the Article 3 of the UNFCCC, for which reason the decision-makers of the GCF were unable to address the fundamental aspects of climate change policy-making. That is, how the global community should manage the global warming problem in the best possible way and how the individuals and communities should adapt to changes in the climate system in an efficient manner (Nordhaus 1994; Seo 2016d, 2017b).

This verdict in turn warns one additional time the policy-makers of global warming and any other global-scale phenomena of the infeasibility of managing such a large-scale policy quandary by the power of a committee of one or two.

References

Adams R, McCarl BA, Segerson K, Rosenzweig C, Bryant K, Dixon BL, Conner R, Evenson RE, Ojima D (1999) The economic effects of climate change on US agriculture. In: Mendelsohn R, Neumann J (eds) The impact of climate change on the United States economy. Cambridge University Press, Cambridge

Ainsworth EA, Long SP (2005) What have we learned from 15 years of free-air CO2 enrichment (FACE)? A meta-analysis of the responses of photosynthesis, canopy properties and plant production to rising CO2. New Phytol 165:351–372

Akasaki I, Amano H, Nakamura S (2014) Blue LEDs: filling the world with new light. Nobel Prize Lecture. The Nobel Foundation, Stockholm. Available at: http://www.nobelprize.org/nobel_prizes/physics/laureates/2014/popular-physicsprize2014.pdf

Baumol WJ, Oates OA (1988) The theory of environmental policy, 2nd edn. Cambridge University Press, Cambridge

Boden TA, Marland G, Andres RJ (2017) Global, regional, and national fossil-fuel CO2 emissions. Oak Ridge National Laboratory, U.S. Department of Energy, Oak Ridge, Tenn., USA. https://doi.org/10.3334/CDIAC/00001_V2017

Bohme R, Christin N, Edelman B, Moore T (2015) Bitcoin: economics, technology, and governance. J Econ Perspect 29:213–238

Brown KM (1973) Welfare implications of congestion in public goods. Rev Soc Econ 31:89–92

Buchanan JM (1965) An economic theory of clubs. Economica 32:1–24

Butt TA, McCarl BA, Angerer J, Dyke PT, Stuth JW (2005) The economic and food security implications of climate change in Mali. Clim Chang 68:355–378

Byerlee D, Eicher CK (1997) Africa's emerging maize revolution. Lynne Rienner Publishers Inc., Boulder

Carlsbad Desalination Plant (2016) DESAL-101. Claude "Bud" Lewis Carlsbad Desalination Plant. San Diego, CA

Carter M, de Janvry A, Sadoulet E, Sarris A (2015) Index-based weather insurance for developing countries: a review of evidence and a set of propositions for up-scaling. Revue d' Economie du Dev 23:5–57

Central Intelligence Agency (CIA) (2018) The world factbook. The CIA, Virginia. Accessed from https://www.cia.gov/library/publications/resources/the-world-factbook/

Church JA, Clark PU, Cazenave A, Gregory JM, Jevrejeva S, Levermann A, Merrifield MA, Milne GA, Nerem RS, Nunn PD, Payne AJ, Pfeffer WT, Stammer D, Unnikrishnan AS (2013) Sea level change. In: Climate change 2013: the physical science basis. Cambridge University Press, Cambridge

Coase R (1960) The problem of social costs. J Law Econ 3:1–44

Commonwealth Scientific and Industrial Research Organization (CSIRO) (2018) Future feed. CSIRO, Canberra. Accessed from https://research.csiro.au/futurefeed/

Conning J, Udry C (2007) Rural financial markets in developing countries. In: Evenson R, Pingali P (2007) Handbook of agricultural economics, vol. 3, pp 2857–2908

Diamond PA, Hausman JA (1994) Contingent valuation: is some number better than no number? J Econ Perspect 8:45–64

Dinar A, Campbell MB, Zilberman D (1992) Adoption of improved irrigation and drainage reduction technologies under limiting environmental conditions. Environ Resour Econ 2:373–398

Ellerman AD, Marcantonini C, Zaklan A (2016) The European union emissions trading system: ten years and counting. Rev Environ Econ Policy 10:89–107

Fischer G, Shah M, Tubiello FN, van Velhuizen H (2005) Socio-economic and climate change impacts on agriculture: an integrated assessment, 1990–2080. Philos Trans R Soc B 360:2067–2083

Freeman AM III, Herriges JA, Cling CL (2014) The measurements of environmental and resource values: theory and practice. RFF Press, New York

Friedman M (1962) Capitalism and freedom. The University of Chicago Press, Chicago
Global Carbon Project (GCP) (2017) Global Carbon Budget 2017. Available at: http://www.globalcarbonproject.org/carbonbudget/index.htm
Global Environment Facility (GEF) (2017) Comparative advantage, adequacy of funding/financing, health of the expanded GEF partnership and governance structure. GEF, UNDP, New York
Goodwin BK, Smith VH (2014) Theme overview: the 2014 farm bill—an economic welfare disaster or triumph? Choices 29(3):1–4
Goswami BN, Venugopal V, Sengupta D, Madhusoodanan MS, Xavier PK (2006) Increasing trend of extreme rain events over India in a warming environment. Science 314:1442–1445
Graziano M, Gillingham K (2015) Spatial patterns of solar photovoltaic system adoption: the influence of neighbors and the built environment. J Econ Geogr 15:815–839
Green Climate Fund (GCF) (2011) Governing instrument for the green climate fund. GCF, Songdo City
Green Climate Fund (GCF) (2018a) About the fund. GCF, Songdo City
Green Climate Fund (GCF) (2018b) Annex III: investment framework. GCF, Songdo City
Green Climate Fund (GCF) (2018c) Annex IX: results management framework. GCF, Songdo City
Green Climate Fund (GCF) (2018d) Status of pledges and contributions made to the green climate fund, status Date: 29 January 2018. GCF, Songdo City
Green Climate Fund (GCF) (2018e) Projects + programmes. GCF, Songdo City. Accessed from https://www.greenclimate.fund/what-we-do/projects-programmes
Green Climate Fund (GCF) (2018f) Green climate fund invests USD 1 billion for developing country climate action, launches first replenishment. Published on October 21, 2018. Accessed from https://www.greenclimate.fund/-/green-climate-fund-invests-usd-1-billion-for-developing-country-climate-action-launches-first-replenishment
Hahn RW, Dudley PM (2007) How well does the U.S. Government do benefit-cost analysis? Rev Environ Econ Policy 1:192–211
Halcrow HG (1949) Actuarial structures for crop insurance. J Farm Econ 31:418–443
Hanemann WM (1994) Valuing the environment through contingent valuation. J Econ Perspect 8:19–43
Hanemann WM (2000) Adaptation and its management. Clim Chang 45:571–581
Hart O, Holmstrom B (1987) The theory of contracts. In: Bewley T (ed) Advances in economics and econometrics. Cambridge University Press, Cambridge
Hristov AN, Joonpyo OH, Fabio Giallongo F, Tyler W, Frederick TW, Michael T, Harper MT, Holley L, Weeks HL, Antonio F, Branco AF, Moate PJ PJ, Matthew H, Deighton MH, SRO W, Kindermann M, Duval S (2015) An inhibitor persistently decreased enteric methane emission from dairy cows with no negative effect on milk production. Proc Natl Acad Sci 112 (34):10663–10668
Hsu A, Rosengarten C, Weinfurter A, Xie Y, Musolino E, Murdock HE (2017) Renewable energy and energy efficiency in developing countries: contributions to reducing global emissions. United Nations Environment Programme, Nairobi
Indian Institute of Tropical Meteorology (IITM) (2012) Homogeneous Indian monthly rainfall data sets & Indian regional monthly surface air temperature data set. IITM, Pune
Intergovernmental Panel on Climate Change (IPCC) (2014) Climate change 2014: the physical science basis. The fifth assessment report of the IPCC. Cambridge University Press, Cambridge
Intergovernmental Panel on Climate Change (IPCC) (2018) Special report on global warming of 1.5 °C. Cambridge University Press, Cambridge
Kafle K, Winter-Nelson A, Goldsmith P (2016) Does 25 cents more per day make a difference? The impact of livestock transfer and development in rural Zambia. Food Policy 63:62–72
Kala N (2015) Ambiguity aversion and learning in a changing world: The potential effects of climate change from Indian agriculture. PhD dissertation, Yale University
Kazianga H, Udry C (2006) Consumption smoothing? livestock, insurance, and drought in rural Burkina Faso. J Dev Econ 79:413–446
Koopmans TC (1965) On the concept of optimal economic growth. Acad Sci Scr Varia 28(1):1–75

Kopp RE, Horton RM, Little CM, Mitrovica JX, Oppenheimer M, Rasmussen DJ, Strauss BH, Tebaldi C (2014) Probabilistic 21st and 22nd century sea-level projections at a global network of tide-gauge sites. Earth's Future 2:383–406

Kurukulasuriya P, Kala N, Mendelsohn R (2011) Adaptation and climate change impacts: a structural Ricardian model of irrigation and farm income in Africa. Climate Change Econ 2:149–174

Laffont JJ, Martimort D (2002) The theory of incentives: the principal-agent model. Princeton University Press, Princeton

Mandelbrot B (2001) Scaling in financial prices: I. Tails and dependence. Quant Finan 1:113–123

Mankiw NG (2014) Principles of economics, 7th edn. Cengage Learning, Stamford

Mas-Colell A, Whinston MD, Green JR (1995) Microeconomic theory. Oxford University Press, Oxford

Meehl GA, Hu A (2006) Megadroughts in the Indian monsoon region and southwest North America and a mechanism for associated multidecadal Pacific sea surface temperature anomalies. J Clim 19:1605–1623

Mendelsohn R (2000) Efficient adaptation to climate change. Clim Chang 45:583–600

Mendelsohn R, Neumann J (1999) The impact of climate change on the United States economy. Cambridge University Press, Cambridge

Mendelsohn R, Olmstead S (2009) The economic valuation of environmental amenities and disamenities: methods and applications. Ann Rev Resour 34:325–347

Mendelsohn R, Nordhaus W, Shaw D (1994) The impact of global warming on agriculture: a Ricardian analysis. Am Econ Rev 84:753–771

Mollick E (2014) The dynamics of crowdfunding: An exploratory study. J Bus Ventur 29:1–16

Muller NZ, Mendelsohn R (2009) Efficient pollution regulation: getting the prices right. Am Econ Rev 99:1714–1739

Nakicenovic N, Davidson O, Davis G, Grubler A, Kram T, La Rovere EL, Metz B, Morita T, Pepper W, Pitcher H, Sankovski A, Shukla P, Swart R, Watson R, Dadi Z (2000) Emissions scenarios. A special report of working group III of the Intergovernmental Panel on Climate Change, Geneva, Switzerland

National Research Council (2008) Desalination: a national perspective. National Academies Press, Washington, DC

New York Times (NYT) (2016) The Obama Era: 'terrifying' path of climate crisis Weighs on Obama. Published on September 8, 2016

Ng N-S, Mendelsohn R (2006) The economic impact of sea-level rise on nonmarket lands in Singapore. Ambio 35:289–296

Nordhaus W (1994) Managing the global commons. MIT Press, Cambridge

Nordhaus WD (2006) Paul Samuelson and global public goods. In: Szenberg M, Ramrattan L, Gottesman AA (eds.) Samuelsonian economics and the twenty-first century. Oxford Scholarship Online

Nordhaus W (2010) Economic aspects of global warming in a Post-Copenhagen environment. Proc Natl Acad Sci U S A 107:11721–11726

Ostrom E (1990) Governing the commons: the evolution of institutions for collective action. Cambridge University Press, Cambridge

Pareto V (1906) Manual for Political Economy. In: Montesano A, Zanni A, Bruni L, Chipman JS, McLure M (eds.) (2014). Oxford University Press, Oxford

Parry ML, Rosenzweig CP, Iglesias A, Livermore M, Fischer G (2004) Effects of climate change on global food production under SRES emissions and socioeconomic scenarios. Glob Environ Chang 14:53–67

Raadschelders JCN (2005) The institutional arrangements for water management in the 19th and 20th centuries. IOS Press, Amsterdam

Reich PB, Hobbie SE, Lee TD, Pastore MA (2018) Unexpected reversal of C3 versus C4 grass response to elevated CO2 during a 20-year field experiment. Science 360:317–320

Reilly J, Baethgen W, Chege F, Van de Geijn S, Enda L, Iglesias A, Kenny G, Patterson D, Rogasik J, Rotter R, Rosenzweig C, Sombroek W, Westbrook J (1996) Agriculture in a

changing climate: impacts and adaptations. In: Climate change 1995: impacts, adaptations, and mitigation of climate change. Cambridge University Press, Cambridge
Reilly J, Paltsev S, Felzer B, Wang X, Kicklighter D, Melillo J, Prinn R, Sarofim M, Sokolov A, Wang C (2007) Global economic effects of changes in crops, pasture, and forests due to changing climate, carbon dioxide, and ozone. Energy Policy 35:5370–5383
Ricardo D (1817) On the principles of political economy and taxation. John Murray, London
Ridker R, Henning J (1967) The determination of residential property values with special reference to air pollution. Rev Econ Stat 48:246–257
Rosenzweig C, Parry M (1994) Potential impact of climate change on world food supply. Nature 367:133–138
Samuelson P (1954) The pure theory of public expenditure. Rev Econ Stat 36:387–389
Samuelson P (1955) Diagrammatic exposition of a theory of public expenditure. Rev Econ Stat 37:350–356
Samuelson P, Nordhaus W (2009) Economics, 19th edn. McGraw-Hill Education, New York
Schlenker W, Roberts M (2009) Nonlinear temperature effects indicate severe damages to crop yields under climate change. Proc Natl Sci Acad US 106(37):15594–15598
Schlenker W, Hanemann M, Fisher A (2005) Will US agriculture really benefit from global warming? accounting for irrigation in the hedonic approach. Am Econ Rev 95:395–406
Schoengold K, Zilberman D (2007) The economics of water, irrigation, and development. In: Evenson R, Pingali P (2007) Handbook of agricultural economics, vol 3, pp 2933–2977
Seo SN (2010a) A microeconometric analysis of adapting portfolios to climate change: adoption of agricultural systems in Latin America. Appl Econ Perspect Policy 32:489–514
Seo SN (2010b) Is an integrated farm more resilient against climate change? a micro-econometric analysis of portfolio diversification in African agriculture? Food Policy 35:32–40
Seo SN (2010c) Managing forests, livestock, and crops under global warming: a microeconometric analysis of land use changes in Africa. Aust J Agric Resour Econ 54(2):239–258
Seo SN (2011) An analysis of public adaptation to climate change using agricultural water schemes in South America. Ecol Econ 70:825–834
Seo SN (2012a) Adapting natural resource enterprises under global warming in South America: a mixed logit analysis. Economia 12:111–135
Seo SN (2012b) Decision making under climate risks: an analysis of sub-Saharan farmers' adaptation behaviors. Weather Climate Soc 4:285–299
Seo SN (2014a) Evaluation of agro-ecological zone methods for the study of climate change with micro farming decisions in sub-Saharan Africa. Eur J Agron 52:157–165
Seo SN (2014b) Adapting sensibly when global warming turns the field brown or blue: a comment on the 2014 IPCC Report. Econ Aff 34:399–401
Seo SN (2015a) Micro-behavioral economics of global warming: modeling adaptation strategies in agricultural and natural resource enterprises. Springer, Cham
Seo SN (2015b) Helping low-latitude, poor countries with climate change. Regulation:6–8
Seo SN (2015c) Adaptation to global warming as an optimal transition process to a greenhouse world. Econ Aff 35:272–284
Seo SN (2016a) Modeling farmer adaptations to climate change in South America: a micro-behavioral economic perspective. Environ Ecol Stat 23:1–21
Seo SN (2016b) Untold tales of goats in deadly Indian monsoons: adapt or rain-retreat under global warming? J Extreme Events. https://doi.org/10.1142/S2345737616500019
Seo SN (2016c) The micro-behavioral framework for estimating total damage of global warming on natural resource enterprises with full adaptations. J Agric Biol Environ Stat 21:328–347
Seo SN (2016d) Microbehavioral econometric methods: theories, models, and applications for the study of environmental and natural resources. Academic Press (Elsevier), Amsterdam
Seo SN (2017a) Beyond the Paris agreement: climate change policy negotiations and future directions. Reg Sci Policy Pract 9:121–140
Seo SN (2017b) The behavioral economics of climate change: Adaptation behaviors, global public goods, breakthrough technologies, and policy-making. Academic Press (Elsevier), Amsterdam

Seo SN (2017c) Measuring policy benefits of the cyclone shelter program in North Indian Ocean: protection from high winds or high storm surges? Climate Change Econ 8(4):1–18

Seo SN (2018a) Natural and man-made catastrophes: theories, economics, and policy designs. Wiley-Blackwell, Oxford

Seo SN (2018b) Will farmers fully adapt to monsoon climate? J Agric Sci (under revision)

Seo SN (2018c) Two tales of super-typhoons and super-wealth in Northwest Pacific: will global-warming-fueled cyclones ravage East and Southeast Asia? J Extreme Events. https://doi.org/10.1142/S2345737618500124

Seo SN, Bakkensen LA (2017) Is tropical cyclone surge, not intensity, what kills so many people in South Asia? Weather Climate Soc 9:71–81

Seo SN, Mendelsohn R (2008) Measuring impacts and adaptations to climate change: a structural Ricardian model of African livestock management. Agric Econ 38:151–165

Seo SN, Mendelsohn R, Munasinghe M (2005) Climate change and agriculture in Sri Lanka: a Ricardian valuation. Environ Dev Econ 10:581–596

Seo SN, Mendelsohn R, Dinar A, Hassan R, Kurukulasuriya P (2009) A Ricardian analysis of the distribution of climate change impacts on agriculture across Agro-Ecological Zones in Africa. Environ Resour Econ 43:313–332

Smith P, Martino D, Cai Z et al (2008) Greenhouse gas mitigation in agriculture. Philos Trans R Soc B 363:789–813

Solow RM (1993) An almost ideal step toward sustainability. Resour Policy 19:162–172

Sumner DA, Zulauf C (2012) Economic & environmental effects of agricultural insurance programs. The Council on Food, Agricultural & Resource Economics (C-FARE), Washington, DC

Thaler R, Rosen S (1976) The value of saving a life: evidence from the market. In: Terleckyj N (ed) Household production and consumption. National Bureau of Economic Research, Cambridge, MA

Thom R (1975) Structural stability and morphogenesis. Benjamin-Addison-Wesley, New York

Tol RSJ (2009) The economic effects of climate change. J Econ Perspect 23:29–51

Udry C (1995) Risk and saving in Northern Nigeria. Am Econ Rev 85:1287–1300

United Nations (1972) Report of the United Nations conference on human environment. Stockholm, Sweden

United Nations (2000) United Nations millennium declaration. UN, New York

United Nations Children's Fund (UNICEF) (2018) Emergency and disaster relief. The UNICEF, New York. Accessed from https://www.unicefusa.org/mission/emergencies

United Nations Development Program (UNDP) (2018) Our funding: top contributors. UNDP, New York. Accessed from http://www.undp.org/content/undp/en/home/funding/top-contributors.html

United Nations Environment Programme (2017) The emissions gap report 2017: a UN environment synthesis report. UNEP, Nairobi

United Nations Framework Convention on Climate Change (UNFCCC) (1992) United Nations framework convention on climate change. UNFCCC, New York

United Nations Framework Convention on Climate Change (UNFCCC) (2009) Copenhagen accord. UNFCCC, New York

United Nations Framework Convention on Climate Change (UNFCCC) (2010) Cancun agreements. UNFCCC, New York

United Nations Framework Convention on Climate Change (UNFCCC) (2011) Report of the transitional committee for the design of Green Climate Fund. UNFCCC, New York

United Nations Framework Convention on Climate Change (UNFCCC) (2012) Decisions adopted by the conference of the parties on its eighteenth session, held in Doha from 26 November to 8 December 2012. UNFCCC, New York

United Nations Framework Convention on Climate Change (UNFCCC) (2015) The Paris agreement. Conference of the parties (COP) 21. UNFCCC, New York

United Nations System Chief Executive Board (UNSCEB) (2018) UN system. UN, New York. Accessed from http://www.unsceb.org/directory

United States Department of Agriculture (USDA) (2016) California drought: farm and food impacts. USDA, Washington, DC. Available at: https://www.ers.usda.gov/topics/in-thenews/california-drought-farm-and-food-impacts/
von Thunen JH (1826) The Isolated State. (trans Carla M. Wartenberg). Pergamon Press, Oxford, New York [1966]
Weitzman ML (2009) On modeling and interpreting the economics of catastrophic climate change. Rev Econ Stat 91:1–19
World Bank (2008) World Development Report 2008: agriculture for development. World Bank, Washington DC
World Bank (2009) Awakening Africa's sleeping giant: prospects for commercial agriculture in the Guinea Savannah Zone and beyond. World Bank and FAO, Washington, DC
World Food Programme (WFP) (2018) Overview. The WFP, Rome. Accessed from http://www1.wfp.org/overview
World Meteorological Organization (WMO) (1985) Report of the international conference on the assessment of the role of carbon dioxide and of other greenhouse gases in climate variations and associated impacts. WMO, Paris
World Resources Institute (WRI) (2005) World resources 2005: the wealth of the poor: managing ecosystems to fight poverty. WRI, Washington, DC
Yohe GW, Schlesinger ME (1998) Sea level change: the expected economic cost of protection or abandonment in the United States. Clim Chang 38:337–342

Chapter 8
Economics of Global Funds: United Nations Specialized Funds and Other Crypto, Crowdfunding, Green Funds

Abstract This final chapter of the book expands the discussion of the Green Climate Fund (GCF) to other global funds. Four important old and new global funds are reviewed: United Nations funds and specialized programs, crowdfunding, Bitcoin and cryptocurrencies, and various green funds such as the European Union's CO2 permit. The review shows that many UN specialized programs share similar characteristics with the GCF and have overlapping areas of funding. The GCF is a type of crowdfunding for the global climate programs in which member nations are donors, but a majority of the crowdfunding projects are politically motivated appealing to a certain segment of the society. The Bitcoin system is transparent in creating and determining the value of a bitcoin through currency exchanges, but the GCF system relies on the GCF Board's implicit determination of the value of a ton of CO2-equivalent emissions. Many green funds such as the EU's emission trading scheme or the sulfur dioxide allowance trading in the US can be adapted to match the GCF's project funding allocations, but the global scope of the GCF works would still remain the biggest challenge. Many important lessons can be learned from the past experiences of the green funds, which would augment the importance of the economic concepts presented in this book.

Keywords United Nations · Specialized funds · Crowdfunding · Bitcoin · Green funds · Emissions trading scheme

8.1 Introduction

This book has, up to this point, clarified and evaluated in depth the economic aspects of the Green Climate Fund (GCF) from many different angles. In particular, the present author provided a rare policy exercise with regard to the GCF by offering a theoretical framework and empirical evaluations of the GCF's allocations of the global money to selected projects from those proposed by developing country parties. The book evaluated each of the GCF fund allocations, about 93 projects in all as of December 2018, that have taken place since the first funding allocation to a batch of projects in November 2015.

S. N. Seo, *The Economics of Global Allocations of the Green Climate Fund*,
https://doi.org/10.1007/978-3-030-18274-8_8

In Chap. 2, I began with a comprehensive review of the GCF, covering the entire spectrum of the organization: its conception, history, governing instrument, institutional elements such as governing board and its secretariat, current pledges and contributions, investment criteria, application procedures, and approved and funded projects. The multifaceted review of the GCF was conducted in tandem with the explanations of major results and agreements from the United Nations Framework Convention on Climate Change (UNFCCC)'s annual conferences of parties since the establishment of the UNFCCC in 1992 (UNFCCC 1992).

From Chaps. 3, 4, 5 and 6, the book elucidated the four scientific traditions and their key predictions that are pertinent to the evaluation works of the GCF's principles and funding decisions and are utilized as such in the succeeding chapter, i.e., Chap. 7. Each of the four traditions is all rooted on a conceptual theory and comprises a family of models, based upon which a considerable basket of empirical results and predictions has been produced and promoted for a policy decision process.

More specifically, Chap. 3 elaborated the tradition of the microbehavioral economics of adaptation to climate change (Seo 2006, 2010a, 2016a). Chapter 4 described the agro-economic modeling tradition of quantifying the impacts of climate change on agriculture (Adams et al. 1990; Parry et al. 2004). Chapter 5 clarified the statistical tradition of estimating yield changes of a particular productivity indicator such as maize or rice yield (Schlenker and Roberts 2009). Chapter 6 explicated the ecosystem impact modeling tradition that integrates the changes in ecosystems induced by climatic shifts with the changes in economic indicators such as agricultural productivity (Fischer et al. 2005; FAO/IAASA 2012).

The reviews of the four traditions were followed by Chap. 7 where each of the GCF projects was evaluated by reflecting upon the salient empirical results and predictions from the four scientific traditions as well as the relevant literature on ocean-based climate events such as hurricanes. The evaluation of each project was conducted with regard to the three criteria which are three fundamental economic concepts: a theory of public goods and public expenditure, a Ricardo-Pareto theory of efficiency in resource allocations, and a theory of public adaptation to climate change (Ricardo 1817; Pareto 1906; Samuelson 1954; Seo 2011).

The ensemble of evaluations has revealed many novel but also perhaps surprising conclusions, some of the essence of which can be summarized presently. When it comes to the theoretical grounds of public goods and public expenditure, there is no quantitative information on climate change-induced damage in any of the GCF project documents. There is neither assessment in terms of avoided damage expressed in US dollars nor assessment of avoided damage in terms of the degree of reduced global temperature increase that may be achievable by a proposed program (UNEP 2017).

With absence of this critical information, the GCF funding by the GCF Board, especially in the category of the energy-generation-access, is not directed to the countries/recipients where such funding is deemed most effective in terms of the reduced damage or the reduced rate of global warming from the allocation of the fund. Rather, the funding is, in general, selectively given to under-developed poor countries in Africa, Latin America, South Asia, and Pacific Islands.

From the vantage point of the theoretical grounds of public adaptation, a careful examination of the GCF funding portfolio reveals that funding recipients, called accredited entities and executing entities by the UNFCCC, are often major international policy organizations such as the United Nations Development Programme (UNDP), the Inter-American Development Bank (IDB), and the Asian Development Bank (ADB). Non-international recipients are national policy organizations: To name but a few, the Agency for Agricultural Development of Morocco (ADA) and the Peruvian Trust Fund for National Parks and Protected Areas (Profonanpe).

When a public adaptation is needed and should be supported, to whom the funding should be directed is a central policy question. That is, a funding recipient should be an appropriate level of the public sector or the community at which level the public project needs to be implemented efficiently. The funding decisions to the GCF projects that were funded since late 2015 have sidelined this key policy question.

The fund-allocated GCF projects are in a mismatch with the efficient level of the public sector interventions for each project. To be more specific, a locally efficient adaptation project is awarded to a national level agency. A district-level efficient adaptation project is awarded to an international-level policy agency. The mismatch between the efficient level of a public adaptation project and the level to which the funding is actually allocated will result in large efficiency losses in the eventual outcomes of the funded projects.

The third evaluation, the Ricardo-Pareto efficiency test, uncovered several intriguing observations. The analysis in Chap. 7 reveals that 12 out of 14 total GCF projects on the project category of food security expressed the ultimate goal of the project as 'resilience' while each of these resilience projects proposes a different way for achieving the resilience. The proposed routes for the attainment of resilience by the funded project include an 'integrated' management, 'diversification' of agriculture, a 'community-based' management, an 'ecosystem-based' management, 'climate information' services, and a 'resilience fund' to climate change (GCF 2018c).

It is conspicuous that none of the 14 food security category projects is proposed on the grounds of efficiency or commensurate concepts such as optimality, efficient adaptation, and the best options. With the absence of a policy goal of efficiency or optimality expressed in the proposal documents, these GCF projects would most certainly fail to pass the efficiency test, barring a magical strike of serendipity.

The present author also pointed to the empirical evidence that the food security projects fail to pass the efficiency test and will even result in creating perverse incentives for the recipients of funding. It was pointed out that many funded projects in the category of food security were awarded by the GCF Board to the currently most vulnerable geographical areas in a vulnerable country to the current weather/climate conditions or to the most fragile of the agricultural systems at the present climate regime. Put differently, the food security funding allocations do not assess whether these vulnerable areas or systems will remain competitive or not in the altered climate regimes (Seo 2015).

Without identifications of the best or efficient practices under the future climate regime either by the GCF Board or the funding recipients, the funding is most likely to be used up on non-related or unintended purposes. More specifically, the funding programs would only serve as a development aid to a country, or an income redistribution to the poor, or even an international aid to weather-related disaster refuges. As such, there would be no climate-change policy purposes to be served directly.

Furthermore, if the funding recipients were to perceive the GCF funding as a foreign aid—a development aid—to the poor or an income redistribution from the rich to the poor, it will force the recipients to stick perversely to the present form of agricultural enterprise, that is, the most vulnerable form of agriculture (Seo and Mendelsohn 2008; Seo 2010b).

The project evaluations in the previous chapter went further to argue that many funding programs by the GCF can actively encourage mal-adaptation of the recipients. A funding to establish an irrigation system in the semi-arid regions, as was the case in the project funding given to Morocco, would force the recipient regions to stick to or switch to an irrigated crop agriculture. The semi-arid climate regime has, however, been a favorite ecosystem for non-crop agriculture such as livestock management owing to its vast grasslands and rangelands as well as low humidity (Dudal 1980; Seo 2014, 2016b). An irrigated form of crop agriculture, if established or expanded through the foreign funds from the GCF, would discourage semi-arid farmers from transforming to a more profitable system such as livestock management or a more profitable species such as cattle and sheep (Seo and Mendelsohn 2008). In this way, the GCF funding could induce mal-adaptation of funding recipients and therefore make semi-arid farmers more vulnerable. Many mal-adaptation events that could be induced by the current GCF funding programs were explained in Chap. 7.

The theories, empirical traditions, and project evaluations of the Green Climate Fund conducted thus far in this book can be extended with some modifications to a generalized analysis of the global funds. From this point on, the present author takes the first step toward the possibility of such a generalized analysis by introducing a host of similar global funds and exploring their relationships with the GCF.

The endeavor in search for a general analysis allows the present author to explain a range of global funds and programs that are many and diverse as well as their economic and political environments, making this book vastly more general and provocative. On the other hand, as will become more evident through this chapter, the examinations of other global funds, old and novel, will also offer readers another opportunity to look at the Green Climate Fund in comparisons with and in the contexts of the other global funds and currencies to be accounted for shortly.

Of the full range of global funds at work today, this chapter takes comprehensively four major global funds and global currencies, each of which offers unique as well as novel features and has attracted much global attention in the recent years. They are United Nations' specialized funds and programs, crowdfunding such as GoFundMe, cryptocurrency such as Bitcoin, and green (environmental) funds such as the European Union's CO2 emission allowance trading.

Of the host of global funds, the most well-known to the general public, especially in the recipient countries, is perhaps a foreign aid fund and program, which we may consider as the fifth type of the global funds. In Chap. 7, the present author alluded to some of the largest foreign aid programs. The foreign aid programs have a long history dating back to the early 1900s. By contrasts, a climate-related funding program began only recently, for example, the first GCF funding allocations in November 2015.

At a finer level, there are many different types of a foreign aid program. By the primary donors, foreign aids can be classified as a civil-group-originated, a religious-group-originated, a national-agency-originated, and an international-organization-originated. One of the largest national-agency-originated foreign aids is the Japan International Cooperation Agency (JICA). At an international organization level, numerous foreign aid programs are operating under the umbrella of the United Nations. A civil-group-originated foreign aid program is the Médecins Sans Frontieres (MSF), also known as Doctors Without Borders, of French origin with an annual expenditure of approximately US$1.9 billion and with more than 30,000 employees in 2017 (MSF 2017). Another civil-origin aid program is the Save the Children Fund of UK origin, established in 1919, with 29 national members as of 2018. One of the largest religious-group-originated foreign aid programs is the Tzu Chi Foundation's international disaster relief, of Taiwanese Buddhism origin, with offices in more than 50 countries (Tzu Chi 2018).

In Chap. 7, the two foreign aid programs by the United Nations were discussed prominently for their similarities to the GCF's climate-related funds. The United Nations' International Children's Emergency Relief Fund (UNICEF)'s numerous relief funds have climate disaster components, albeit indirectly: earthquake relief, children in conflict, food crises, flood relief, hurricane relief, tsunami relief, child refugee and migrant crisis (UNICEF 2018a, b). The World Food Programme (WFP) is another United Nations' aid program focusing on the humanitarian food-assistance. The WFP provides food assistance to about 80 million people in about 80 countries annually (WFP 2018a, b).

The large array of foreign aid programs has a century-long history, therefore many of them are well-known to the general public in both donor countries and recipient countries. In the previous chapter, I emphasized that the GCF-funded programs share many similarities with the existing foreign aid programs on multiple facets. The foreign aids are one of the oldest international fund transfers, the primary concern of this book.

Of the range of foreign aid or development aid programs, many are operating under the structure of the United Nations funds and programs. The aforementioned UNICEF and the WFP are considered specialized funds and programs of the United Nations. Other UN specialized funds include the United Nations Development Programme (UNDP), United Nations Enviromnent Programme (UNEP), Food and Agriculture Organization (FAO), United Nations International Strategy for Disaster Reduction (UNISDR), International Fund for Agricultural Development (IFAD), World Meteorological Organization (WMO) (UNSCEB 2018). These programs have separate budgets from those of the United Nations, that is, the United Nations general budget and the United Nations peacekeeping/security budget (UN 1945).

The United Nations Framework Convention on Climate Change (UNFCCC) is also considered a United Nations specialized program. The Green Climate Fund has a separate budget and expenditure from either the United Nations or the UNFCCC.

This chapter will explain in detail the United Nations' numerous funds and programs with a different legal status and relationships with the United Nations. For several major UN funds, the present author will highlight their budgets, expenditures, how contributions to their budgets are made, and how the outcomes of their expenditures to certain projects are assessed, attempting to approximate the evaluations in Chap. 7.

Thenceforth, it is followed by the examinations of the three other global funds' transfers that are relatively novel therefore not well grasped by the public and researchers: crowdfunding, cryptocurrency, and a green fund. Among the three, there is a difference in the scale of circulation at the present time. The cryptocurrency such as the Bitcoin is global in its generation and circulation, at least in the conceptual level. That is, anyone in the world who can access the internet can create the bitcoins according to the protocol proposed by its founder (Nakamoto 2008). The CO2 emissions allowance is implemented at a regional level, for example, like the European Union's Emission Trading Scheme (ETS) (EC 2016). The crowdfunding such as Kickstarter and GoFundMe is used predominantly for supporting various personal and business projects within a country of its origin, but projects intended for international recipients outside the country of the fund creator are increasing.

Through the three novel global currencies and funds, along with the evaluations of the Green Climate Fund, we hope to gain generalized insights into the globally transferred funds. All of the three funds are not issued by the central bank of any sovereign nation. For the Bitcoin, there is no central bank that is responsible for printing the currency, but can be traded globally for exchanges with any major international currency. The EU's CO2 emissions permit is not issued by the European Central Bank, but the permit is nearly equivalent to the Euro paper money within the Euro Zone. For the crowdfunding which is more a financing tool, a project creator generates the fund that will be directed to specified works and recipients of the project.

These non-traditional funds that are transferred internationally are of themselves much interest to academic economists as well as financial investors (Ellerman and Buchner 2007; Freeman and Nutting 2015; Böhme et al. 2015; Liu and Tsyvinski 2018; Shiller 2018; CNBC 2018). The experts agree broadly that these funds are a 'new' type of currency and are debating about the opportunities and risks that these funds could unleash. However, there is still little attention paid to these strange funds as a global currency with their unique potential and disruptions. This chapter will bring forth some of the prominent characteristics of a global fund/currency via these new and emerging funds.

From another viewpoint, a review and analysis of the novel global funds will offer additional insights to the economic evaluations of the GCF funding projects provided in this book up to Chap. 7. As will be made clearer, the GCF shares many similarities with the crowdfunding organized for a particular cause or event of concern to the project creator. Like the CO2-equivalent emissions permit, the GCF funding is a sort of a green permit given to the fund recipients which commands a dollar value. The GCF funding also takes on the aspect of the cryptocurrency in that the bitcoin is created without borders and allocated to a participant (miner) who successfully performs a desired task, that is, updating a block chain record.

This chapter is composed of the following sections. In the next section, the United Nations funds and programs are explained, with the details on their budgets, expenditures, and contributions. Section three introduces the development of crowdfunding, along with major crowdfunding platforms and major notable projects. In Sect. 8.4, the internet protocol that gave birth to the Bitcoin system is presented with analyses of currency exchanges, bitcoin price history, comparisons with the GCF. Section 8.5 provides a review of the theory and historical developments of the CO2 emission permit and examines the possibility of the price of the permit being utilized as a price signal for the GCF funding decisions. The chapter is concluded with Sect. 8.6 where the present author outlines a set of guidelines for the GCF investment decisions, drawing from the lessons learned from the global funds and currencies.

8.2 United Nations: Funds, Programs, and Specialized Agencies

As far as the international funds and their allocations are concerned, the United Nations is by far the largest organization for such monetary transfers. The Green Climate Fund (GCF) is also one of the specialized agencies of the United Nations Framework Convention on Climate Change (UNFCCC) which belongs to the United Nations System. The GCF is a special program of the UN, whose budget is not included in the UN's regular budget but is supported by voluntary contributions by member countries and corporations, which is also true of many UN's specialized programs such as the UNICEF and the World Food Programme.

For the book on the economics of allocations of global funds, it is inevitable for the present author to provide a comprehensive exposition on the UN's numerous funds, their revenues and distributions. Further, it is not an exaggeration to say that the GCF's framework is laid out after the UN's existing funds and programs (UN 1972).

The United Nations was created by the Charter of the United Nations signed in 1945 after the two catastrophic wars in the first half of the twentieth century that the world had never seen. The foremost purpose of the United Nations declared by Article 1 of the Charter was the maintenance of international peace and security (UN 1945):

> To maintain international peace and security, and to that end: to take effective collective measures for the prevention and removal of threats to the peace, and for the suppression of acts of aggression or other breaches of the peace, and to bring about by peaceful means, and in conformity with the principles of justice and international law, adjustment or settlement of international disputes or situations which might lead to a breach of the peace;

The Article 1 also declared international cooperation on numerous international economic, social, cultural, and humanitarian issues as one of the purposes for the establishment of the United Nations (UN 1945):

> To achieve international cooperation in solving international problems of an economic, social, cultural, or humanitarian character, ...;

The activities of the United Nations are financially supported by financial contributions by the member nations who are obligated under the United Nations Charter. Without financial contributions, member countries are excluded from the benefits, including security, that the United Nations provide. Member contributions are determined by the United Nations committee, considering many factors such as the size of economy.

The United Nations principal organs declared by the UN Charter are the following: General Assembly, Security Council, Economic and Social Council, Secretariat, International Court of Justice, and Trusteeship Council (UN 1945).

The United Nations' budgets are maintained separately for the general activities on the one hand and the security/peacekeeping activities on the other. The biennial budget for the UN general activities was set most recently at US$ 5.4 billion for the period starting July 1, 2018 at the conclusion of the 72nd meeting of the United Nations General Assembly. The approved budget for the security/peacekeeping activities for the fiscal year starting July 1, 2018 was US$ 6.7 billion (UN 2018).

Figure 8.1 shows percentage contributions of members to the UN regular budget for the aforementioned fiscal year. The United States paid 22% of the UN budget (US$ 591 million per annum and US$ 1.2 billion per biennium), the maximum threshold set by the UN for the individual member's contribution, which was followed by Japan (9.7%) and China (7.9%). They are followed by 10 other largest contributors to the UN's regular budget.

The present author added two developing countries' contributions to the figure: India and Kenya. Both countries, the former in South Asia and the latter in Africa, are one of the largest recipients of the UN's numerous funds and programs, which will be explained shortly. An India's contribution amounted to 0.7% while a Kenya's contribution amounted to 0.01% of the UN general budget.

The security and peacekeeping budget of the UN is separately maintained. Figure 8.2 shows percentage contributions of 10 largest contributors to the UN security/peacekeeping budget in the 2018–19 fiscal year. The United States contributed 28.5% of the budget (US$ 1.9 billion), which was followed by China (10.3%) and Japan (9.7%). They are followed by Germany, France, UK, Russia, Italy, Canada, Spain. The permanent members of the UN security council are the US, China, France, UK, Russia.

In addition to the two UN budgets, specialized agencies of the United Nations have their own budgets, which are separate from the aforementioned UN budgets and are sustained through voluntary contributions by various actors. According to the United Nations System Chief Executives Board (CEB), there are four different types of programs and organizations: UN funds and programs, specialized agencies,

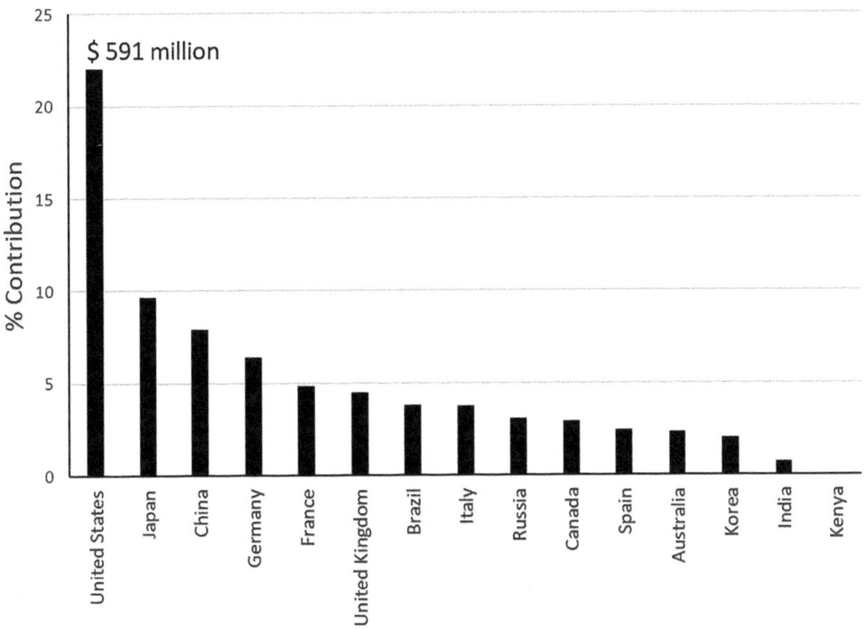

Fig. 8.1 Percentage contributions of members to UN general budget

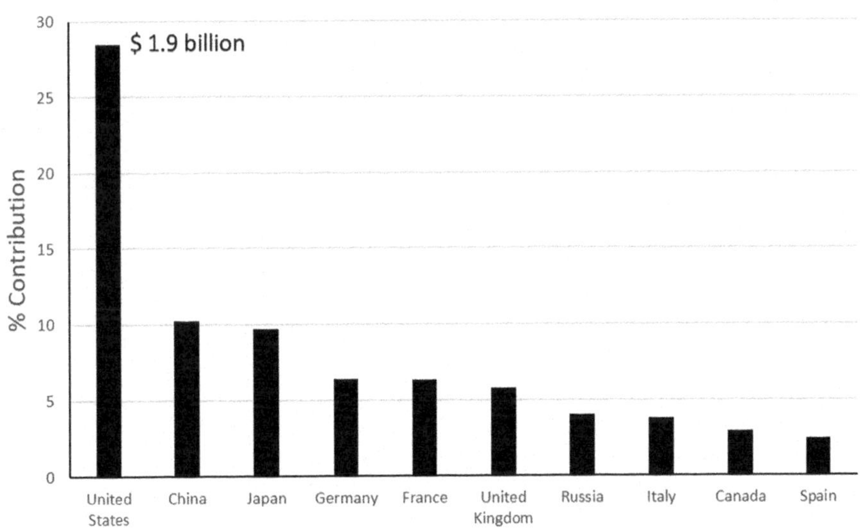

Fig. 8.2 Percentage contributions of members to UN peacekeeping budget

other UN entities, and related organizations (UNSCEB 2018). The UN funds, programs, specialized agencies, and related agencies are summarized in Table 8.1.

The UN funds and programs are created by the UN to meet the needs not envisaged at the meeting in San Francisco in 1945. Of the four types of organizations, the specialized agencies and other UN entities are mostly similar: They are legally independent international organizations which are in relationship with the UN through international agreements. Related organizations are also international organizations that are closely related to the UN through negotiated agreements. In addition to these specialized agencies, there are five regional commissions of the United Nations, for example, the United Nations Economic Council for Africa (UNECA), and the UN research and training institutes such as the United Nations University.

All organizations listed in Table 8.1 derive their financial resources independently, that is, not from the UN budgets. The table shows annual contributions/revenues, annual budgets, and annual expenses of the five specialized programs of the United Nations: UNDP, UNEP, UNFPA, UNICEF, and WFP. According to the most recent reports from these programs, the annual revenue in the most recent year amounted to US$ 5.2 billion for the UNICEF, US$6.7 billion for the WFP, US$ 1.06 billion for the UN Population Fund (UNFPA), and US$889 million for the UNEP. The annual expenses amounted to US$ 5.4 billion for the UNICEF, US$ 2.6 billion for the UNDP, and US$ 953 million for the UNEP. The annual budget is equivalent to US$5.6 billion for the UNDP, US$ 683 million for the UNEP (UNDP 2018; UNEP 2018; WFP 2018b; UNICEF 2018b).

For these specialized programs and agencies, funds are raised from non-governmental sources as well. For example, the World Food Programme's major donors are corporations, both private and non-profit, as well as individuals. In addition, governmental contributions are the principal source of funding for the WFP, but the WFP receives 'no dues or portions of the UN assessed contributions (WFP 2018b).'

The UN's regular budget and the security/peacekeeping budget are determined by the UN's committee by considering many factors. These budgets can be set higher or lower than the existing budgets. In recent years, Trump administration sought to cut the US contributions to the UN budgets. He announced a reduction of US$ 285 million annually to the UN operating budget for the fiscal 2018–2019 year (Fox News 2018). This was in addition to the annual US$ 200 million reduction proposed by the UN Secretary-General in its regular budget. The UN peacekeeping budget was slashed by US$ 600 million under the pressure from Trump administration (UN 2018).

President Trump argued that although the US is the largest financial contributor to the United Nations, the decisions made by the UN or the UN Security Council often contradict the interests of the United States. He argued that other rich countries should contribute more to the UN budgets.

Table 8.1 does not yet list the Green Climate Fund as one of the UN System organizations, but list the UN Framework Convention on Climate Change (UNFCCC) as other entities and bodies (UNSCEB 2018). The GCF is an independent international organization separate from either the UNFCCC or the UN.

Table 8.1 UN specialized programs and agencies: revenues and expenses

	Total annual contributions/ revenues	Annual budget	Annual expenses
UN funds and programs: Created by the UN to meet the needs not envisaged at San Francisco in 1945			
United Nations development Programme (UNDP)		$5.664 billion (2018)	$2.736 billion (2018)
United Nations Environment Programme (UNEP)	$889 million (2017)	$683 million	$953 million
UN Population Fund (UNFPA)	$1.06 billion (2017)		
UN Children's Fund (UNICEF)	$5.2 billion (2018)		$5.4 billion (2018)
World Food Programme (WFP)	$6.8 billion (in 2017)		
UN Human Settlements Programme (UN-Habitat)			
United Nations Conference on Trade and Development (UNCTAD)			
UN High Commissioner for Refugees (UNHCR)			
United Nations Relief and Works Agency for Palestine Refugees (UNRWA)			
United Nations Office for Project Services (UNOPS)			
United Nations Office on Drugs and Crime (UNODC)			
UN Women			
Specialized agencies: Legally independent international organizations, brought into by international agreements			
	Food and Agriculture Organization (FAO)		
	International Civil Aviation Organization (ICAO)		
	International Fund for Agricultural Development (IFAD)		
	International Labor Organization (ILO)		
	International Monetary Fund (IMF)		
	International Maritime Organization (IMO)		
	International Telecommunication Union (ITU)		
	UN Educational, Scientific and Cultural Organization (UNESCO)		
	United Nations International Development Organization (UNIDO)		
	World Health Organization (WHO)		
	UNWTO (World Tourism Organization)		
	Universal Postal Union (UPU)		
	World Intellectual Property Organization (WIPO)		
	World Meteorological Organization (WMO)		
	World Bank		

(continued)

Table 8.1 (continued)

	Total annual contributions/ revenues	Annual budget	Annual expenses
Other UN entities and bodies: not members of the Chief Executives Board for coordination of the UN system			
	Joint United Nations Programme on HIV/AIDS (UNAIDS)		
	International Trade Centre (ITC)		
	Office of the United Nations High Commissioner for Human Rights (OHCHR)		
	United Nations Framework Convention on Climate Change (UNFCCC)		
	United Nations International Strategy for Disaster Reduction (UNISDR)		
	Organization for the Prohibition of Chemical Weapons (OPCW)		
	Preparatory Commission for the Comprehensive Nuclear-Test-Ban Treaty Organization (CTBTO)		
Related organizations: similar to specialized agencies			
	International Atomic Energy Agency (IAEA)		
	World Trade Organization (WTO)		
	International Organization for Migration (IOM)		

The table indicates that the GCF's activities would overlap the activities of the UN System organizations listed in Table 8.1. The UNDP, UNEP, Food and Agriculture Organization (FAO), World Bank, United Nations International Strategy for Disaster Reduction (UNISDR), International Fund for Agricultural Development (IFAD), World Meteorological Organization (WMO), International Civil Aviation Organization (ICAO), and International Maritime Organization (IMO) are all heavily involved in the works of the UN Framework Convention on Climate Change while many activities of these organizations are also found in the funded projects by the Green Climate Fund in a similar form (UNFCCC 2015; UNEP 2016; ICAO 2016).

The UN funds listed in Table 8.1 also share with the GCF that the funds are generated from voluntary contributions of member countries, but also from corporations' and individuals' contributions. The contributions by corporations and individuals are particularly larger in the Green Climate Fund.

The UN funds listed in Table 8.1 and the GCF also share the same criticism that the funding (investment) decisions as well as project outcome evaluations are not rigorously defined for these funds and programs. When presented with multiple possible projects, there is no scientific guideline on which project should be ultimately selected and given the funding. Further, there is rarely a scientific quantitative evaluation of a completed/funded project supported by these funds and programs. Detailed analyses with regard to these points were offered in Chap. 7 in the contexts of the Green Climate Fund.

8.3 Crowdfunding

Crowdfunding is a novel method utilized to finance a project, an individual, or an event through small contributions made through an online platform by a large number of donors, that is, the crowd. The concept, albeit ancient, became a powerful financing tool after a personal computer (PC) such as Microsoft made revolutionary changes in the public data exchanges. It expanded exponentially with the remarkable success of smartphones such as iPhone.

The Green Climate Fund (GCF) can be interpreted as a form of crowdfunding because it relies on voluntary pledges by individual countries and donors and its allocations of the funds to selected projects are determined by the originator, that is, the GCF Board.

The history of crowdfunding may be as long as the major human civilizations. It is not hard to find a crowdfunding project in the early civilizations, that is, a project voluntarily funded by small contributions by a large group of donors. But, a crowdfunding project through an online platform emerged only recently.

There are three actors in the crowdfunding: a crowdfunding platform, a project campaign, donors. A project campaigner seeks donations from potential donors through the crowdfunding platform. The crowdfunding platform charges the project campaigner a service fee. In the Kickstarter platform to be described presently, a successful campaign is required to pay a fee of 5% of the total donation.

In the United States, a widely-publicized fan-funding website, named ArtistShare.com, was launched in 2003 by a Boston musician and computer programmer, Brian Camelio. The ArtistShare is the first internet crowdfunding platform. The ArtistShare platform was set up so that fans could finance production costs of an album by their artist. Through the platform, musicians were allowed to seek donations from their fans with a promise of producing digital recordings using the fan donations (Freeman and Nutting 2015).

The first crowdfunding project by the ArtistShare was American composer Maria Schneider's jazz album "Concert in the Garden" in 2003. The campaign raised about US$ 130,000. The donations were used to cover production expenses such as composing the music, paying her musicians, renting a recording studio, and producing and marketing the album through the ArtistShare website. The album won a 2005 Grammy Award for best large jazz ensemble album. Since the launch, the ArtistShare projects have won 10 Grammy Awards.

8.3.1 *Rewards-Based Crowdfunding*

The crowdfunding model set by the ArtistShare is a rewards-based crowdfunding. In the Schneider's album project, fans who donated to the project were rewarded according to the sizes of their contributions. The fans who contributed US$ 9.95 were given the right to first download the album upon its release before the public's downloads. The donors who contributed US$ 250 or more were given the recognition in the album booklet as participants who helped to make the recording possible. The fan who contributed US$ 10,000 was listed as executive producer of the album (Freeman and Nutting 2015).

Owing to the success of the ArtistShare, many rewards-based crowdfunding platforms followed suits. Besides the crowdfunding projects for arts, the new platforms allowed crowdfunding campaigns on social causes in numerous areas such as animals, community, education, environment, health, politics, and religion. Further, crowdfunding campaigns on entrepreneurs and small businesses in the areas of food, sports, gaming, publishing, and technology were soon followed.

Among the most prominent rewards-based crowdfunding platforms is Kickstarter which launched in 2009. Key performance statistics of the Kickstarter platform are presented in Table 8.2 (Statistica 2018). In the Kickstarter platform, there were 265,000 campaigns during the period from its launch to October 2015, of which 95,200 (36%) campaigns were successful campaigns, that is, the campaigns that achieved their stated funding goals.

Table 8.2 Key statistics of Kickstarter campaigns, as of october 2015

Total campaigns (N)	Successful campaigns (N)	Total size of donations	Total size of backers (N)	Percent of campaigns that raised more than x	Top ten business categories (by April 10, 2018)
265,000	95,200 (36%)	$1.76 billion (by October 2015);	9.7 million (by October 2015);	29% ($x = \$10{,}000$);	Music (27,284)
		$3.6 billion (by April 2018)	14.45 million (by April 2018)	2.6% ($x = \$100{,}000$)	
					Film & video (24,612)
					Games (13,330)
					Publishing (12,740)
					Art (12,018)
					Design (11,057)
					Technology (6732)
					Theater (6596)
					Food (6303)
					Comics (6185)

Note: Data are from Statistica (2018)

During this period, there were 9.7 million donors to various Kickstarter projects, whose contributions added up to about US$ 1.76 billion. Of the successful Kickstarter campaigns, 29% of them raised more than US$ 10,000 and about 3% of the successful campaigns raised more than US$ 100,000. By April 2018, the total size of donations reached US$ 3.6 billion and the total number of donors reached 14.5 million.

During the six-year period, the music and film/video category was by far the most successful business category as measured by the number of a successful campaign, which was followed by the business categories in art, publishing, games, design, and theater (Statistica 2018). However, some of the Kickstarter campaigns not in the categories of music, art, film, games were widely successful in raising donations as well as promoting the campaigns to the public.

One of the most successful Kickstarter campaigns was the Pebble smart watch campaign in 2012 by a group of entrepreneurs from Palo Alto, California. The smart watch was a digital wristwatch that runs downloadable sports and fitness apps, connecting wirelessly to a smart phone. The Palo Alto team set the fundraising goal of US$ 100,000, but eventually raised more than US$ 10 million from about 70,000 contributors during the funding period from April to May in 2012.

Other than the Kickstarter, there are other rewards-based crowdfunding platforms with a narrow product focus or with a focus on a niche market. To note some of such platforms, Experiment.com is a crowdfunding website for scientific research projects. The rewards to the donors to the Experiment.com campaigns are described as "insight behind the science" offered by the sponsored science project teams. For another example, Plum Alley is a crowdfunding platform for women entrepreneurs (Freeman and Nutting 2015).

Of the rewards-based crowdfunding campaigns, there are two types of funding model: An all-or-nothing model, a keep-it-all model. In the all-or-nothing funding model, only the campaigns that meet the stated funding goal within the preset deadline are funded. The campaigns that fail to meet the funding target should walk away with nothing and donors' credit cards are not charged in those campaigns.

In the keep-it-all funding model, on the other hand, a project campaigner is allowed to keep all the donations even if the predetermined funding target is not reached within the deadline set by the campaign.

8.3.2 Donors' Motivations, Risks, and Fraud

Why do people donate to a crowdfunding project? One of the primary motivations is to receive the rewards from the completion of the project. The more attractive the reward is, the more people would contribute to the campaign.

Some backers, however, may simply want to support a particular project without expecting a financial or material reward. In the rewards-based crowdfunding campaign, a backer should donate at least the minimum amount set by the project to receive a reward after the project's completion. Empirical data show that many backers donated less than the minimum threshold, which indicates that people donate not just for a material/financial reward.

For the second type of donors who do not seek a reward in return for the giving, it is still important for them to keep track of the project's progress and witness the project's coming into fruition. Therefore, it can be more properly said that they are placing the value, or satisfaction, on the project itself when they donate, regardless of whether the benefit from the project falls upon them or not.

In donating to a crowdfunding campaign, a donor assumes several risks, including the risk of a complete fraud by a project campaigner. First, a very common risk to the crowdfunding donor is that there is no certainty that the funding campaign would be successful in reaching the funding goal. Even if the campaign should be successful, there would be no guarantee that the project deliver the promises and do so on time. In the Kickstarter platform, at least 70% of the successful campaigns failed to deliver the promised products within the project deadlines (Freeman and Nutting 2015).

Second, there is a higher probability of a fraud in the crowdfunding than in other investments in a company's product, say, purchasing a bundle of stocks. In an outright fraud scenario, a startup company may ask donations through a crowdfunding platform and receive hundreds of thousands of dollars from a group of donors, but run away with the donation without even attempting to deliver a product. In addition, there can be other more obscure forms of fraudulent behaviors by the crowdfunding campaigner.

A study by Mollick gathers empirical data and offers early insights on the risk of fraud in crowdfunding projects. Based on his 2013 survey of about 48,000 Kickstarter projects, he reported that a failure to deliver promised rewards was about 9% of all projects (Mollick 2015). In the category of technology and design, the fraud rate measured by undelivered pledges was less than 1%.

Mollick attributes the low fraud rate found in Kickstarter projects, among other things, to a sort of crowd effect, that is, influence of the online community. In an online crowdfunding campaign, donors interact each other closely and also interact with the campaign creator, forming a strong community unlike in other traditional investments into a company. One of the crowd effects is that the presence of the crowd could deter its members from acting fraudulently.

The crowd effect also plays a key role in the Paris Agreements and the Green Climate Fund when the member submitted their Intended Nationally Determined Contributions (INDCs) and the voluntary pledges to the GCF (Seo 2017a, b). The strategy is often referred to as a 'name and shame' strategy.

However, an online-based crowdfunding campaign may also increase the risk of fraud through other opaquer routes. In an online platform, a crowd can be more easily manipulated by a smaller group, say, a sub-crowd within the crowd, which is another type of crowd effect. Also, a crowdfunding campaign is built on the premise of voluntary giving, which makes it harder for donors to enforce contract details strictly unlike the traditional financing tools. As such, a vast majority of funded projects do not deliver promises within the deadlines or the promised products.

8.3.3 Other Types of Crowdfunding

Other than the rewards-based crowdfunding, other types of crowdfunding have subsequently emerged: debt-based crowdfunding, donation-based crowdfunding, and equity-based crowdfunding platforms (Freeman and Nutting 2015). We need to ask whether these alternative crowdfunding methods can mitigate donors' risks described in the above.

The debt-based crowdfunding is Peer-to-Peer (P2P) lending or marketplace lending done through an online platform. It is regarded as an alternative to a bank loan. Each donation to the crowdfunding campaign is rewarded with the principle (donation) and an interest to the donation.

The donation-based crowdfunding is a crowdfunding based on pure donations, that is, without rewards expected. It is a traditional funding model used by charitable organizations but is done through an online platform in the crowdfunding.

The equity-based crowdfunding is done through sales of an equity which is purchased by donors. Considering the success of the rewards-based and the donation-based crowdfunding, financial intermediaries whose profession is to raise capital soon followed up by offering an online platform to startup companies from which equities can be offered to potential investors. Such trades are regulated by the Regulation D of the Federal Securities Act of 1933 (Douglas and Bates 1933).

The most successful equity-based crowdfunding platforms were MicroVentures in 2011 focusing on technology companies such as Twitter, Facebook, Yelp, and CircleUp in 2012 concentrating on computer products and retail.

8.3.4 GoFundMe

Another stage in the evolution of crowdfunding is reached by the GoFundMe platform for the reasons to be explained shortly. As of September 2018, the GoFundMe is the largest crowdfunding platform, raising more than US$ 5 billion since its first campaign in May 2010. The platform hosted more than 2 million campaigns and received donations from more than 50 million individual donors (GoFundMe 2018).

The GoFundMe shares many similarities with the other crowdfunding platforms explained above. There are several unique features that separate it from the others. First, it is a for-profit platform. The creator of a campaign should pay multiple fees: platform fee (5%), payment processing fee (2.9% + 0.3$ per donation). For the US, Canada, UK project creators, the platform fee of 5% is exempted.

Second, the platform is far more international. As of September 2018, the GoFundMe is available to the citizens of North America, Europe, and Australasia: Australia, Austria, Belgium, Canada, Denmark, Finland, France, Germany, Ireland, Italy, Luxembourg, Netherlands, Norway, Portugal, Spain, Sweden, Switzerland, United Kingdom, and the United States. It is planned that the platform be extended further to other countries in the near future.

Third, the GoFundMe is neither an incentive-based campaign in which donors are rewarded upon the success of a campaign nor a commercial campaign whose ultimate goal is commercial success. Instead, the platform supports a campaign for personal causes as well as a campaign for life's most important events. For example, it can be a campaign for donations for her/his medical expenses in treating a rare disease or for someone's memorial event (GoFundMe 2018).

The most successful or noticeable GoFundMe projects are summarized in Table 8.3. All-time top five projects raised donations that range from US$ 6.8 million to US$ 21.5 million. The all-time top five fundraising projects are dominated by the campaigns to help mass shooting victims in the United States. The other projects are about the sexual assault victims and the bus crash victims.

As shown in Table 8.3, the GoFundMe campaigns are often controversial politically-focused campaigns. Three politically motivated campaigns are listed in the Table: Support Officer Wilson from Ferguson, Missouri in 2014, March for Our Lives by Stoneman Douglas High School students in 2018, Support for FBI Veteran Pete Strzok in 2018.

Of particular interest to the readers of this book is that many GoFundMe campaigns were started as a response to various natural disasters such as hurricanes, floods, earthquakes, and fires. The Vieques Love, Puerto Rico Fund raised about US $1 million for the victims of Hurricane Maria that hit the small island in 2017. The Louisiana Floods in 2016 started multiple GoFundMe projects that raised more than US$ 11 million. The Ecuador earthquakes initiated multiple funding campaigns that generate US$ 1.5 million (GoFundMe 2017).

Also, many GoFundMe projects are related to an event of poverty, famine, refugees, and rare/fatal diseases. Further, some of these projects are directed to international recipients, as in the campaign entitled "Love Army for Somalian famine" (2017) and "Love Army for Rohingya refugees" (2017). As explained previously, these events are also the events to which development and foreign aids have been historically directed. In fact, in addition to the aids they receive through the GoFundMe projects, they receive the aids from other development and international aid programs.

8.4 Bitcoin and Cryptocurrencies

In the same way that the crowdfunding was made possible by the internet revolution exemplified by the world wide web, personal computers (PC), and smart phones, the bitcoin emerged from the internet revolution but in a more dramatic fashion. The bitcoin is the currency given birth purely by an internet protocol, that is, not by a national central bank of any nation and is becoming a truly international currency. It was reported only a week ago, at the time of this writing, that the most influential academic endowment manager, David Swensen from Yale University who manages almost US$ 30 billion university endowment, made investments into two cryptocurrency funds (CNBC 2018).

Table 8.3 The most successful and noticeable GoFundMe campaigns

Projects	Amount raised	Notes	Date
Time's Up Legal Defense Fund	US $21.51 million	The fund connects people who have experienced sexual harassment or assault in the workplace with public relations and legal teams	All-time top five
Las Vegas Victims' Fund	US $11.87 million	The fund was set up for the victims of the Oct. 1, 2017 shooting at a country music festival in Las Vegas where 58 people were killed	All-time top five
Support Victims of Pulse Shooting	US $7.85 million	The campaign set up to support the victims of the June 2016 shooting at Pulse Nightclub in Orlando, Fla., where 49 people were killed	All-time top five
Funds for Humboldt Broncos	US $7.3 million	The online fundraiser for the victims of the April 6 bus crash in Saskatchewan, Canada	All-time top five
Stoneman Douglas Victims' Fund	US $6.82 million	The fund was set up to support the victims of a Feb. 14, 2018 shooting at Stoneman Douglas High School in Parkland, Fla., which left 17 people dead	All-time top five
Love Army for Somalia	US $2.7 million	Campaign to provide famine relief for Somalis	2017 Famine
Love Army for Rohingya	US $2.0 million	Campaign to help Rohingya refugees from Myanmar	2017 Refugees
Charlie Gard, UK	£1.3 million	Charlie's parents started this campaign to try to get experimental treatment to save their son's life from a rare disease	2017 Diseases
Nabeel Qureshi Cancer Fund	US$959 K		2017 Diseases
Vieques Love, Puerto Rico	US$923 K	Help the tiny island of Vieques, off Puerto Rico's eastern coast, recover from Hurricane Maria	2017 Hurricane
Louisiana Floods	US $11.2 million		2016 Floods
Hurricane Matthew	US$3 million		2016 Hurricane
Ecuador earthquakes	US $1.5 million		2016 Earthquakes
West Virginia floods	US $1.3 million		2016 Floods
Fort McMurry fires	US $1.1 million		2016 Fires
Tennessee fires	US$1 million		2016 Natural disasters
Support Officer Wilson	US$183 K	In support of Ferguson, Missouri, police officer Darren Wilson, who shot and killed Michael Brown	2014 Political
March for Our Lives	US $3.5 million	Support the demonstration, created following the Stoneman Douglas High School Shooting by student Cameron Kasky	2018 Political
Support for FBI Veteran Pete Strzok	US$461 K	Support fired FBI agent Peter Strzok for his active engagements to stop Trump's 2016 election victory	2018 Political

As will be clarified in this section, the Green Climate Fund can become a global crypto currency that is automatically generated to the person who makes a certain contribution to GHG emissions reduction and that can be transferred from one to another beyond national boundaries. The 'GCF currency' will then become exchangeable with any national currency such as the US dollar or Euro. The system can be functional if there can be a process established to verify the emissions reduction by an individual. Without doubt, the 'GCF currency' will also give birth to similar side-effects to those witnessed in the transactions of the cryptocurrency.

The Bitcoin, the first cryptocurrency, is a virtual currency created and exchanged through an online communication protocol since 2009 when an unknown identity, which can be a person or a group, named Satoshi Nakamoto first uploaded the Bitcoin protocol entitled "Bitcoin: A Peer-to-Peer Electronic Cash System (Nakamoto 2008)." The Bitcoin is a global currency because it is issued by a certain internet protocol called a block chain, that is, not issued nor limited by a national government or a central bank.

A bitcoin can be exchanged with other currencies. Also, it can used to make electronic payments for goods and services. As of 2015, a daily transaction volume was US$50 million at market exchange rates and the total value of the bitcoins in circulation was $3.5 billion (Böhme et al. 2015). According to one analysis, the total value of bitcoins by market capitalization is predicted to pass US$1.2 trillion, as of October 2018 (Investopedia 2018).

8.4.1 Block Chain: Creation of Bitcoins

The basic transaction in the Bitcoin system is simple. The Bitcoin core software can be downloaded freely from its website bitcoin.org. Each user is given a public key which guarantees anonymity of the user as well as a wallet in which bitcoins are stored and spent. Consider the following three users—Alice, Brian, and Charlie—and transactions among them. When Alice sends five bitcoins electronically to Brian, five bitcoins are added to the Brian's wallet and subtracted from the Alice's wallet. If Brian sends five bitcoins electronically to Charlie, the same changes occur in the two persons' wallets.

The Bitcoin system is made possible through the block chain technology. The block chain data is the entire record of the Bitcoin transactions from the creation of the first bitcoin transaction. In the Bitcoin system, every user is given the access to the block chain data. Therefore, every user in the world that are in the Bitcoin system can see the above transactions are made among Alice, Brian, and Charlie. However, no one in the Bitcoin system can see who made those transactions. That is, the identities of the Bitcoin users are hidden in the system. Only public keys of the users can be seen.

How is the bitcoin created at all in the first place? The block chain technology is the key to the creation of the bitcoin and the record-keeping of the bitcoin transactions. A block is a set of transactions. It is a record of a set of transactions in the

Bitcoin system. When a set of transactions occur, a new block is created by a Bitcoin miner. A new block is then added to the historical chain of existing blocks, that is, the current block chain. The record-keeping of the entire history of bitcoin transactions is done by this block chain technology, solely by the Bitcoin software users.

A new block is updated every 10 min in the Bitcoin system. A block is created and added, instead of an individual transaction, because multiple online transactions can occur at the same time, that is, simultaneously. Bitcoin miners compete for the next block which is the history of newly submitted transactions in the Bitcoin system. Miners are required to solve a puzzle and whoever solves the puzzle first gets to append the new block to the existing chain of blocks.

Updating a new block is done through computer intensive methods, therefore, not any computer user can compete. Miners should purchase a specialized computing facility in order to compete at all. Further, it is time and energy consuming to solve a puzzle. Sine the size of the block chain is big, as big as 30 Giga Bites as of 2015, miners should purchase a storage device. Mining also requires heavy electric power consumption, 173 MW continuously. Further, the difficulty of a puzzle has gradually increased since the first Bitcoin transaction (Böhme et al. 2015).

The Bitcoin puzzle is an image with words, numbers, and symbols, within which are clues to the puzzle. The clues may reveal a private key or a seed key of the Bitcoin wallet which contains the bitcoins as the prize. The puzzle can be also a sequence of numbers within which secrete words are hidden.

To get the readers acquainted with how the bitcoin puzzle works, a Bitcoin puzzle is presented in Table 8.4 selected from the Bitcoin puzzles made public online (Reddit 2018). The puzzle is a sequence of numbers. The poster of the puzzle says that 12 words are hidden beneath the sequence of numbers. The reward to the miner who solves the puzzle is 0.18665328 unit of a bitcoin which is worth $1200.

Bitcoin miners are the bankers in the Bitcoin system in that they keep the record of Bitcoin transactions by competing for the next block chain. The Bitcoin system rewards a miner for the mining efforts with 25 bitcoins, having been reduced periodically from the reward of 500 bitcoins in the beginning of the Bitcoin system, every time a miner is successful in updating the block chain.

8.4.2 No Governance, Decentralization, Globalization

The uniqueness of the Bitcoin lies in that, other than the Bitcoin software, there is no governance structure in the Bitcoin monetary system. There is no central government or central bank that manages the creation of the money and its transactions. Instead, the system is maintained by the verification of the participants in the system while the entire history is verifiable from the block chain data. Anybody in the system can verify whether a specific transaction occurred or not.

This unique governance structure has resulted in several salient features of bitcoin transactions. First, there is no obligation in the Bitcoin system for a financial institution or an intermediary to verify the identities of its users. Nor is there any

Table 8.4 A bitcoin puzzle

Introduction	Beneath the sequence of numbers in the following are hidden 12 words. Find them
Puzzle	5199109198198809256200212281219820512
	2201501061199304251200208141220130429
	520010407220101204820091005119851024
	1196701214196111021120030620220120622
	6200408077198108171320060723720171106
	1219480106220051213419710509319950420
	120150106920120619319970706720080930
	119961004820011028419960107320110910
	419500701619870126919860320519560124
	1119580402119500701319550403119440526
	6194707019196702277195106211219820512
	4199304151420021228520050730519550403
	1019861122219921231820151203619890808
	220120619420030414219810817819390405
	8194403027193509131219480106419310811
	12196805121119730403219501209220070726
	319880520719840105119900429
Rewards	With the hidden words you can open a blockchain wallet that contains 0.18665328 Bitcoin which is equal, at the moment of writing, to US$1200.
Answer	?

structure in the Bitcoin system to prohibit sales of particular items. As such, the countries in the international watch-lists or embargoed countries may find bitcoin transactions more attractive. Further, illegal and prohibited items can be sold and bought by these countries and others through the Bitcoin system.

Second, the block chain technology requires that any transaction in the Bitcoin system is final and irreversible. The block chain protocol provides no way for a payer to reverse accidental or unwanted purchases.

Third, the Bitcoin system is global. There is no limitation of participation in the Bitcoin system on the basis of nationalities. As long as there is a successful miner who has access to the internet, bitcoins can flow to anywhere on the Planet. As long as there is a Bitcoin system user, the bitcoins can be transferred from a user from a country to another user in any other country. There is no central bank or national government who controls its issues and transactions. It is a global currency, in which sense it has bearing to the host of global funds explained in this book.

8.4.3 Bitcoin Exchanges

One of the fundamental questions with regard to the Bitcoin is why some people would want to own the bitcoin by investing personal resources and time. An immediate answer to the question is that a bitcoin can be traded for, e.g., the US dollars at a certain exchange rate. There are multiple currency exchanges that allow bitcoin users to trade bitcoins for traditional currencies such as US dollar, Euro, and Japanese Yen.

As of 7 May 2018, the US$/bitcoin exchange rate was 9353.88. That is, one bitcoin was traded for US$ 9353.88. As in the other financial markets, most bitcoin exchanges operate based on double auctions with bids and asks. For a bitcoin exchange, a commission is charged, which is at the range of 0.2–2%. Most bitcoin transactions are accompanied by conversions to traditional currencies (Böhme et al. 2015).

What are these bitcoin exchanges? In Spring 2012, the Mt. Gox, a Japan-based exchange, served over 80% of all bitcoin transactions. The Mt. Gox collaspsed early 2014 and filed for bankruptcy, reportedly losing 754,000 bitcoins equivalent to US$ 450 million. As of May 2018, the largest Bitcoin exchanges were, in order by monthly trading volume, Bitfinex (31%), Bitflyer (14%), Coinbase (13%), Bitstampy (12%), Kraken (11%), Hitbtc (5%), Bit-x (4%), Gemini (3%), Itbit (2%), and others (Bitcoinity 2018).

The history of bitcoin exchange rates recorded in these exchanges is drawn in Fig. 8.3 for the period from July 2010 to May 2018, using the markets data provided by the Bitcoinity (2018). The exchange rates at the two largest exchanges, Bitfinex and Coinbase, and the exchange rate compiled by other exchanges (than the above-mentioned large exchanges) are put together in the figure.

The figure shows that the US dollar value of a unit of bitcoin started from almost zero, 0.049 US$ per bitcoin in July 2010. By February 1, 2017, the exchange rate surpassed US$1000 per bitcoin. Through the rest of 2017, the dollar price of a bitcoin climbed up steeply to reach a peak at $16,000 by December 1, 2017. The price then fell precipitously. By April, 2018, the exchange rate fell to about $7000 per bitcoin. By January 2019, the exchange rate stood at around US$ 3700.

8.4.4 *Bitcoin Uses and Users*

With regard to the question of whom would trade for the ownership of bitcoins, it is noteworthy that the first adopters of bitcoins were the businesses that wanted to exploit (1) anonymity given to the users and transactions of bitcoins; and (2) absence of the rules that govern the bitcoin transactions equivalent to those applied to other financial transactions. A prime example of such businesses was the online sellers of narcotics such as marijuana, prescription drugs, and benzodiazepines (Böhme et al. 2015).

The first online market place that supported bitcoins exclusively as the means of payment was Silk Road. Designed as a free market outside the control of the government in February 2011, over US$ 1 billion worth of goods had been traded before the FBI seizure of the website in October 2013. The ten most popular product categories listed by Silk Road were dominated by narcotics. From January 2012 to July 2012, the top ten categories were weed (13.7%), drugs (9.0%), prescription

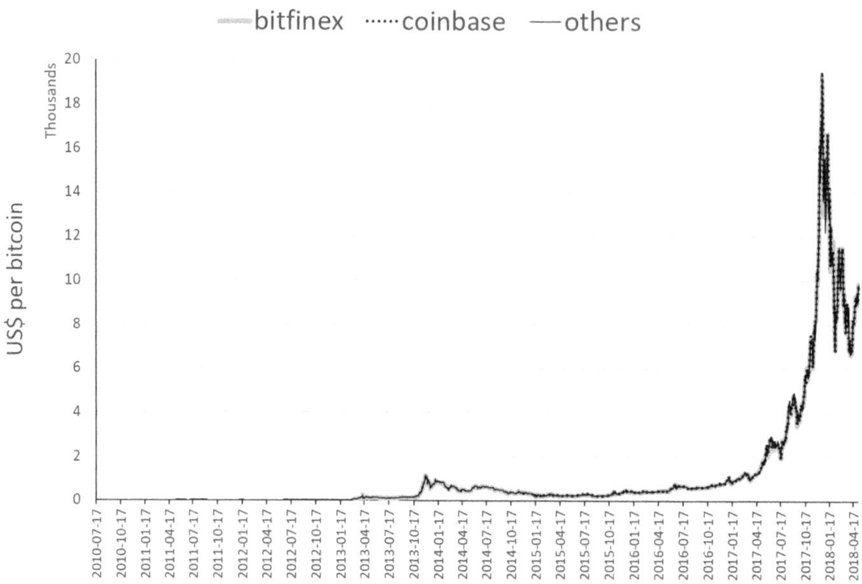

Fig. 8.3 History of the bitcoin exchange rates. (Note: The source data are from Bitcoinity (2018))

(7.3%), benzodiazepines (4.9%), books (3.9%), cannabis (3.6%), hash (3.4%), cocaine (2.6%), pills (1.9%) (Christin 2013). The Silk Road operated as a Tor hidden service in the darknet and relied on bitcoins for payments, both of which ensured almost complete anonymity of users and transactions.

Early Bitcoin adopters were also gambling sites and the individuals who sought to avoid international capital controls. Gambling sites wanted to protect their customers' identities and receive funds from the customers who cannot make payments using other methods. Similarly, interests in the Bitcoin are expected to be higher in the country whose governmental policy strictly limits monetary transfers to other countries.

Notwithstanding the dark aspects of bitcoin transactions, the cryptocurrency has been gaining wider acceptance in the society. On the one hand, some retailers started to accept bitcoins as an alternative form of payments in light of the criticism of high fees charged by credit and debit cards. Such prominent retailers include Overstock.com (online retailer), Expedia (travel), Newegg (electronics), Foodler (restaurant delivery), TigerDirect (electronics), and Gyft (gift cards) (Böhme et al. 2015). A downside to the customers who pay by bitcoins, however, is that the credit card payments often offer a rebate or other benefits such as frequent flyer credits and merchandise credits which the bitcoin payments do not offer.

On the other, many bitcoin users do not spend the coins for payments for goods and services but hold them in expectation of future appreciation in the coin value. According to one study, more than 60% of the bitcoins mined in 2009–2010 remained unspent by early 2013 (Meiklejohn et al. 2013). A Yale economic study of investment portfolios shows that an investor should hold 6% of her portfolio in the Bitcoin if she believes that the Bitcoin will perform as well as it has in the past. If the investor believes that the Bitcoin will perform half as well, then she should invest 4% of her portfolio in the Bitcoin (Liu and Tsyvinski 2018). As mentioned earlier in this section, it was reported that Yale endowment, the world's second largest university endowment and the most influential, invested in two cryptocurrency funds in 2018 (CNBC 2018).

8.4.5 Risk in Bitcoin

Some prominent businessmen called Bitcoin a fraud while others predicted that it will soon become worthless. Is it a truly global currency or just a global fraud that happens to be very smart? Including the risk of owning bitcoins explained in the above, we can summarize the risk as the following five components: market risk, counterparty risk, operational risk, privacy risk, and regulatory risk.

A market risk is the risk of holding the bitcoin that arises from fluctuations in the exchange rate of the Bitcoin, which has been highly volatile as shown in Fig. 8.3. Another market risk arises when, because of a shallow market, a person who wants to trade a large volume of bitcoins for either purchases or exchanges cannot do so quickly.

A counterparty risk refers to the risk that Bitcoin exchanges may end operations abruptly. A Bitcoin exchange that holds a large volume of bitcoins may default owing to an internet security breach or may simply abscond without explanations. Another type of counterparty risk is that a bitcoin transaction may be made owing to an error or a fraud. But, in the block chain technology, the transaction is irreversible.

Third, an operational risk refers to the risk that the Bitcoin system and its security can be attacked. An individual's private key and wallet can be stolen because of security flaws or by malware that specifically targets them. A distributed denial-of-service (DDoS) attack is one form of the operational risk, in which attackers swamp a particular target firm with a large number of messages to make it inaccessible or unstable.

Fourth, a privacy risk of the Bitcoin system stems from the block chain technology which reports the entire history of bitcoin transactions from the first bitcoin ever created. If a Bitcoin user's public key can be divulged, the entire transaction history of the user can be traced by others. For example, when a user converts bitcoins to national currencies or makes purchases using bitcoins, her/his identity can be revealed by the name used in the transactions or the mailing address used for the purchases.

Fifth, a legal and regulatory risk. At its current stage of developments, the legal and regulatory risk of the Bitcoin system is certainly large, although it seems to have decreased in the past several years. As in the Silk Road takedown by the FBI, there are many avenues that legal and regulatory troubles shock or even cripple the Bitcoin system. For example, a national government may decide to ban the Bitcoin system on the grounds of crime fighting or consumer protection (Böhme et al. 2015).

8.4.6 Bitcoin and the GCF

Because of the unusual origin and cryptic characteristics of the Bitcoin, readers may wonder whether there is any direct relationship between the cryptocurrency and the GCF funds. However, when the present author describes another unusual currency not in the form of a traditional currency in the next section, that is, an international green currency such as the European Union's CO2 emissions permit, readers will be able to better appreciate the importance of this currency.

Is the GCF fund a cryptocurrency? Both the Bitcoin and the GCF fund are a global currency in that they can be issued to anywhere in the world and can be converted to major international currencies. We can make intriguing comparisons between the two global funds. First, the Bitcoin seems to have a more transparent mechanism in the generation of the currency than the Green Climate Fund. The bitcoin is created only when someone is successful in appending a new block chain record, regardless of his nationality, race, status, etc. On the other hand, the GCF is created through an opaque mechanism, that is, a voluntary pledge and contribution. Therefore, the size of the GCF fund depends on many political, international, and emotional factors.

Second, the distribution rule is transparent in the Bitcoin system in which the bitcoins flow to the miners who succeed in updating the block chain and a transfer from one user to another in the system is done electronically and openly to the public. By contrasts, the distribution rule, so called an investment rule, of the Green Climate Fund is not 'firmly' established, as highlighted in Chap. 7 of this book. The investment rule is loosely tied to the expected contribution in terms of an expected ton of CO2-equivalent emissions avoided or an expected number of beneficiaries from a funded project.

Further, there is no central authority that determines the distribution and redistribution of the bitcoins. Neither a national government nor any non-governmental organization is responsible for the distribution decisions. On the other hand, there is a clearly defined authority that is responsible for the distribution decisions of the GCF. The GCF Board comprising 24 members determines which project among multiple proposals gets to be funded or not. The review in Chap. 7 of the funded GCF projects revealed that such decisions are done deliberately considering geopolitics of global warming.

Third, the market value of a bitcoin was extremely volatile (Fig. 8.3), but is determined in the currency exchanges through bids and asks. The true value of a bitcoin may approximate the amount of efforts that a miner exerts in order to earn the coin, including equipment costs, electricity cost, and time cost, but could deviate from the true value owing to such factors as speculative market elements and regulatory risk (Shiller 2018).

On the other hand, the value of a GCF project is determined by the GCF Board. As explained in Chap. 7, the size of a grant to a particular GCF project is determined by the Board considering, among other things, the expected contribution through avoided CO2-equivalent emissions. This means that the GCF Board dictates the price of a ton of CO2 emissions, that is, the carbon dioxide price. There is no formal international conference on the determination of the carbon dioxide price, therefore, the GCF Board likely picks one point from the range of the CO2 price distribution (Tol 2009; Nordhaus 2014).

On this last point, we can establish a direct link between the GCF and other green funds such as the EU's CO2 emissions permit. In the latter, the price of carbon dioxide is determined through the exchanges in the CO2 emissions permit markets. But, even in the CO2 markets, the price hinges fundamentally on the emissions limit set by the European Council (Nordhaus 2007; Ellerman and Buchner 2007).

The comparisons between the Bitcoin and the GCF and other green funds can be made graphically as in Fig. 8.4. Four global funds are put together: Bitcoin, GCF fund, CO2 permit by the EU ETS (Emission Trading Scheme), and carbon tax. All four funds are international currencies. The main difference between the GCF and the ETS is that there is a market price in the latter. There is a market price in the carbon tax as well in the sense that it creates the price of carbon dioxide that should be embedded in all product prices. The main difference between the ETS and carbon tax is that there is no volume limit set by a central authority in the latter. In the Bitcoin, all four blocks in the figure appear. The main difference between carbon tax and the Bitcoin is that there is no central authority that determines either its volume or price.

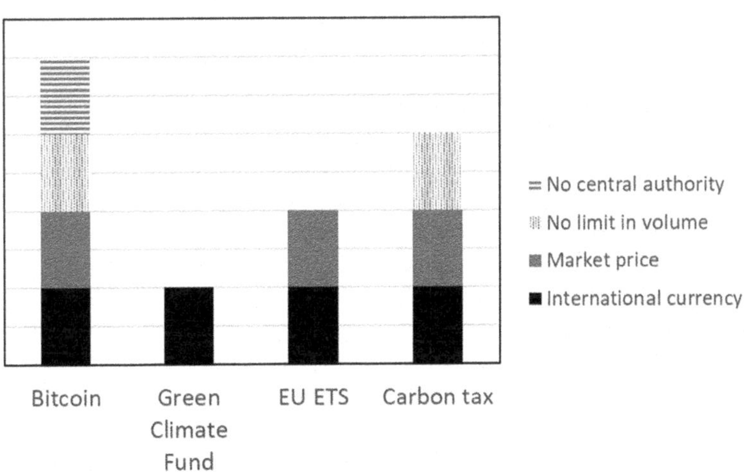

Fig. 8.4 Bitcoin versus GCF and other green funds

8.5 Green Funds: A CO2 Permit

The concept of a green fund or a green money has long been lurking throughout the international environmental policy dialogues and negotiations such as the Stockholm Conference, the Kyoto Protocol, and the Paris Agreements. A green fund can be defined as a fund created for the purpose of environmental protection against a specific pollutant.

A green fund is a new type of money that is circulated presently across the globe but only in an implicit manner. That is, it can be exchanged with a major world currency, as is the case with the bitcoins. It is not, however, used to pay directly for goods and services in the market. There are many different green funds in use in the world.

It would be much easier for the readers to understand the green funds by way of visualizing an emissions permit as a twenty-dollar note. At the national level, an SO2 allowance has been circulated for two decades in the United States to control sulfur dioxide emissions from coal-fired power plants (Schmalensee and Stavins 2013). At the international level, the EU's Emissions Trading System (EU ETS) has been implemented during the past decade as a way for cutting CO2 emissions in the EU in compliance with the Kyoto Protocol (UNFCCC 1997; EC 2016).

A unit of the emissions permit for, say, carbon dioxide, is a paper money just like the national currencies. A unit of the emissions permit gives the holder the right to emit a one ton of CO2 from its production activities. The holder can sell the permit to another party in exchange for an international currency, say, Euro. The negotiation between the seller of the permit and the buyer leads eventually to the monetary market value of the permit through the emissions permit market.

Unlike the national currencies, however, the CO2 permit is not issued by a national Central Bank. For the EU ETS, the permit is issued by the European Commission's Department of Environment, the agency which decides the stringency of the climate policy in the region. It is therefore an environmentalist's currency. That is, the issue and value of the CO2 permit is controlled by environmentalists, more concretely, an environmental organization.

The attempts to make the EU ETS a truly global trading system, a success of which would have made the CO2 permit a global currency, have failed multiple times. In the Copenhagen Conference held in 2009, the EU negotiators failed to achieve its goal of extending the Kyoto Protocol over the first implementation phase at which period the EU ETS became the cornerstone policy instrument for the European Union (UNFCCC 2009).

EU and global negotiators launched 2 years later in Durban, South Africa their renewed efforts to accomplish a legal framework in which all parties of the convention would take legal responsibility for cutting carbon dioxide emissions, for example, through the purchase of the emissions permit, which should be in circulation across the globe legally, for each unit of CO2 emissions each party releases (UNFCCC 2011). Negotiators convened in Paris in December 2015 for a breakthrough deal, but fell short of reaching a legally-binding agreement or treaty (UNFCCC 2015; Seo 2017a).

Although unsuccessful up to this point, a successful global Protocol that was envisioned in Durban, South Africa would be similar to the EU ETS framework, but implemented at the global level and across all emission behaviors. More specifically, a green fund or a CO2 emissions permit would be created and issued by the United Nations, which would then be accepted and exchanged across all nations in the world. It would be a legally binding global policy via the permit's monetary power exchangeable anywhere in the world that disallows a free release of carbon dioxide.

The monetary value of the permit is created by allocating the total number of CO2 permits to each nation smaller than the total current unmitigated quantity of emissions of CO2. Each nation should reduce the quantity of CO2 emissions by employing costly means or purchase additional permits from another polluter, through which processes the monetary value of the emissions permit is born.

If global negotiators would be successful in agreeing on such a legally binding policy framework in the future, the green CO2 permit—a green fund—could provide an ultimate guideline for the GCF's investments, a fascinating propect for accomplishing the central task of this book. That is to say, numerous GCF investment decisions can be guided by the market price of the CO2 permit globally determined and circulated. We will come back to this point in the final section of this book.

What is the rule that determines the total number of CO2 permits globally, which eventually settles the market equilibrium price of the permit (Montgomery 1972)? The global environmental policy negotiators who would gather at the United Nations Conferences at some point in the future would determine this number. Therefore, any number in principle can be picked by them. However, that the number chosen by the negotiators turns out to be the globally most desirable number cannot be guaranteed at all. In the language of this book, a random choice by the policy negotiators does not guarantee a Pareto optimal decision for the globe.

Then, one of the most rational choice rules is to choose the Pareto optimal number (Stavins 2007). A choice by policy-makers and enforcements in the market of the Pareto optimal total number of permits would create the market equilibrium price of the carbon dioxide which is equal to the Pareto efficient carbon tax (Nordhaus 1991, 2007). At this equilibrium point, the carbon tax per ton equals the marginal benefit of a ton of carbon dioxide reduced as well as the marginal cost of a ton of carbon dioxide abated.

Since it is individual firms and individuals that emit CO2-equivalent gases, the total number of permits should be divided to the individual actors, which is not a handy task at all. A global policy body should have the full knowledge of the current level of emissions from all potential emitters of CO2-equivalent pollutants in the global economy. Further, nearly all individuals and firms are the emitters of CO2-equivalent gases but some of them are also 'sinkers' of those gases.

An allocation rule may be established as follows: In the first stage, each polluter is given the number of permits that is equal to the current level of emissions. In the second stage, the number of permits allocated to the individual is reduced by a certain percentage periodically, say, every 5 year. Another big question with regard to the initial allocation of carbon dioxide permits is whether the permits should be given away to the polluters or whether polluters should be forced to pay for the initial number of permits (Stavins 2007).

Once the emissions permits are given to individual polluters, the permits become an asset to the holders. A holder can sell the permits at a market price if the company plans to trim its production activities. A company which intends an expansion of its production activities would have to purchase the permits from the market.

With the emissions permit system in place, environmentalists will be able to sell excess permits in the market thereby earn profits by engaging in 'environmentally friendly' practices: burning dried cow-dungs for heating, paper-free toilets, choosing public transportations over private vehicles, purchasing electric vehicles, switching to natural gas-fired electricity, installing solar panels, purchasing wind energy, replacing incandescent bulbs with Light Emitting Diodes (LED). By engaging in these pollution reduction activities, environmentalists will have spare permits that can be sold in the market.

Non-environmentalists, on the other hand, will have to purchase additional permits from the market because of engaging in pollution increasing activities. They rely, among other things, on the electricity produced from coal-fired power plants, private gasoline-powered vehicles with a low mileage per gallon, red-meat based diets with high methane emissions from production activities. By having to purchase additional permits for these activities under the emissions permit system, non-environmentalists will pay for the social cost of the pollution generating activities.

How much should they pay for a unit of CO2-equivalent permit? Given the total number of emissions permits determined by the global negotiators, the price of a permit will reflect and be a shadow price of the social cost of a ton of carbon dioxide equivalent emissions (Nordhaus 2014). That is, it is the marginal cost of a ton of emissions on the global society.

The European Union adopted the ETS (Emissions Trading Scheme) from the first implementation phase of the Kyoto Protocol in 2008. Since then, the market price of the CO2-equivalent permit showed high volatility. In Fig. 8.5, a trajectory of the historical prices of the CO2-equivalent emissions allowance is presented by the author using the closing prices of the EUA Futures traded in the Intercontinental Exchange (ICE) (ICE 2018).

The price history shows that initial enthusiasm before the Copenhagen Conference at the end of 2009 kept the price float, but since then the price had gradually fallen near to zero by the first months of 2014 as a 'global warming hiatus' hypothesis was being debated. Since that point, the permit price started to incrementally rise up to the time of the Paris Conference at the end of 2015, which then tumbled again after the Paris Conference. Since the early months of 2018, the CO2 allowance price has increased steeply to reach more than 25 Euros by the end of July of the year, but sharply fell again during the latter half of 2018. The high volatility during 2018 has raised concerns among the traders.

A rough calculation shows that the average price during the time period in Fig. 8.5 is 8.8 Euros and the standard deviation is 4.3 Euros. The volatility of the CO2 allowance price, measured in percentage standard deviations, is more than 10 times larger than that of the price of the Dow Jones Industrial Average Index or the S&P500 Index traded in the New York Stock Exchange (Shiller 2006). The EUA futures has so far exhibited very large volatility.

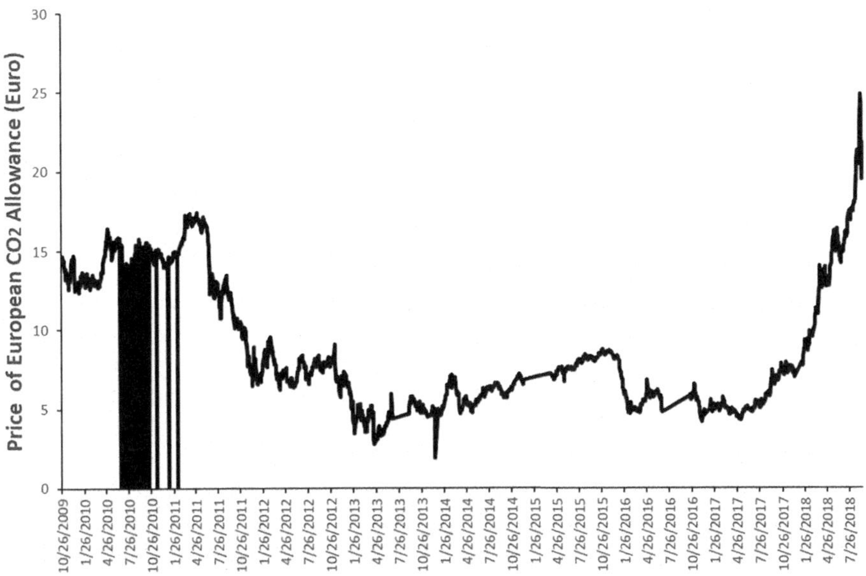

Note: The data are from the Intercontinental Exchange (ICE 2018).

Fig. 8.5 Price volatility of the CO2 allowance. (Note: The data are from the Intercontinental Exchange (ICE 2018))

The volatility of the CO2 allowance price can be compared to that of the bitcoin price shown in Fig. 8.3. A visual inspection shows that the former shows a high degree of volatility, but not as high as that of the latter, that is, the bitcoin price.

8.6 Guidelines on the Green Climate Fund's Investment Decisions

How can the CO2 allowance price such as the EUA futures provide useful guidelines for the Green Climate Fund's investment activities? In a comprehensive market of CO2-equivalent gases, the allowance price settled in the market provides a signal on the true price of carbon dioxide emissions. If the market existed, individuals and agencies will make investment decisions upon seeing the allowance price in the market. With the allowance price embedded, their decisions will achieve an optimal allocation of carbon dioxide intensive resources.

Let's suppose that a group in Mali submits a proposal to the GCF with project details and a funding request. The GCF Board would consider the proposal and, if they decide to give a grant to the project, would request the group to submit the total number of CO2-equivalent emissions from the host of economic activities it operates at the time of submission. The GCF Board would then award the number of permits equal to the total number of CO2-equivalent emissions to the group for free, after review and corrections by the GCF-affiliated research centers.

At the end of the suggested project period, the GCF would reassess the total number of CO2-equivalent emissions from the project group. If the project were to yield a reduction in the CO2-equivalent emissions, the GCF Board would purchase the spare CO2 permits, that is, the difference between the CO2-equilvanet emissions at the time of proposal and that at the time of project completion, from the project group. The GCF would pay the CO2 allowance price determined at the CO2 allowance exchanges to each unit of spare permits.

Having obtained the residual permits from the participating group, the GCF Board will participate in the CO2 allowance exchanges in order to sell the acquired permits, the revenue from which can be used in the future for funding other projects.

In this coupled GCF-CO2 allowance approach, an essential task of the GCF Board is to assess the level of CO2-equivalent emissions at the time of project initiation and at the time of project completion. The difference in the levels of CO2-equivalent emissions is the total dollar value that the project undertaking agency ultimately receives.

The CO2 allowance price determined at the, e.g., the Intercontinental Exchange, provides a signal to the potential project initiators. The higher the price of CO2 allowance, the stronger the incentive to participate in the GCF funding programs. The lower the price of the CO2 allowance, the weaker the incentive for the potential participants.

The incentive created by the CO2 allowance price goes beyond this. Given the monetary incentive, a proposal maker would consider carefully a range of CO2-equivalent emission reduction activities and submit a proposal with the project that yields the largest amount of carbon dioxide equivalent emissions reduction. Put differently, heavier emitters of the CO2-equivalent emissions will have stronger incentive to submit a proposal to the GCF Board than lighter emitters.

The GCF Board should be given the authority to issue the CO2 allowances to any organization in the globe after a rigorous review process. This authority could be bestowed by the international agreement achieved at a future UNFCCC Conference. The authority given should also be specific with regard to the ability of the GCF Board to exchange the excess permits at international CO2 allowance exchanges.

The GCF Board then relies on the revenue from selling the excess permits at the international markets in order to fund various projects submitted to the international organization. The GCF Board may continue to have the option to solicit monetary contributions from the world's major CO2 emitting and developed countries such as the US, EU, and Japan, which is, however, no longer necessary.

In this GCF-CO2 allowance price framework, an individual can be motivated to participate in the GCF program. The present GCF program awards funding to only an international organization or a major national-level agency. In this alternative policy framework, an individual can have an incentive to submit a proposal if a proposed activity can lead to sufficiently large revenue, that is, a sufficiently large number of emissions reductions.

What underlies such a strong incentive created by the CO2 allowance market? The underlying motivation is a green money that is created by the CO2-equivalent allowance market. Once the green currency, that is, a green permit, is created and traded, people will make everyday decisions in ways to obtain the currency.

Under the GCF-CO2 allowance framework, the CO2 market does the magical work of guiding the GCF project activities. The GCF does not any more choose a particular project arbitrarily from a large pool of submitted proposals. Neither does the GCF committee any more pick a specific political area, e.g., sustainable development, over other priority areas for funding. Nor does the GCF decide any longer on the size of monetary aids to be provided to selected project undertakers. The heavy burdens of the GCF Board, which were elucidated throughout this book, will be lifted off from the shoulders of the exclusive group of 24 board members of the Green Climate Fund.

Under the coupling of the GCF-CO2 allowance market, everyone makes one's own decisions for CO2 abating and emitting activities, knowing that s/he can be rewarded with monetary compensation for the CO2-equivalent emissions reductions. The heaviest emitters will come out voluntarily to participate in the GCF-CO2 allowance program since they know they are the largest monetary beneficiaries of participation.

Why hasn't this magical framework been adopted by the international green community or, more narrowly, the United Nations Framework Convention on Climate Change? Needless to say, there are quite a few hurdles to overcome to make this framework functional. Taking aside those hurdles to be explained shortly, however, there is ignorance among the UNFCCC negotiators about the carbon price and CO2 allowance market, which is one of the factors that has prevented a wider application of this kind of market-based climate policy.

Let me make clear some of the hurdles that face the GCF-CO2 allowance framework. First, the CO2 allowance price framework such as the EU ETS has been treated by many to be useless as a policy instrument because, as we have seen in Fig. 8.5, the price of the allowance crashed after the Copenhagen Conference and stayed at less than 5 Euros for most of the time period until the beginning of 2018. Such a low price of the CO2 permit would fail to provide a strong incentive to the heaviest polluters to cut their emissions, neither to the policy-makers of the UN.

Second, a steeply rising price of the CO2 allowance since the dawn of 2018, noticeable in Fig. 8.5, would certainly give UN policy-makers a motivation to have a second look at the possibility of the CO2 allowance market linked to the GCF Board's works. On the other hand, the large volatility of the CO2 allowance price would be interpreted by the UN policy-makers to be another source of concern for the mandated works of the GCF and the UN. What if a large volatility of the CO2 allowance price results in a world-wide economic turmoil at some point in time?

The risk of the CO2 permit system has not been realized at a global scale up to this point. If materialized, it may resemble the US subprime mortgage defaults in 2008 which had rippled into a global financial crisis (Shiller 2009). The 'subprime mortgage' was a financial asset which was owned by the public and its defaults *en masse* led to an economy-wide meltdown of the financial sector in the US, which struck other countries by linked assets and investors in the following years.

The CO2 permit price would crash at some point in the future if the Fig. 8.5 is any predictor of future events. For example, a ton of CO2-equivalent emissions permit may decline from 50 Euros to 5 Euros within a week of turmoil. In such an event, the firms that own a large number of CO2 permits and use the permit sales as one way of debt payments would face severe difficulties owing to the sudden drop in the CO2 permit value. These firms may default, which then may lead to an economy-wide crisis through linked assets among the firms in the economy.

This asset value crash scenario, referred to as a bubble-burst event in financial economics, is one area that the policy-makers of the UNFCCC and the GCF need to figure out how to prevent the event and minimize the impacts. A crash of the SO2 allowance price in the United States that occurred in 2012 is a historical lesson to CO2 policy-makers (Schmalensee and Stavins 2013).

The third concern is a more theoretical one: In what sense does the price shown in Fig. 8.5 represent the 'correct' price of a ton of carbon dioxide emissions? Does it reflect the true social cost of a ton of carbon dioxide equivalent emissions? (Mendelsohn and Williams 2007; Tol 2009; Nordhaus 2014; NASEM 2017).

At present, the CO2-equivalent emissions allowance is traded in the European Union and other limited national markets. The price trajectory shown in Fig. 8.5 captures the price determined from the EU ETS program (EC 2016). The EU ETS is limited to the EU emitters of the CO2-equivalent gases. As such, it does not include the effects of CO2 emissions from outside the EU such as the United States, China, and India. More specifically, the price of a ton of CO2 emissions is zero in these countries.

Further, the EU ETS is not designed to be inclusive of all the emitters within the European Union, but rather focused on power plants. The program excludes a large number of emitters of carbon dioxide and other equivalent gases. In total, it is argued that the EU ETS covers about half of the total emissions of CO2 in the EU (EC 2016):

> The EU ETS covers more than 11,000 power stations and manufacturing plants in the 28 EU member states as well as Iceland, Liechtenstein and Norway. Aviation operators flying within and between most of these countries are also covered. In total, around 45% of total EU emissions are limited by the EU ETS.

Moreover, the pollutants covered are not inclusive of all greenhouse gases. Specifically, the EU ETS covers carbon dioxide from (1) power and heat generation; (2) energy-intensive industry sectors including oil refineries, steel works, and production of iron, aluminum, metals, cement, lime, glass, ceramics, pulp, paper, cardboard, acids and bulk organic chemicals; (3) civil aviation. The other gases that are covered by the EU ETS are (1) Nitrous Oxide (N2O) from production of nitric, adipic, glyoxal and glyoxylic acids; (2) Perfluorocarbons (PFCs) from aluminum production (EC 2016).

Considering these limitations, what is the true price, namely, shadow price, of a ton of carbon dioxide emissions and how is it related to the realized CO2 allowance prices in the EU ETS shown in Fig. 8.5? In fact, an efficient CO2 allowance market cannot exist without a careful account of the truce price of a ton of carbon dioxide emissions beforehand (Nordhaus 2007). This is because the CO2 permit system gets into motion only after the CO2 allowance limit is set by the policy body, say, the European Council.

The EU sets the total number of tradable CO2 permit in accordance with its commitment to the Kyoto Protocol (UNFCCC 1997). That is, the goal declared at the Kyoto Conference of reducing the level of CO2-equivalent emissions by 5% from the baseline of the 1990 level of emissions was the guiding principle for designing and determining the quantity of CO2 permits to be allocated by the EU. Of course, there is no reason that the Kyoto Protocol declaration should be the only fixed point for designing the allowance market in the EU.

A theoretically sound approach for setting the CO2 allowance limit is, if it were to be an optimal policy for the globe, to set the allowance limit at the level that accords with the true price of carbon dioxide equivalent emissions (Nordhaus 2007). That is, a carbon price provides an ultimate guidance for setting the CO2 allowance limit, that is, the emissions cap.

How should the policy-makers obtain efficient carbon prices then? The carbon price (tax) can only be obtained from an integrated assessment model (IAM) which accounts for all carbon dioxide emitting and abating decisions that take place globally. It is calculated through the IAM from considerations of all emitters (including tailpipes and smokestacks), impacts of carbon dioxide emissions on numerous economic sectors and natural ecosystems, costs of reducing the emissions, an array of behavioral adaptations, and monetary valuations.

The description in this section thus far has provided another vantage point from which readers can evaluate the GCF's numerous investment decisions by analyzing it through the lens of another global currency, that is, the CO2 permit (GCF 2011, 2018a, b). The two global environmental currencies are indeed quite intricately related. Although to a lesser degree, the GCF funding decisions are also entangled with the other global funds introduced in this chapter, that is, the UN specialized funds, crowdfunding, and cryptocurrency.

References

Adams R, Rosenzweig C, Peart RM, Ritchie JT, McCarl BA, Glyer JD, Curry RB, Jones JW, Boote KJ, Allen LH (1990) Global climate change and US agriculture. Nature 345:219–224

Bitcoinity.org (2018) Markets. Available at https://bitcoinity.org/markets/list. Accessed 8 May 2018

Böhme R, Christin N, Edelman B, Moore T (2015) Bitcoin: economics, technology, and governance. J Econ Perspect 29:213–238

Christin N (2013) Traveling the Silk Road: a measurement analysis of a large online anonymous marketplace. In: Proceedings of the 22nd international world wide web conference (WWW'13), pp 213–24. Rio de Janeiro, Brazil, May 2013

CNBC (2018) The most influential endowment manager just jumped into crypto with bets on two Silicon Valley funds. Oct 5, 2018, CNBC. Accessed from https://www.cnbc.com/2018/10/05/yale-investment-chief-david-swensen-jumps-into-crypto-with-bets-on-two-silicon-valley-funds.html
Douglas WO, Bates GE (1933) The Federal Securities Act of 1933. Yale Law J 43(2):171–217
Dudal R (1980) Soil-related constraints to agricultural development in the tropics. International Rice Research Institute, Los Banos
Ellerman AD, Buchner BK (2007) The European Union emissions trading scheme: origins, allocations, and early results. Rev Environ Econ Policy 1:66–87
Ellerman AD, Marcantonini C, Zaklan A (2016) The European Union emissions trading system: ten years and counting. Rev Environ Econ Policy 10:89–107
European Commission (EC) (2016) EU ETS Handbook. European Commission, Brussels
Fischer G, Shah M, Tubiello FN, van Velhuizen H (2005) Socio-economic and climate change impacts on agriculture: an integrated assessment, 1990–2080. Philos Trans R Soc B 360:2067–2083
Food and Agriculture Organization (FAO)/International Institute of Applied Systems Analysis (IIASA) (2012) Global agro-ecological zones (GAEZ v3. 0) model documentation. FAO, Rome
Fox News (2018) Haley announces $285M cut in 2018–19 UN operating budget. Accessed from https://www.foxnews.com/world/haley-announces-285m-cut-in-2018-19-un-operating-budget
Freeman DM, Nutting MR (2015) A brief history of crowdfunding including rewards, donation, debt, and equity platforms in the USA
GoFundMe (2017) GoFundMe 2017: a year in giving. The GoFundMe, Red Wood City, California. Accessed from https://www.gofundme.com/2017
GoFundMe (2018) How GoFundMe works. The GoFundMe, Red Wood City, California. Accessed from https://support.gofundme.com/hc/en-us/articles/115011182268-How-GoFundMe-Works-
Green Climate Fund (GCF) (2011) Governing instrument for the green climate fund. GCF, Songdo City
Green Climate Fund (GCF) (2018a) Annex III: investment framework. GCF, Songdo City
Green Climate Fund (GCF) (2018b) Annex IX: results management framework. GCF, Songdo City
Green Climate Fund (GCF) (2018c) Projects + programmes. GCF, Songdo City. Accessed from https://www.greenclimate.fund/what-we-do/projects-programmes
Intercontinental Exchange (ICE) (2018) ICE futures Europe: EUA futures. The ICE, Atlanta. Accessed from https://www.theice.com/products/197/EUA-Futures/data?marketId=400185&span=3
Intergovernmental Panel on Climate Change (IPCC) (2014) Climate change 2014: the physical science basis. The fifth assessment report of the IPCC. Cambridge University Press, Cambridge
International Civil Aviation Organization (ICAO) (2016) Carbon offsetting and reduction scheme for international aviation (CORSIA). ICAO, Montreal
Investopedia (2018) Bitcoin total value to hit $1.2T in'18: Fundstrat. Accessed from www.investopedia.com
Liu Y, Tsyvinski A (2018) Risks and returns of cryptocurrency. NBER working paper no. 24877. NBER, Boston
Medecins Sans Frontieres (MSF) (2017) International financial report. The MSF, Geneva, p 2017
Meiklejohn S, Pomarole M, Jordan G, Levchenko K, McCoy D, Voelker GM, Savage S (2013) A fistful of bitcoins: characterizing payments among men with no names. In: Proceedings of the 2013 conference on Internet Measurement Conference, Barcelona, Spain, pp 127–140
Mendelsohn R, Williams L (2007) Dynamic forecasts of the sectoral impacts of climate change. In: Schlesinger M, Kheshgi H, Smith J, de la Chesnaye F, Reilly JM, Wilson T, Kolstad C (eds) Human-induced climate change: an interdisciplinary assessment. Cambridge University Press, Cambridge
Mollick ER (2015) Delivery rates on Kickstarter (December 4, 2015). Available at SSRN: https://ssrn.com/abstract=2699251

Montgomery WD (1972) Markets in licenses and efficient pollution control programs. J Econ Theory 5:395–418
Nakamoto S (2008) Bitcoin: a Peer-to-Peer electronic cash system. Available at https://bitcoin.org/bitcoin.pdf. Accessed May 5 2018
National Academies of Sciences, Engineering, and Medicine (NASEM) (2017) Valuing climate damages: updating estimation of the social cost of carbon dioxide. The National Academies Press, Washington, DC. https://doi.org/10.17226/24651
National Post (2018) Here are the top five GoFundMe campaigns. Published on April 11, 2018. Accessed from https://nationalpost.com/pmn/sports-pmn/hockey-sports-pmn/here-are-the-top-five-gofundme-campaigns
Nordhaus W (1991) To slow or not to slow: the economics of the greenhouse effects. Econ J 101:920–937
Nordhaus W (2007) To tax or not to tax: alternative approaches to slowing global warming. Rev Environ Econ Policy 1:26–44
Nordhaus W (2014) Estimates of the social cost of carbon: concepts and results from the DICE-2013R model and alternative approaches. J Assoc Environ Resour Econ 1:273–312
Pareto V (1906) In: Montesano A, Zanni A, Bruni L, Chipman JS, McLure M (eds) Manual for political economy. Oxford University Press, Oxford, (2014)
Parry ML, Rosenzweig CP, Iglesias A, Livermore M, Fischer G (2004) Effects of climate change on global food production under SRES emissions and socioeconomic scenarios. Glob Environ Chang 14:53–67
Reddit (2018) Bitcoin puzzles. Accessed from https://www.reddit.com/r/bitcoinpuzzles/
Ricardo D (1817) On the principles of political economy and taxation. John Murray, London
Samuelson P (1954) The pure theory of public expenditure. Rev Econ Stat 36:387–389
Schlenker W, Roberts M (2009) Nonlinear temperature effects indicate severe damages to crop yields under climate change. Proc Natl Acad Sci U S A 106(37):15594–15598
Schmalensee R, Stavins RN (2013) The SO2 allowance trading system: the ironic history of a grand policy experiment. J Econ Perspect 27:103–122
Seo SN (2006) Modeling farmer responses to climate change: climate change impacts and adaptations in livestock management in Africa. PhD Dissertation, Yale University, New Haven
Seo SN (2010a) A microeconometric analysis of adapting portfolios to climate change: adoption of agricultural systems in Latin America. Appl Econ Perspect Policy 32:489–514
Seo SN (2010b) Is an integrated farm more resilient against climate change? A micro-econometric analysis of portfolio diversification in African agriculture. Food Policy 35:32–40
Seo SN (2011) An analysis of public adaptation to climate change using agricultural water schemes in South America. Ecol Econ 70:825–834
Seo SN (2014) Evaluation of Agro-Ecological Zone methods for the study of climate change with micro farming decisions in sub-Saharan Africa. Eur J Agron 52:157–165
Seo SN (2015) Helping low-latitude poor countries with climate change. Regulation 38(4):6–8. Winter 2015–2016
Seo SN (2016a) Microbehavioral econometric methods: theories, models, and applications for the study of environmental and natural resources. Academic/Elsevier, Amsterdam
Seo SN (2016b) Modeling farmer adaptations to climate change in South America: a micro-behavioral economic perspective. Environ Ecol Stat 23:1–21
Seo SN (2017a) Beyond the Paris agreement: climate change policy negotiations and future directions. Reg Sci Policy Pract 9:121–140
Seo SN (2017b) The behavioral economics of climate change: adaptation behaviors, global public goods, breakthrough technologies, and policy-making. Academic/Elsevier, Amsterdam
Seo SN, Mendelsohn R (2008) Measuring impacts and adaptations to climate change: a structural Ricardian model of African livestock management. Agric Econ 38:151–165
Shiller RJ (2006) Irrational exuberance, 2nd edn. Princeton University Press, Princeton
Shiller RJ (2009) The subprime solution: how today's global financial crisis happened, and what to do about it. Princeton University Press, Princeton

Shiller RJ (2018) The old allure of new money. Project Syndcate. May 21, 2018
Statistica (2018) Number of successfully funded projects on Kickstarter.com as of April 2018, by project category. Statistica, Hamburg. Available at https://www.statista.com/statistics/222460/number-of-successfully-pledged-kickstarter-projects-by-category/
Stavins R (2007) A US Cap-and-Trade system to address global climate change. Hamilton project discussion paper 2007–13
Tol RSJ (2009) The economic effects of climate change. J Econ Perspect 23:29–51
Tzu Chi (2018) About Tzu Chi. Accessed from http://tw.tzuchi.org/en/
United Nations (1945) Charter of the United Nations and statue of the international court of justice. UN, San Francisco
United Nations (1972) Report of the United Nations conference on human environment. Stockholm, Sweden
United Nations (2017) Assessment of Member States' advances to the Working Capital Fund for the biennium 2018–2019 and contributions to the United Nations regular budget for 2018. Secretariat, United Nations, New York
United Nations (2018) United Nations peacekeeping: how we are funded. Accessed from https://peacekeeping.un.org/en/how-we-are-funded
United Nations Children's Fund (UNICEF) (2018a) Emergency and disaster relief. The UNICEF, New York. Accessed from https://www.unicefusa.org/mission/emergencies
United Nations Children's Fund (UNICEF) (2018b) UNICEF integrated budget 2018–2021. The UNICEF, New York
United Nations Development Program (UNDP) (2018) Our funding: top contributors. UNDP, New York. Accessed from http://www.undp.org/content/undp/en/home/funding/top-contributors.html
United Nations Environment Programme (2017) The emissions gap report 2017: a UN environment synthesis report. UNEP, Nairobi
United Nations Environment Programme (UNEP) (2018) Annual Report 2017. UNEP, Nairobi
United Nations Environmental Programme (UNEP) (2016) The Kigali amendment: the Montreal protocol on substances that deplete the ozone layer. UNEP, Kigali
United Nations Framework Convention on Climate Change (UNFCCC) (1992) United Nations framework convention on climate change. UNFCCC, New York
United Nations Framework Convention on Climate Change (UNFCCC) (1997) Kyoto protocol to the United Nations framework convention on climate change. UNFCCC, New York
United Nations Framework Convention on Climate Change (UNFCCC) (2009) Copenhagen Accord. UNFCCC, New York
United Nations Framework Convention on Climate Change (UNFCCC) (2011) Report of the transitional committee for the design of Green Climate Fund. UNFCCC, New York
United Nations Framework Convention on Climate Change (UNFCCC) (2015) The Paris agreement. In: Conference of the Parties (COP) 21. UNFCCC, New York
United Nations System Chief Executive Board (UNSCEB) (2018) UN system. UN, New York. Accessed from http://www.unsceb.org/directory
World Food Programme (WFP) (2018a) Overview. The WFP, Rome, Italy. Accessed from http://www1.wfp.org/overview
World Food Programme (WFP) (2018b) Contributions to WFP in 2017. The WFP, Rome, Italy. Accessed from https://www.wfp.org/funding/year/2017

Index

S. N. Seo, *The Economics of Global Allocations of the Green Climate Fund*,
https://doi.org/10.1007/978-3-030-18274-8

D

E

Printed by Printforce, the Netherlands